Felix Alba-Juez

Felix was born in Burgos (Spain) in 1948. In 1949, his parents settled in Necochea (Argentina) where he completed his elementary and high school education. In 1966 he moved to Bahía Blanca (Argentina) where he graduated in Electrical Engineering at the 'Universidad Nacional del Sur' (UNS). In 1971, he started his academic life as a Teaching Assistant of Mathematics at the UNS and, from 1974 until to 1983, he was Adjunct, Associate, and Full Professor at the 'Universidad Nacional de San Juan' in San Juan, Argentina.

In 1983, he moved to Salt Lake City, USA as Postdoctoral Fellow and soon after became a Research Associate for the Department of Metallurgy at the University of Utah, conducting basic Research and Development on Optimal Control of Mineral Grinding Operations.

In 1987, he left Academia and founded his own consulting company working over the years with private companies and governmental entities as DuPont, ALCOA, US Department of Transportation, NASA, and Dow Chemical. His first patent was granted in 1992 in USA, United Kingdom, France, Germany, and Japan, protecting a technology based on ultrasonic spectroscopy for measuring particle size in industrial suspensions and emulsions.

In the period 1997-2001, Felix developed a fundamental theory for the generic mathematical modeling of multiple scattering of optical and acoustical waves interacting with highly-concentrated suspensions and emulsions.

During 2001-2007, he developed a particle size analyzer based on optical spectroscopy, and commercialized a generic simulation software tool connectable to acoustic and optical spectrometers, so as to convert them into particle size analyzers. The patent for this generic technology was granted in 2007.

In 2008, Felix sold all his intellectual property to Agilent Technologies, Inc, and currently is a scientific consultant, and writes Popular Science books on Epistemology and Philosophy of Science in English and Spanish. The Spanish Edition of this book was published in 2009 by the 'Ciudad de las Artes y las Ciencias, S.A.' in Valencia (Spain). He is currently working on his next book entitled *What is Reality? - Einstein, Quantum Physics, and Folklore.*

Praise for *Galloping with Light*

I'm impressed at your pedagogical capability … with words (almost) only. Physics is a very intere… nuclear energy, space exploration, astronomy an… a few fields. But to understand physics requires m… are legions, even many books are best-sellers. Th…

D1158599

Although most readers probably think that th… what is meant by mass, weight, force, and energy, perhaps also by inertia, gravitation, and momentum, FAJ devotes his first chapter to elucidate these concepts. Here and in every following chapter he first discusses the epistemology and history of the concepts, clarifying the semantics and fighting like a Don Quixote against misconceptions. Surprisingly, this approach makes an absolutely fascinating reading, a cultural Odyssey through the roots of physics.

Without mathematical equations, but with clever examples from everyday life, the reader rapidly gets used to moving in different frames of reference. Chapter 2 on time is mainly a fascinating review of the human perception of time, vividly illustrated by fictitious and amusing encounters with Mrs. Chaplin and film stars, and exemplified by the perceived duration of a kiss. Soon one discovers that one has learned effortlessly about topological simultaneity, and that the cigarette smoke of Humphrey Bogart in Casablanca illustrates an irreversible process.

Altogether this book is written on a fascinating subject in a captivating style, so I can warmly recommend it to anyone interested in science.

Matts Roos
Emeritus professor of Physics, University of Helsinki, Finland
Author of *Introduction to Cosmology*

...an introduction, intellectually accessible and literarily enchanting, to one of the principal components of the scaffold scientists have built for the current understanding of the Universe: the Theory of Relativity. This initiative is the more commendable if we consider that the immense majority of the Popular Science books on the subject normally limit themselves to inform the reader of some of the prominent consequences of Einstein's discoveries, but they are silent, or overly unclear, as to why those discoveries and consequences are valid. ...Given this scenario, Galloping with Light is an illuminating exception, and I hope the readers will enjoy it as much as I have.

Jesús Zamora Bonilla
Professor of Philosophy of Science at the UNED (Madrid, Spain)

...By reading this book anyone can comprehend the immensity of Einstein's concepts and implications. And not only that: it is read with gusto and at the end one would like to start all over, in part just for the mere pleasure of it, but mostly because one is urged to review ideas, concepts, examples, reflections... In sum, to review everything one already read but did not pay enough attention to capture its depth, even after having surmised there was profound philosophical content in most of the sentences read. I am therefore proud to recommend this book as a potentially giant step to demolish the barriers that impede non-professionals having access to scientific knowledge.

Manuel Toharia Cortés
Scientific Director of the 'Ciudad de las Artes y las Ciencias' (Valencia, Spain)

By reviewing the best of 2010 I found that reading this book was one of the most rewarding experiences of the year. Rewarding because it showed me how Science that is explained with examples drawn from everyday life, becomes a fascinating subject. Thanks to this book I discovered that YES! I can (finally) understand the basic laws of the universe and in particular, a lot more about "time" (which has always been of special interest to me). I had always liked mathematics, but I struggled with my Physics I and Physics II at college, so I went into languages and linguistics instead. Now after being a teacher for many years, I can see the potential in this book for Science teachers, it should actually become a compulsory resource!! how much value, richness and depth could be added to their lectures if some of this book examples (taken from everyday life) were used? This book is a jewel!

Read this book ...if only to discover how much more clever than you thought YOU are! simply delicious experience.

Liliana Brogan
Language Teacher (Sydney, Australia)

I bought the book in December, 2010. I could not have had a better Christmas present. I was devouring every page of it throughout the holidays. What a treat. At first glance, the 325-page book looked very dense to start reading, but in the first few pages I found myself sucked in by the author's passion to make the reader understand common physics terminology and hence inspire him or her. In each chapter Felix gives the historical perspective before arriving at the currently relevant concept. That is very much appreciated. I almost felt like that I had read Ptolemy, Copernicus, Brahe and Kepler, da Vinci, and Galileo before reading about Newtonian mechanics.

For me this book clarified a number of misconceptions I had in physics as taught by my college teachers. After all, many a college teachers learned physics by rote from books and they spit out a partial but often misconstrued logic in the class room. E = MC^2 is another misconception we all carry. We have been told that mass is converted to energy as in a nuclear reaction. The author categorically states that in a nuclear blast every elementary particle prior to the blast ended up as is in the post blast. I will let you read the details.

What is more interesting is Felix lays bare his biography throughout the book. You see the young author falling in love with his 1st grade teacher (surprise), wandering through his home town of Necochea, later Madrid and Salt Lake City. Everywhere his hunger to understand the Theory of Relativity grows like a fire. Each chapter inspires your curiosity about things you know already!

Oh, by the way the book does not bring up a single algebraic equation until very late and that too a minimal set is used. Well, I thought this must be a qualitative book without rigor. On the contrary, through epistemology of terms in physics and his skill of putting in words his own understanding, Felix makes it just as rigorous as a book with tens of equations in each chapter.

I highly recommend this book to anyone who has had a couple of college level physics course. It is a delightful treat for the inquisitive mind. I am eagerly looking forward to the author's upcoming book on Quantum mechanics!

Raj Rajamani
Professor of Metallurgy, University of Utah, Salt Lake City, Utah, USA.

In this book, the author is convinced that you can "know very much but comprehend very little", thereby he must have dedicated innumerable hours to comprehend with so much depth and precision the Special and General Relativity theories. Besides, he has explained them with an astonishing clarity and in non-scientific terms (without mathematics)... I must confess that, after 40 years of teaching, this book ended up demolishing part of my knowledge scaffold but, at the same time, it has the virtue of providing me with a new much more solid foundation to rebuild it... In summary: a different book that I recommend to all lovers of knowledge and, in particular, of physical sciences.

José L. Pombo
Professor of Mechanical Engineering, Universidad Nacional del Sur (Argentina)

Understanding the Universe and its Laws was always a great interest of mine. This book has made possible for me to understand the Theory of Relativity and its implications without any background in mathematical physics... The author had the ability to transform a 'studying' book into a 'reading' book -- certainly not easy reading, but the intellectual revenue is more than worth the effort. It is remarkable the clarity and depth with which all subject-matter is presented. Information, deep understanding, and demystification are the goals achieved by this book. It is a great work which I highly recommend.

Héctor Goldenberg
Architect (Belgium)

Excellent book! After 30 years of reading only literature, I felt a natural 'inertia' that gradually disappeared with an increscent reading pace until I felt the pleasure of resuming a subject-matter with which I was in love when young. Your references to your youth in Necochea add to this book an incredible plus of emotions and human content. I admit that I was surprised to find out that -besides being technically competent- you are a great writer in a genre normally arid and unfriendly to the general public. You have the virtue of leading the reader -in the most pleasant way- through intellectual labyrinths, so as to intuitively enlighten -at the very outset- very complex concepts, without sacrificing humor and humility. Anything else I could say would not reflect all that I have felt and learned reading this book. Congratulations; it is a great work that shows your concern for depth in the amplest meaning of the word, from the strict and elegant use of the language, to the didactic discourse on physical phenomena. I look forward to your next book.

Héctor Boido
Industrial Engineer (Mexico City, Mexico)

Galloping with Light

Einstein, Relativity, and Folklore

Felix Alba-Juez

Felix Alba-Juez, Publisher/Editor
Salt Lake City, Utah - USA

AUTHOR

© Felix Alba-Juez

GRAMMAR AND STYLE ADVISOR

Laura Alba-Juez, Ph.D.
Associate Professor of Linguistics at the UNED (Madrid, Spain)

ILLUSTRATIONS

Joaquín Armijo
NASA Public Domain Images
Front cover: "Albert Einstein in Viena, 1921" of Ferdinand Schmutzer (public domain)
Back cover: NASA picture taken by the Apollo 11 Mission (public domain)

PUBLISHER/EDITOR

Felix Alba-Juez
1159 Sunset Dunes Way
Draper, Utah 84020, USA
felixalba@q.com
Phone: (801) 501-8379; Fax: (801) 501-8392

First English Edition (updated), March 2011 **(printed by CreateSpace)**
First Spanish Edition by 'Ciudad de las Artes y las Ciencias, S.A.', Valencia (Spain) Nov. 2009

ISBN-13: 978-1456373856

ISBN-10: 1456373854

Copyright © 2010-2011 by Felix Alba-Juez

ALL RIGHTS RESERVED. Printed in the United States of America. No part of this work may be used or reproduced, transmitted, stored, or used in any form or by any means graphic, electronic, or mechanical, including but not limited to photocopying, recording, scanning, digitizing, taping, Web distribution, information networks, or information storage and retrieval systems, or in any manner whatsoever without prior written permission.

I dedicate this book to Women...

Grandmother, Mother, Teacher, Sister, Friend, Lover,
Professional, Wife, Daughter... the order being chronological...

All of whom have taught me much more than
what I can include in this brief account of my life

Prologue by *Manuel Toharia Cortés*

Teaching Science to the layperson in the twenty-first century is a challenge of incalculable proportions because science has been progressing at such a fast pace that the vast majority of the population is incapable of comprehending the variety and depth of the achievements in the field. This is a serious matter, in part because we lose our capacity of incorporating this accumulated and ever-increasing knowledge into our cultural baggage, but mainly because of its dramatic effects in our daily lives.

Almost everything we use, learn, or talk about, those things which interest or simply impress us... all those have a relation with a world as fascinating as unknown which we could call "Science". And due to our lack of knowledge, most of those achievements, all of them fascinating -though not always innocuous or altruistic- are not only ignored by us but also make us feel uncomfortable. At the same time, it is obvious that never before in the history of humanity have we lived so long and so comfortably -- at least those of us who live in the rich countries.

What we call "Science" can, most approximately, be divided into Basic Science and Technology (or Applied Science). Technology is the product of the application of the principles of Basic Science in order to attain a particular benefit for society.

In today's world we are submerged in Applied Science; almost everything around us has to do with some technological development. And this is evident not only in events of historical importance, like the recent 40th anniversary of man's first landing on the Moon, but also in the large majority of the appliances and gadgets we use every day. From the TV screen made of plasma, LCD or, more recently LEDs -three terms who nobody understands in depth, even though commercial advertisements employ them profusely- through mobile telephony, transgenic organisms, Internet, the ABS in cars, vaccines or antibiotics, or even antiques like radio or TV, the fundamentals of all these technologies remain hugely unknown for most of humanity.

Basic Science is essential; i.e. it is in the essence of everything we know and utilize. It is in all we have discovered and continue discovering through a complex process that, at its core, attempts to respond questions that humanity, and scientists in particular, continuously ask themselves: why and how things are the way they are? To each attained answer, many others pop up, creating a tree of knowledge which branches *ad infinitum*. In vain shall be any attempt of knowing everything! We will never achieve such a target because there will always be new questions to answer...

In this scenario of lack of understanding -many speak of scientific illiteracy- of what we think we know and utilize on a daily basis, books on Popular Science have a singular relevance with respect to Applied Science, as we can learn why a plane flies or an artificial satellite stays in orbit, how to reduce cholesterol in our arteries, and many other things. But, regarding Basic Science, books for the layperson are even more significant because without comprehending the fundamentals behind what we think we know, it is impossible to understand, utilize, or simply accept the devices we possess and use.

Before we can achieve a practical application, we should be capable of intuiting how to do it by establishing some kind of model, natural law, or simple natural regularity. Upon that, we can devise a practical application to improve our lives -or, in the worst case, deteriorate it as with sophisticated weapons that kill more and better-... Of course there exist those strikes of genius that lead to inventions without an *a priori* theoretical basis; but, even in those rare cases, this basis appears quite rapidly as a need to extend the idea, improve it, and generalize it so it can lead to other inventions and technologies.

A theoretical concept as powerful as Einstein's Relativity, already older than a century, remains being misused and abused by most mortals -the stereotyped phrase "everything is relative" is attributed to 'don Alberto', who would laugh in his grave to a new death if he could listen to it-, because very few understand the most basic of its premises, namely, that time, even though it seems to us as absolute and totally independent of everything else, in reality is relative and depends upon our position in the Universe and the velocity with which we move from that position at each instant.

All this generated an authentic revolution in Physics. And this is not only a theoretical revolution, but also an eminently pragmatic one. Today, we cannot live without the applications that, thanks to the relativistic science and its complementary Quantum Physics which -as a matter of fact- have both been equally ignored by the general public, inundate most of what we use daily, be it in information technology, electronics, medical and space technologies, or even in our kitchens, offices, etc.

Aware of the great significance of teaching to the big public what Science is all about, many authors have attempted the difficult task of explaining in layperson terms the fundamentals of those revolutionary concepts. This attempt had -in all truthfulness- slim success, though I suppose that some people of course must have benefited from those commendable efforts.

With this book I believe another step is made, and I hope it is a giant step, in the good direction. It is obvious, against what many think, that those Einstein's ideas which shook the solemn edifice built by Galileo and specially by Newton, are not mere theoretical speculations -more philosophical than anything else-, but that they have also had and still have innumerous practical and theoretical implications, helping to consolidate and further stimulate our basic and technological knowledge.

As a good scientist who decided to become a Popular Science writer, Felix Alba-Juez takes on the folklore associated with Relativity Theory and Science in general. He is well aware that it is the mythical, the romantic seduction of the pseudoknowledge, i.e. the folklore -both popular and scientific- that propagates quickly and easily through society, hiding and diminishing the powerful reality of what the new ideas and technologies can offer to humanity.

But, at the same time, the author goes to great lengths in trying to explain things in such a way that any person with a minimum education, even not having any dexterity in mathematics or physics, can comprehend the immensity of the ideas conceived at the beginning of the twentieth century by a young engineer employed by the patent office in Bern -- ideas which caused a commotion in the academic world of his time.

I believe the book achieves the objective the author envisioned. And not only that: it is read with gusto and at the end one would like to start all over, in part just for the mere pleasure of it (which is a very good sign when we finish a book) but mostly because one is urged to review ideas, concepts, examples, reflections... In sum, to review everything one already read but did not pay enough attention to capture its depth, even after having surmised there was profound philosophical content in most of the sentences read.

I am therefore proud to recommend this book in a new attempt to demolish the barriers that impede non-professionals having access to scientific knowledge. Comprehending and knowing better and deeper are the best guarantees we can have to attain ideas and criteria of our own; i.e. to stop depending on what other people say. In summary, to be freer to choose our own path in life.

Manuel Toharia Cortés
Scientific Director of the 'Ciudad de las Artes y las Ciencias'
President of the 'Asociación Española de Comunicacion Científica'

Valencia (Spain), Fall of 2010

Prologue by *Jesús Zamora Bonilla*

In one of the worst nightmares created by human imagination, the Athenian Aristocles (better known as "Plato", a nickname given by his wrestling trainer in honor of his broad back and shoulders) conceived a subterranean prison in which various children grew up without any direct access to the external world, from where they could only detect sporadic shadows. Based on their perception of those shadows or, better, holes floating on the dimmed light illuminating one of the walls in the cave, the children, or adults had they survived such a torture, were condemned to imagine what those phantasmagoric visions to which their life had been reduced could be. Plato, or Aristocles, employed this image of the unfortunate troglodytes as an impressive tool to provide verisimilitude to his theory affirming that, besides our eyes, skin, and the other senses through which we receive the shadows of the exterior reality, we have a 'mental eye' (intelligence) with which we can perceive reality as it is.

Almost two thousand years later, another philosopher, this time the French René Descartes, envisioned another version -perhaps even crueler- of this terror movie, in which the troglodyte paraphernalia is replaced by a more naked environment. And this environment was so empty as to be reduced to the solitude of a conscience incapable of ascertaining whether what he is observing is real or simply a dream orchestrated by a malevolent genius (or an evil scientist, we would say a few centuries later) who produces with masterful skill the sharp images and vivid sensations that Descartes firmly believed he himself was experimenting at the very same moment. It is known that the Cartesian film happily ends confirming that our intellect can -if not perceive a firm and substantial reality as was the case imagined by Plato- at least elaborate a powerful argument with the purpose of convincing the spectator of the existence of that reality, with the cloying Supreme Being at the top, or at the bottom, of everything.

This prologue is not, obviously, the place to deepen into Plato and Descartes' philosophical stands or into their consequences or weaknesses. What I do want to pinpoint is the enormous load of anthropological truth both allegories convey in, at least, one important point: it is rigorously true that we are something like minds imprisoned in a cave reached only by shadows of the external world, a cave whose limits are nothing more than our skin and the fortunate hardness of our skull. The brain... I was about to say the human brain but in reality we are talking about all living brains, those incredible machines blindly designed by natural selection (or by the hand of Whoever we believe in -- it does not matter for this discussion), executes the incommensurable artifice of allowing us to live in a 'world' we experience as perfectly natural, despite this experience consisting purely and plainly in the manner that billions of neuronal connections respond to physical stimuli reaching our sensorial organs.

We are, like Plato's prisoners, isolated in the interior of our brain, in our cranial cave, and the trip to the exterior imagined by the wrestler Aristocles, where we would see things as they are thanks to the enlightenment of a purely noetic faculty, is an impossible trip. We can only imagine, conjecture how that 'exterior world' is, and contrast those hypothetical entities that, according to our imagination, populate that world, with the experiences we should have if those hypotheses were correct. It was another philosopher, this time the Scottish David Hume, a century after Descartes, who said that when cogitating about these matters (though he knew nothing about neurons and synapses) he felt falling in an unbearable abyss, avoidable only by returning to the daily sub-consciousness of chats, business, and games with friends (and even with enemies I would venture to say). But the good man Hume ignored that what allows us to disconnect from these terrifying philosophical truths, and come back to the naïve experience of the every-day life, in which what appears real is real, and in which our peers are enjoyably or painfully real with their own -and our- happiness and grief, is also a trick of that benign prestidigitator, our brain, which is programmed so that the every-day life, the apparent reality of things and people, appeals instinctively to us captivating our soul with its extraordinary verisimilitude.

For this purpose, our brain has an ample set of neuronal circuits which come, so to speak, as 'factory defaults': we are born with them (though, we should more appropriately say that we develop them during our first years, without more help from the exterior than a regular nutrition and a healthy environment). These circuits are the ones which allow us to walk, hold things, learn how to speak, distinguish shapes and colors, and a thousand other skills necessary for our smooth displacement so we can develop an efficient picture of the 'living world' in our conscience -- like a vehicle dashboard with integrated GPS- for navigating within our surroundings. However, there are millions of other things we need to learn about the world around us which are not genetically built in the program which controls the development of our brain, starting from the fertilized ovum from which each one of us came. Those other things are to be discovered by us (in general, -but fortunately not always- with the help of others who already discovered them), and we thus build with them a superstructure of neuronal connections which add to the default ones, in order to integrate the "world we live in". This set of added connections is called our culture.

So the culture we have is nothing more and nothing less than the world in which we live (aside from the most basic and indispensable cognitive software). There exist many arguments, more or less effective, about the usefulness of possessing a "certain" culture or having a "vast" culture (usually associated with knowing a lot of literature, philosophy, history, and some other of the wrongly-called "humanities", when there is nothing more humane than -or at most, equally humane as- botany or astronomy). But for me, the principal argument is that, prisoners as we are in our cranial caves (isolated though not incommunicado), the culture that each one of us possesses is the world... and I, personally, prefer to live in a much richer world, full of fascinating realities and in which, at the same time, I can be capable of comprehending it in its whole complexity, than in a world full of the silliness surrounding the soap-operas and soccer TV programs.

Likewise, I also prefer to have the maximum possible assurance that, when

I build in my brain a specific scaffold to represent a piece of the structure of the "external world", that representative fragment is the best possible... i.e. not just a random invention (I already cheerfully fulfill this need with novels, cinema, TV, etc.) but a solid theoretical framework possessing good arguments to assume that, quite probably, reality is close to what our theory says it is. In other words, I prefer for "my world" to have the highest certification of being a faithful representation of the exterior world. Descartes and Plato searched for absolute, indubitable, unbreakable, benevolent warranties, but after many failures we suspect, if not know, that such guarantees do not exist, that knowledge and business are alike in that, for instance, if the TV set does not work, the promise of getting the money back cannot resist the bankruptcy of the company who sold it to you and, ergo, it is not an absolutely reliable warranty.

In summary, we have to content ourselves with relative warranties: the more different phenomena our theoretical framework can sustain (explain) without collapsing, especially because of new experimental facts, the more probable will be for our theoretical construct to be on the right track. However, our scaffold is not assembled on a solid-rock foundation but, instead (as another of my favorite philosophers, the Austrian Otto Neurath, said in the first half of the twentieth century), is more like a ship which keel has to be constantly repaired while we stay in high seas. The best ship, the best culture, the best knowledge, is the one which allows us to go farther, explore more territories or oceans of reality, and have the least damaging leaks possible.

Naturally, modern science is a critical element of this scaffold because it is based on the quest for theories which could explain and integrate the greatest possible variety of phenomena, as well as predict the largest possible set of experimental data with precision. And, since our culture is the scenario within and from which we interact with others, I am ready to assert that we all have the responsibility of including in our culture those lines of argumentation that Science has built along the centuries, describing what the world in which we live is like. Certainly, mandatory education is one of the main ways to implement such a responsibility, but we would be blind (or, better, would see visions) if we thought that the education our children receive, or we have received in school, is sufficient for us to grasp a minimally acceptable comprehension of what Science teaches about reality. But due to the fact that -fortunately- it is not possible to force people to remain at school after a reasonable age, we have to appeal to individual responsibility and, for this purpose, the best tool is a good book on Science purposely written for the non-professional, like the one the reader has now in his hands (since media, particularly TV, give very coarse glances, if not distorted, and museums and expositions are more appropriate for impressive spectacles than for argumentative comprehension).

Unfortunately, in Spain and the Spanish-speaking countries, few people read books on Popular Science (go to a bookstore and compare the section dedicated to this literature with the volume of books on the so-called "pseudosciences" -ridiculous name, because they do not even look like sciences- or with books on the various forms of mysticism), and this fact greatly discourage the writers so that the already slim demand is normally met with translations. Besides, the institutionalized system for the 'carrera investigadora' (researcher career) existing in our countries stimulates researchers very little, if not nothing, to steal time from their research projects, teaching, and administrative chores so they can sit down and try to explain to the "general public" (who are, after all, the ones who pay the salaries of the scientists) as clearly as possible what those wonderful things they are working-on are.

It is therefore welcome that authors from academic fields or, like in this case, professionals in the field of technological research and development and, at the same time, aficionados to knowing-and-understanding, contribute to improve the scientific culture of our society with an introduction, intellectually accessible and literarily enchanting, to one of the principal components of the scaffold scientists have built for the current understanding of the Universe: the Theory of Relativity. This initiative is the more commendable if we consider that the immense majority of the Popular Science books on the subject normally limit themselves to inform the reader of some of the prominent consequences of Einstein's discoveries, but they are silent, or overly unclear, as to why those discoveries and consequences are valid. In sum, those books inform about the theory, but do not explain it, aggravating and perpetuating a poor comprehension of what doubtlessly is (together with quantum mechanics and the theory of evolution by natural selection) the most strange contribution that Science has made to the understanding of Nature -- despite its rationality to the core.

Given this scenario, *Galloping with Light* is an illuminating exception, and I hope the readers will enjoy it as much as I have. Its reading is pleasant, but it does require some intellectual effort to comprehend and follow the arguments from time to time, even though such an effort should be reasonably easy to attain for anyone who has passed high school. In those instances, the reader should have the conviction that those efforts constitute an indispensable step in order to build -with a solid foundation- the scaffold upon which our neurons maintain a reasonable image of the world afloat -- as reasonable as the most sagacious and fortunate minds of history (and Albert Einstein's was certainly one of them) have been able to build for us.

Jesús Zamora Bonilla

Full Professor of Philosophy of Science at the UNED (Madrid, Spain)
Director of the Master in Scientific Journalism (UNED)
Author of the blog entitled 'A bordo del Otto Neurath'

Madrid (Spain), Fall of 2010

Acknowledgments

This book was first written in Spanish and published in November of 2009 by the 'Ciudad de las Artes y las Ciencias, S.A.' (Valencia, Spain). Many people were behind that Spanish Edition with the only unselfish objective of serving, helping me improve its grammatical precision and literary quality, constructively criticizing its pedagogical letdowns, emphasizing didactic successes to exploit them even more, stimulating me with questions and suggestions, assisting me to redefine its market, detecting and correcting errors, etc. and -not least important- helping me to maintain afloat the good spirits, inner peace, and optimism necessary to finish it. The Old and the New World got together without knowing each other to collaborate with me. Doubtlessly, I am a privileged person.

From the Old World (to where they arrived from the New one), I start with my little sister Laura (Madrid, Spain), who thinks more highly of me than I deserve, and whose specialized knowledge in Linguistics met my needs perfectly for the Spanish project, and even better during my writing of this English Edition: not many scientists are as lucky as to have a sister with a Ph.D. degree in English Linguistics, and who was willing to spend innumerable hours in order to polish my grammar and style[1], distorting neither the scientific meaning nor my pedagogical approach. If you think this book is well written from the grammatical point of view, please give much credit to her; if not, blame it on me -- as I did not follow her advice on many occasions. You will also find that, in a few instances, I chose to use some slightly colorful (non-academic) idioms with the purpose of forcefully expressing my passion for understanding, as well as the frustration I felt while reading most of the existing popular science books on Relativity Theory. These linguistic escapades were also made against Laura's advice.

During the Spanish project, besides her invaluable and meticulous contribution, Laura -together with her husband Gustavo Armijo- introduced me to the intellectual world in my native Spain, (which I left when I could barely utter my first cry) and gave me the privilege of knowing and interacting with Manuel Toharia Cortés (Valencia) and Jesús Zamora Bonilla (Madrid). I wish to express my gratefulness to Jesús for his detailed reading and critique, which helped me produce a better quality book, as well as for his consideration in writing one of the two prologues of this book. Likewise, I am in debt with Manuel for his valuable comments, advice, and thoughtful prologue.

In addition, I am grateful to the following people from the Old World who helped me during the Spanish Edition: Enrique Higalgo (Madrid), Jorge Melluso (Canary Islands), María Teresa Gibert (Madrid), Santiago and Gonzalo Madero (Madrid), Jorge Juez (Burgos, Spain), Héctor Goldenberg (Belgium), Francisco Alba (Madrid), Julio Juez (Burgos), Mónica Aragonés (Madrid), and Raymond Golle (Madrid). My special thanks go to my nephew Joaquín Armijo (Madrid), the artist who combined the body of the 'Master of the Military Arts' (The Cid) with the face of the 'Master of the Scientific Arts' (Einstein), galloping on 'Babieca' throughout *space-time*, represented by the magnificent cathedral of my

1 Needless to say, all the quotations in this book were reproduced verbatim.

beloved Burgos.

From the New World, I had the invaluable assistance of the following persons: my brother Francisco Alba-Juez (San Juan, Argentina), my sister María Alba-Juez (Villa Gesell, Argentina), Oscar Cosatti (Trelew, Argentina), Zulema de Barcala (La Plata, Argentina), Eduardo Casals (Necochea, Argentina), Noemí Pineda (Buenos Aires, Argentina), Sergio Storniolo (San Juan, Argentina), and Roberto Saroglia (La Plata, Argentina).

Last, but not least, I owe much to my friend, lover, and wife Susana, who has been tolerating my passion for studying during 34 years, and whose dedication to the family and limitless work in the prevention and cure of cancer affecting indigent women are a continuous source of inspiration and respect for me, and for everyone who knows her. Undoubtedly, I am a privileged person.

Most certainly, I may have forgotten to explicitly mention many other persons who have assisted me in producing this work, and, if so, I apologize for it. Besides, as it should always be acknowledged (though my attorney wouldn't like it but... what the heck, this is not technology), every mistake or omission are entirely and exclusively my responsibility.

Felix Alba-Juez

Salt Lake City, Fall of 2010

Table of Contents

Introduction

Teaching Science to the Layperson is not Distorting but Educating

Divulgare Tergiversare (teaching the uninitiated is distorting the subject) -- I was always convinced of the veracity of this Latin aphorism. However, after having travelled the arduous and specialized path of Mathematical Physics for over 40 years, here I am writing a Popular Science book on one of the most difficult and detached-from-the-prestigious-common-sense subject that Science has produced in the history of humanity and, what is worse, doing it without the use of mathematics.

To introduce myself, I was born in Spain and grew up in Argentina, where I graduated in Electrical Engineering in 1974. I started my academic career as a Teaching Assistant in Mathematics, and I became a Professor in an Argentinean University for almost a decade, teaching courses on Electrical Circuits, Digital Systems, Industrial Control, and others. In 1983, I resigned from my academic post and was hired by the Department of Metallurgy at the University of Utah, USA, as a Postdoctoral Fellow and later on as a Research Associate.

In 1987, I resigned again from my academic post and founded my own company for the development of novel technologies for measurement and control of particle size in suspensions and emulsions. The significance of nanotechnologies on the future of society was clear by then. In the last 27 years I have worked in collaboration with private companies and governmental institutions like DuPont, ALCOA, US Department of Transportation, Dow Chemical, and NASA, which partially financed/cooperated on the development of the necessary prototypes to prove the new concepts and to patent the novel instruments.

Finally, I have recently transferred all my patents and intellectual property to Agilent Technologies, Inc. As a result, my scientific and commercial website was closed in 2008. At the beginning of the same year I started my dream of finishing my life explaining the most profound concepts in Science to the big public, and decided to start doing so by writing this book about Einstein, Relativity, and *Folklore*. By *Folklore*, I mean the set of popular (and scientific) beliefs, mostly erroneous, associated with Relativity Theory and with our scientific activity in general. If you are reading this book in its electronic version, you have at your disposal -via Wikipedia and the Internet in general- a virtually infinite world of **objective** knowledge but... I remind you that, by the same token, you are exposed to a cornucopia of *folklore*. The epilogue of this book is dedicated to help the reader in distinguishing between **Science** and *pseudoscience*.

In this new project as a writer of Popular Science, my objective is to reach the mass reader, that non-scientific person with the curiosity of understanding the Universe in which we live, who has the habit of reading and thinking and the respect for his/her own capacity of logical analysis and comprehension, and who is conscious of the relevance Science carries in current society, and the essential role it plays as a modeler of our destiny as a biological species.

Ever since I developed an interest in Science during my adolescent years, I have always had an incontrollable desire for understanding or, better, for convincing myself that I could understand something -- as it took me not very long to realize it was impossible to understand everything. Around my 14 years of age I avidly read a little book on Relativity Theory without understanding a thing. However, I was perplexed, because the three concepts about which the author elaborated upon over and over were precisely those which every 14-year kid had to be already intimately familiarized with, by the sheer fact of growing up: *time, distance,* and *speed.* Even so, what I was reading was impossible for me to understand and, even though the author avoided any mathematics, he constantly was saying that "there was a formula with which, had the reader known mathematics, what he was saying would have been much easier to understand."(?)

What did I learn from that frustrating juvenile experience? That the author expected me to employ what every one of us (even adults) understands by *time, distance,* and *speed* to reach conclusions which only could be considered untrue and preposterous precisely because, had we considered them credible and sensible, we would have been compelled to throw away our millenary understanding of those three cherished concepts. What a peculiar approach, I said to myself: if the conclusion is going to destroy our deepest convictions, why not start analyzing those convictions, attempting to grasp the experiences, basic principles, and hypotheses that made us erroneously though vehemently believe in them for thousands of years?

Despite my virgin intellect in those years, it seemed evident to me that Einstein had achieved something more than a revolutionary physical theory and that I was very far from understanding both the theory and that 'something more' -- whatever it was. This is how my compulsive interest in epistemology[1] and the philosophy of science started. At the same time, I became convinced that I needed to acquire a solid command of Mathematics and theoretical and experimental Physics, and while traveling the arduous and specialized path of acquiring those skills and developing new technologies for over 40 years, my absolutist appreciation of the meaning of the adage *Divulgare Tergiversare* was little by little inconspicuously evolving.

I had started in full agreement with Ernesto Sabato, who masterfully expressed the meaning of the proverb in his excellent book entitled '*Uno y el Universo*' (*One and the Universe*) on its segment regarding Popular Science:

> *Somebody asks me to explain the Theory of Einstein. With much enthusiasm, I talk about tetra-dimensional tensors and geodesics.*
>
> *- I have not understood a word - he tells me stupefied.*
>
> *I reflect for a moment and then, with less enthusiasm, I give him a less technical explanation, preserving some geodesics, but including aviators and revolver shots.*
>
> *- I understand almost everything -says my friend, partially content-. But there is something I still do not understand: those geodesics, those coordinates...*
>
> *Depressed, I fall into a long and deep cogitation and end up abandoning forever those geodesics and coordinates; with real ferocity, I exclusively focus on aviators that smoke while traveling at the speed of light, manag-*

1 Epistemology is the discipline which studies the nature, foundation, and limitations of *knowledge*.

ers of railroad stations shooting guns with their right hand while checking time with chronometers on their left hand, trains, and bells.

- Now yes, now I understand Relativity! - shouts my friend in happiness.

- Yes -I respond, bitterly disappointed-, but now, what I told you is not Relativity any longer.

With time, from being a passionate defender of the idea that teaching Science to the big public necessarily leads to distortion, I was gradually reaffirming my childhood's suspicion that one can **know** very much but **comprehend** very little and, besides, that different objectives require different levels of knowledge -- though always with the maximum possible comprehension suited to the purpose. The importance of Popular Science clearly revealed before me when reading the correspondence between Faraday and Maxwell. The ability Faraday had to conceive and execute experiments was as magnificent as his lack of dexterity in Mathematics and Theoretical Physics; his objective was to understand the Universe through experimentation. Being 66 years old, he wrote to his disciple Maxwell (the greatest theoretical physicist of the nineteenth century who was then only 26) the following:

I would like to ask you something. When a mathematician involved in the investigation of physical actions and their effects issues conclusions, would it not be possible for him to explain them in common language; in a way as clear, complete, and definitive as when she does it with mathematical equations? If that was the case, would it not be an advantage for people like me? -- translating them from their hieroglyphic form, so we also can work with them in an experimental way. I believe that is the way it should be, because I have always thought that you could transmit clearly your conclusions and, even though I might not understand all the steps of your process, I would be allowed to work and think about it. If this were possible, would it not be a good thing that mathematicians who work on these subjects gave us their results in a popular and useful form, in the same way they do it with their own mathematical language?

Along my professional life, I confirmed that the command of our language is crucial to focusing our thoughts and communicating them with precision to others. I fully agree with Ludwig Wittgenstein in that many of the 'great philosophical problems' do not really exist simply because they are semantic confusions created by the inappropriate use of language -- which is patently visible in the false dichotomy between *theory* and *practice* and in the fallacy of likening the observable **behavior** of the electron **as** a particle or **as** a wave to the ontological[2] assertion that the electron **is** a particle or **is** a wave.

Finally, Henri Poincaré in his book *Science and Hypothesis* triggered my starting the dream of finishing my life explaining deep scientific and epistemological concepts to the big public. Poincaré said:

...But much advantage will accrue if men of science become their own epistemologists, and show to the world by critical exposition in non-technical terms of the results and methods of their constructive work, that more than mere instinct is involved in it: the community has indeed a

2 Ontology is the discipline which studies the concept of *being* or *existing*.

right to expect as much as this.

Aware of the transformation taking place within my soul, in 1996 I had the opportunity of reading the updated and expanded version of the famous *A Brief History of Time* written by the prestigious theoretical physicist Stephen Hawking, and published in commemoration of the tenth anniversary of his original work. In a few words: I was disappointed. The book appeared to me as an excellent revision of the most important theories in Physics for the layperson, but in an exclusively *informative* fashion: not even with my scientific professional background was I able to find a route to improve my conceptual understanding of the ideas and theories the book described. As soon as in its *Foreword*, the author referred to a proposal affirming that "...the universe has no boundaries or edges in the *imaginary time* direction" and, after checking the book glossary for the meaning of *imaginary time* (in a desperate attempt to make sense out of such a statement) I found: "time measured using *imaginary* numbers". After this early experience, very little doubt remained in my mind that, even though I would read the book with interest, my understanding of the Universe would not increase substantially. I merely updated myself on what had been going on in Theoretical Physics while I had been busy developing new technologies. But the way Hawking ended his book did impress me:

However, if we discover a complete theory, it should in time be understandable in broad principle by everyone, not just a few scientists. Then we shall all, philosophers, scientists, and just ordinary people, be able to take part in the discussion of the question of why it is that we and the universe exist. If we find the answer to that, it would be the ultimate triumph of human reason - for then we would know the mind of God.

That was quite a statement in 1986; in 1980 -in his lecture entitled "Is the end in Sight for Theoretical Physics?"- he had said: "we may see a complete theory within the lifetime of some of those present here", and had claimed that there was a fifty-fifty chance of such a *theory of everything* being found in the next twenty years.[3] From my humble point of view, after reading such an ending assertion, I could have made another instinctive prediction: that there was a fifty-fifty chance that in 500 more years such a statement would still be as remarkable as it was in the 1980s.

Regardless of my personal impressions, when his tenth anniversary edition was published, the original one had sold one copy per every 750 human beings on this planet! The sale of more than nine million copies, leaving aside the indisputable caliber of Hawking as a physicist and writer, could only indicate another equally or more undisputable fact: the big public is thirsty of understanding the Universe, and begs for help from scientists to achieve a sense of existence and transcendence in this our mysterious habitat.

When those twenty years passed in the year 2000, Hawking changed his mind and gave another twenty years chance to his fifty-fifty prediction. However, barely two years later, in 2002 -well before the deadline for his new prediction- he radically changed his position acknowledging that the famous Gödel's incompleteness theorem[4] could impose a severe limitation to achieve a *com-*

3 As cited in *The Physics of the Impossible* by Michio Kaku, Anchor Books, 2009.

4 Gödel's incompleteness theorem asserts that every mathematical system or theory will always contain true

plete theory. In 2005, Hawking published his *A briefer History of Time* in which he eliminated the most obscure parts of the previous book, and included plenty of illustrations -- though he preserved the sheer *informative* character of his work. In addition, despite his dramatic change in philosophical position, he left his above-cited ending statement intact.

Reaching my fifties (1998), I had demonstrated to myself I could understand everything I wanted and, combining Theoretical Physics with my training as an engineer, I had developed three novel technologies for measuring particle size in suspensions and emulsions -- a problem of great importance in a wide variety of industries. At the same time, my interest in the development of technologies had been gradually fading, and my past experiences -as an adolescent with the little book on Relativity and as an adult with Hawking's- were painfully present in my mind as witnesses of an unfinished task. I thus started reading Popular Science books on Relativity as well as monumental works on the philosophy of *space* and *time*, like those of Hans Reichenbach and Adolf Grünbaum and, of course, the original writings of Einstein (technical as well as popular). With some astonishment, I discovered that, forty years after my adolescence experience and almost a hundred years after the *annus mirabilis* of Einstein (1905), the state of affairs had not changed much: 90% of the Popular Science books I read were plagued with erroneous folklore and full of apparent paradoxes. They were purely and simply an offense to the reader's intellect. I also found with disgust another class of literature in which the authors, taking advantage of the confusion between *relativity* and *subjectivity*, and Einstein's popularity and prestige, attempted to validate preposterous assertions in fields like psychology, morality, spirituality, sociology, literature, art, etc.

In one of the popular lectures on General Relativity that Einstein gave in Berlin after becoming a celebrity due to the favorable results on the deflection of light during the eclipse in 1919, a person in the audience asked him: "What is the meaning of *potential*, *invariant*, *contravariant*, *energy tensor*, *scalar*, and *inertial system*; could you explain those terms quickly?" to which Einstein responded: "Of course!, all those are technical terms". There could not be a more rapid response! And Einstein gave it with his habitual sardonic craftiness because he knew very well that swiftness is the enemy of comprehension and he had made the mistake of using those highly specialized terms while talking to the general public. After reading this anecdotic tale, I said to myself: doubtlessly, what the non-scientific reader needs (as well as the scientist) is a book giving more importance to the intuitive physical ***meaning*** of the *words* than to the unmeasured erudition an author can display with them.

And, now, while writing the English edition for this book in 2010, a new Popular Science book by Hawking appeared in the market with a very appealing and intriguing title: *The Grand Design*. As expected, it soon became one of the ten science best-sellers in Amazon.com -- though many of the on-line reviews are less than favorable. Having renounced to his dream of a *complete and ultimate theory* of the Universe, he presents the so-called *M-Theory* as "not a theory in the usual sense" but "a whole family of theories", and promises at the outset

statements which cannot be proved as such within the system, i.e. a new more general theory is required for such a proof, but the new theory will -in turn- contain other non-provable true statements requiring another theory... and so forth *ad infinitum*.

that -because this theory asserts that "many universes[5] were created out of nothing"-, it "may offer answers to the question of creation". He continues: "Their creation [of the universes] does not require the intervention of some supernatural being or god". He goes on: "...bodies such as stars and black holes cannot just appear out of nothing. But a whole universe can..."

Unfortunately, I finished its reading with the sensation that not even the authors believe in their guts that our Universe (and the many others they talk about) was "created out of nothing" -- not because I do or don't believe that, but because they do not explain what they mean by the words 'creation', 'nothing', and 'god'. All those three terms are deeply philosophical and, as such, every soul in this Universe has a different idea about what their ultimate meaning is. In short: another big disappointment -- this time in my sixties.

My instinctive conclusion when I read my first little book on Relativity at 14, kept resonating in my brain at 55: the fact that the reader may not have a scientific education, does not mean s/he does not have the capacity to understand profound concepts, as long as they are discussed from the epistemological point of view and we start analyzing the semantics of the words employed when referring to them. After all, it was Einstein who said that Science is simply the refinement of our intuition and daily experiences. My objective thus is **not** to turn the reader into an expert on Relativity; on the contrary: that would require a much bigger effort than simply reading this book. My purpose is to show that the Theory of Relativity, experimentally confirmed in the last hundred years, regardless of how strange and opposed to our prejudices -disguised in the mask of 'common sense'- may seem to be, is rational, consistent, and intelligible for the layperson -- if, and only if, s/he has the audacity of accepting the unfounded nature of those preconceptions.

It is my desire and the honor of stimulating in the non-specialized person such a necessary intellectual boldness that have motivated me to write this book. These intellectual strength and courage have nothing to do with our academic or professional credentials; even more: had I believed they are strictly necessary to seize a concrete and positive message out of this book, I would have not gone through the trouble of writing it. I thus dream for the reader finishing this book with the sensation that it is possible to achieve an acceptable understanding of the Theory of Relativity based on the **objective** *truth* and free of *folklore* (popular as well as scientific). And... if, after some time, the reader feels the urge to read my book again so as to strengthen his/her understanding, my ephemeral stay in Plato's cave will have had the sense and transcendence that all of us seek for our existence.

As I clearly see it now, teaching the layperson (*divulgare*) is not distorting (*tergiversare*) the subject, but educating the public; and it is our duty as scientists to **educate** without *distorting* the essence of the scientific knowledge attained by humanity. The future of our society depends upon this premise.

Felix Alba-Juez

Salt Lake City, Fall of 2010

5 I thought the semantics of the word *universe* did not admit the existence of more than one...

Chapter 1

$E = m.c^2$
The Most Famous Equation in History... and its Folklore

Sir: ...This new phenomenon would also lead to the construction of bombs, and it is conceivable -though much less certain- that extremely powerful bombs of a new type may thus be constructed[1]. Albert Einstein (1939).

The most famous equation in history (though not in its current form) was published by Einstein in his *annus mirabilis* (miraculous year) of 1905[2] as a kind of corollary to the famous previous publication[3] where he presented what later would be known as the *Special Theory of Relativity.*[4]

The "new phenomenon" alluded in the letter to the President of the United States was a nuclear chain reaction conceived by Leó Szilárd, and the only reason why Einstein signed the letter (and not Szilárd) was because of his scientific stature and popularity, and also because he was concerned about solid evidence implying that Germany was working on the atomic bomb. This and a second letter were the only contributions of Einstein to the history of the atomic bomb -- since he never participated either in nuclear fission basic research or in its posterior technological development.

Five years before this letter, in 1934, while giving a lecture to the American Association for the Advancement of Science in Pittsburgh, Einstein was asked whether it would be possible to liberate the tremendous energy predicted by his famous equation, to what Einstein responded: "NO; for practical reasons". Even more, when Szilárd suggested to Einstein the possibility of a nuclear chain reaction, he said: "It never occurred to me something like that could be possible". Notwithstanding these historical events and the scientific truths we will discuss in this chapter, the most famous equation in history is intimately linked to the atomic bomb and nuclear energy in general.

Even though by the end of the 1930-1940 decade the equation $E = m.c^2$ had been confirmed experimentally, it only went from the scientific field to the public domain in 1945 when two atomic bombs were dropped in Japan by

1 Fragment of the first letter to President Franklin D. Roosevelt signed by Einstein and written by Leó Szilárd on August 2, 1939.

2 A. Einstein, *Does the Inertia of a Body depend on its Energetic Content?* Annalen der Physik, 18:639-641, September 27, 1905.

3 A. Einstein, *On the Electrodynamics of Moving Bodies*, Annalen der Physik, 17:891-921, June 30, 1905.

4 The adjective *special* is used with the meaning of *restricted* (as opposed to *general*).

the United States, ending the Second World War. The press and all news media incorrectly referred to the equation as the foundation, archetype, reason, and synthesis of the immense power of nuclear energy -- initiating a body of erroneous folklore that remains among us to this very day.

The Folklore:

1) The atomic bomb converted matter (tangible) into energy (intangible) and this was possible due to the secret that Nature had zealously kept away from us but was finally discovered by Einstein and expressed in the equation $E = m.c^2$; and 2) The equation was the key for the technological development of the atomic bomb.

The reader may be surprised to know this folklore does not exclusively belong to the layperson: it finds itself pretty healthy among scientists, including Nobel Prize winners, as can be confirmed in http://www.pbs.org/wgbh/nova/einstein/experts.html.

The Reality:

1) Matter was not converted into energy: not even a single elementary particle in the 50 kg of Uranium-235 wrapped up in 5 tons of cordite[5] and steel, aboard the "Enola Gay" on August 6 of 1945, disappeared of this world to become energy; 2) the equation $E = m.c^2$ is as pertinent to describe the physics behind the public display of destructive power in 1945, as to describe the phenomenon that took place during the equally-transcendent though solitary moment when man discovered fire 1.5 million years ago; and 3) Neither the scientists of multiple nationalities who participated in the successful development of the atomic bomb here in the United States, nor those scientists who attempted the same in Germany, needed to know the existence and validity of the most famous equation in history in order to reach technological success or failure.

The hype given by the Press to this transcendent and public event as well as its consequent popular interest in 1945 was because the *nuclear reaction* behind the atomic bomb releases (pound per pound) **100 million times** more energy than the *chemical reaction* in the process of combustion does, and therefore the former displayed destructive and constructive possibilities beyond comprehension for the *lay-Homo Sapiens* of those days.

Likewise, it is not difficult to imagine the perplexity of the *Homo Erectus* when s/he discovered fire with all its destructive and constructive possibilities, both completely unintelligible for the average *Homo Erectus*. Since then, all the power of fire has been developed and controlled by humanity much before the most famous equation in history was published. The Press did not exist 1.5 million years ago and the fortunate and/or sage discoverer of fire never in her/his life had to get defensive for the destructive features of the discovery. Neither the Hiroshima bomb nor the Nagasaki one, dropped on August 9, 1945, could go unnoticed by humanity and much less ignored by the Japanese people; the Japanese government surrendered to the Allies on August 15, 1945.

---

5 Smokeless powder comprised of nitroglycerine and wool powder, mixed with acetone into a paste and finished to look as a rope.

In spite of the aura of mystery and veneration around it, the equation is extremely simple mathematically, containing three symbols for three physical concepts intimately related to our intuitive and daily world: *mass*, *energy*, and *velocity*. As we shall see, the first one (from *massa* in Latin) is the main culprit behind the erroneous folklore associated with the equation, due to the various meanings that science and popular language alike have given to the word throughout history. The second (from the Greek *energos* meaning 'activity') has a much simpler semantic history. As for the third, the concept of *velocity* is so engraved in our daily routines that it does not require (so far!) further commentary. It is also taken for granted that the reader understands without ambiguity (at least intuitively) the concept of *acceleration* as the rate of change of *velocity* in time.

Finally, we accept the intuitive and classical notions of *space* and *time*; notions that we will change gradually, in subsequent chapters, to the point of believing we have lost our proud common sense and precious mental health. To be able to communicate with each other, I will begin by agreeing in the language we will use to analyze in depth the most celebrated equation in history.

1.The term 'mass' and its semantics throughout history

During Roman hegemony, the Latin word *massa* referred to something material and palpable. In Newton's days (1643-1727) the term *mass* evolved to meaning the **quantity** of that material something. Both Newton and Euler (1707-1783), who defined the concept of *inertial mass* for the first time, clearly understood the concept of *inertia* of a body, and today all of us, scientists or not, understand it without difficulty and daily perceive it as the reluctance of a body to change its speed. For instance, if the body is *at rest*, it resists starting to *move* and, if it is *moving*, it resists being *stopped*. Because of that reluctance, we hold tight while standing on a bus since upon its sudden stop we would be violently and mysteriously pushed in the direction the bus had before the unexpected brake. Contrariwise, while the vehicle is moving at a ***constant** velocity* on a road without bumps, we can freely move without danger of falling and even hold a glass of water without spilling it.[6]

Nonetheless, the idea that a body with more *mass* has **necessarily** more *matter* (besides having more *inertia*) comes naturally from the referred meaning of 'quantity of matter' for the term *mass* and also from our limited daily experience, and therefore sounds as natural as correct -- not only for the layperson but also for the not-very-rigorous scientist.

In Einstein's days, there already existed five different meanings for the word *mass*; one crude, non-scientific meaning: *something palpable*, and four scientific ones: *quantity of matter*, *inertial mass* (reluctance to change its *speed*), *active gravitational mass* (attraction of other bodies), and *passive gravitational mass* (attraction by other bodies). Newton already knew that the last three types of *mass* were all related by a universal constant, i.e. the numerical ratio between any two of them is always the same number and, by the beginning of the twentieth century, this curious relation used to be expressed by saying

6 Even so, every bus driver or responsible flight attendant will advise us to apply caution while being out of our seats and, when seated, keep our safety belt fasten to protect us from unexpected bumps on the road or turbulence in the air.

that the three *masses*, though conceptually different, had the same numerical value[7]. The meaning of this mysterious numerical equality will be discussed later in this chapter.

What is the situation while writing this book in 2010? If we look for the word *mass* in the Merriam-Webster on-line dictionary (www.merriam-webster.com), we find as acceptation[8] '1a': "A **quantity** or aggregate of matter usually of considerable size" and, as meaning '1c' (most suited for our purposes): "the property of a body that is a measure of its inertia and that is commonly taken as a measure of the **amount** of material it contains and causes it to have weight in a gravitational field".

The meaning 'amount of material' is the major contributor to the erroneous folklore accompanying the most famous equation in history. For instance, it is difficult to understand why a theoretical physicist of the caliber of Stephen Hawking, in his renowned *A Brief History of Time* (Chapter 2), added 'quantity of matter' between parentheses right after the word *mass*, for the sake of clarification. Even more: in the glossaries of both his original book and his latest *A Briefer History of Time*, the first acceptation for the word *mass* is 'quantity of matter'. As we shall see, this piece of folklore is as interesting as subtle, and it persists among the collective mind because it is more than the mere result of a semantic confusion[9]. Finally, the dictionary gives multiple other acceptations of a domestic, political, sociological, or technical nature with no relation to its meaning in Einstein's equation.

We need thus to understand the meaning of the word *matter*, for which we find in the dictionary the meaning '2a': "The *substance* of which a physical object is composed". Acceptation '2b' instead reads: "**material** *substance* that occupies space, has mass, and is composed predominantly of atoms consisting of protons, neutrons, and electrons, that constitutes the observable universe, and that is **interconvertible** with energy". The alert reader is probably already sensing how hard it is to define concepts as basic as these: regarding '2a' meaning, search for *substance* and you will be disappointed, because it is **circularly** defined in terms of *matter*. As for '2b', it is logically vitiated because it uses the word *material* to define *matter* as well as the word *mass* (which forced us to look for *matter* in the first place). Even worse, '2b' affirms that *matter* and *energy* are **interconvertible**, which is one of the erroneous folkloric beliefs I have already denounced.

In order for us not to lose our mental sanity before starting this book, let us forget the dictionary for this case and agree that *matter* distinguishes itself by its **tangibility**, i.e. by occupying *space* and being directly perceivable by our senses. When the occupied *space* becomes too small for our senses to perceive it directly (i.e. when dealing with the microscopic world), we will have to use other criteria based on the knowledge acquired within our macroscopic world.

Returning now to the word *mass*, we ask ourselves which of all these scientific uses of the term (quantity of matter, inertial, gravitational) is the one that Einstein had in mind when he developed his equation and why he included the

7 The proper selection of the measurement unit makes the three masses numerically equal.
8 An acceptation is an accepted meaning of a word or understanding of a concept (www.merriam-webster.com).
9 Einstein himself contributed to the folklore, though inadvertently, when he referred sometimes to the "Principle of Equivalence between Matter and Energy" in his writings.

term *energy* but not the term *mass* in the title of his famous publication *Does the Inertia of a Body depend on its Energetic Content?*[10]

If the equation presented by Einstein for the first time since the discovery of fire established an unknown relation between two basic **attributes** of a physical system -*mass* and *energy*- (so relevant for combustion and many other known and to-be-known phenomena) why did he include the word *energy* but avoided using the word *mass* in the title? Beyond doubt, Einstein knew of the chameleonic character of the word *mass* and, ergo, relegated its appearance to the paper's content where the specialized reader would supposedly know its precise meaning. But then... which was the meaning he had in mind? Well, by now, even our respected *Homo Erectus* would agree that Einstein was talking of the *inertial mass* because the word *inertia* (Trägheit in German) was meticulously selected on the title together with the term *energetic content*.

Notice the term **attributes** I chose above to emphasize that both *mass* as well as *energy* are not independent entities but two crucial **properties** of *matter*. In proper English: it is correct to say that something **has** *mass* and/or *energy* but it is incorrect to say that something **is** *mass* and/or *energy*. Now that we know what Einstein meant by *mass* in his paper (inertia), let us elaborate a little about what is intuitively understood by *inertial* and *gravitational masses*.

Inertial mass is the physical **property** that expresses the *force* necessary to impart a specific **change** of *motion* to a body. This is the verbal expression of Newton's well-known Second Law: *force* is equal to *mass* times *acceleration*, which is in accord with our intuition that the bigger the *inertia* of a body, the more *force* we need to apply, so as to modify its *velocity* (accelerate or decelerate it). As for the concept of *force*, the intuition associated with our daily experience of *pushing* and *pulling* is enough for now. Let us also define the concept of *field of force* or simply *field*, as a region of *space* in which a body would experiment *forces* which may vary with *position* and *time*.

Gravitational mass is the physical **property** upon which the *force* with which a body attracts or is attracted by other bodies depends. This is the verbal expression of *Newton's Law of Universal Gravitation*: the *force* of attraction between two bodies is proportional to the product of their *masses* and inversely proportional to the square of their *distance* (*distance* multiplied by itself). As an example: if any of the *masses* is **doubled**, the *force* of attraction is **doubled**; if one of the *masses* is **three** times **smaller**, the *force* is **three** times **smaller**; if the *distance* is **tripled** and the *masses* do not change, the *force* of attraction is **nine** times **as small**.

1.1. The Difference between Mass and Weight

A concept closely related to *gravitational mass* is that of *weight*, to the point that even the careful scientist uses both terms indistinguishably during his daily work. The pertinent acceptation given by the Merriam-Webster dictionary for weight is '6b': "The *force* with which a body is attracted towards the earth or a celestial body by gravitation and which is equal to the product of the mass and the local gravitation *acceleration*".

10 A. Einstein, *Does the Inertia of a Body depend on its Energetic Content?* Annalen der Physik, 18:639-641, September 27, 1905.

It is clear then that *weight* **is** a *force* and, ergo, cannot be confused with the concept of *mass* (an attribute of *matter*). However, on our planet, as long as we do not consider large regions (laterally or vertically), the *force* (*weight*) exerted by the planet on a body is proportional to its *gravitational mass*. As an example, if the *mass* **quadruples**, the *weight* **quadruples** with the local gravity defining the proportionality factor. For this reason, the use of *mass* or *weight* indistinctly, although wrong, has no pernicious consequences. To aggravate the semantic confusion, the unit 'kilogram (kg)' is used to measure *gravitational mass* (kg-mass) as well as to measure *weight* (kg-weight). Given that the words after the hyphen (mass and weight) are simply modifiers of the noun 'Kilogram', they are rarely used -- making the difference between both concepts even more obscure.

To understand the abysmal difference between both concepts, notice that a given body has the same *gravitational mass* here as on the Moon, while its *weight* here is greater than over there. And, why is that? Because, even though the body has the same *gravitational mass* here and there, the Moon's *gravitational mass* is much lower than the Earth's and the *force* of attraction (*weight*) between the object and each one of the mentioned celestial bodies is proportional to the product of their *gravitational masses* (those of the small body and the celestial one). For instance: if the *mass* of any one of the two bodies **quintuples**, and their *distance* does not change, the *force* of attraction **quintuples** as well. Consequently, given that the Earth gravitational mass is about 81 times greater than that of the Moon, the weight of a body located at the same distance from the center of the Moon than it was from the center of the Earth will be 81 times smaller over there than over here.

I can not help thinking of the curious reader who would be wondering whether what the balance (scale) in a grocery store measures is the *weight* or the *mass* of our food. Looking up *balance* in the dictionary we find as its first acceptation: "Instrument for *weighing*: as a: a beam that is supported freely in the center and has two pans of equal weight suspended from its ends and as b: a device that uses the elasticity of a spiral spring for measuring *weight* or *force*".

Etymologically, the word *balance* comes from the combination of two Latin words: *bis*, meaning *two*, and *planx* meaning *plate*. This means the original instrument had two pans on one of which we put the object to be measured, and on the other of which we add or take away known 'weights' until the *forces* exerted by our planet on both pans are equal, thereby declaring the object's *weight* as equal to the sum of all the standard 'weights' we used to attain equilibrium. Thus, according to this operational description and the dictionary, a *balance* measures *weight*, but... this would imply -as explained before- that if we took the balance, object, and standard 'weights' to the Moon we would obtain a *weight* much lower than here on Earth. True? False!

It is true that the Moon *gravitational mass* is 81 times smaller than the Earth's one, thereby producing a *force* of attraction 81 times smaller on the to-be-measured body (at the same distance). But this smaller *gravitational mass* of the Moon equally and proportionally affects both the *force* of attraction on that body, as well as on the standard weights (and the whole instrument!), with the final result that the equilibrium is achieved with exactly the same number and type of standard weights and, ergo, delivering on the Moon exactly the same result for the **property** we are trying to measure than on Earth! From this, we have to conclude, against common belief, that what this type of *balance* is measuring is the *gravitational mass* of the body and not its *weight*. For the same reason, what we call 'standard *weights*' are, in reality, 'standard *masses*'.

To complete our comprehension and be fair to the dictionary (acceptation 1b), the 'spring scale' is instead an old instrument that really measures *weight* because it is based on the *expansion* experienced by a spring when we hang the to-be-measured body from it, and this *expansion* is proportional to the *force* exerted on it by the celestial body. Given this operational principle, the spring extension will be about six times shorter[11] on the surface of the Moon than on Earth's and therefore the result for the measurement will be the *weight* and not the *mass*. In fact, there is no need to go to the Moon; if a spring scale is sufficiently sensitive, it will deliver -for the same object- a **greater** *weight* on the North Pole than on the Equator of our planet, because the *force of gravity* on the former is **greater** than on the latter[12]. The *force of gravity* for a given object is also different on the top of a mountain than at sea-level or inside a deep mine[13]; it does even depend upon the geological composition of Earth.[14]

1.2. The Equality between Gravitational Mass and Inertial Mass

Newton knew that the three *masses* (active/passive *gravitational* and *inertial*) were numerically the same, a fact considered quite curious and inexplicable. The equality between active and passive *gravitational masses* is easy to understand given the symmetry of *Newton's Law of Universal Gravitation*, which states that the *force* between two bodies is proportional to the product of the two *masses* and inversely proportional to the square of the *distance*. Given that multiplication is a commutative operation (i.e. A.B=B.A) and *distance* is naturally reciprocal (i.e. d(A,B)=d(B,A)), it does not matter which *mass* is active and which is passive.

The equality between *gravitational* and *inertial masses* is much less obvious. In spite of Aristotle's negative influence (he used to teach that bodies with **greater** *mass* fall **faster**), there is historic evidence that Ioannes Philiponos (490-570 AD) had noticed that bodies whose weights were very **different** reached the floor in *time intervals* which were very **close** to each other. Giambattista Benedetti (1530-1590) proposed in 1553 the **equality** between *inertial* and *gravitational masses* in order to explain the **equality** of the arrival *times*, and Simon Stevin put it to the test in 1586.

Let us imagine we let two bodies with different *gravitational masses* fall from the same height and that the air does not impose any friction[15] (or imagine we are on the Moon where there is no atmosphere). The initial attractive *force* exerted by the Earth (or Moon) on both bodies will be proportional to their *gravitational masses* with the proportional factor being the same (*dis-*

11 The radius of the Moon is about 3.67 times shorter than the Earth's, so that 81/(3.67x3.67)=6.

12 In Newtonian terms, it can be explained because, due to our planet rotation, the centrifugal force is greater on the Equator than on the poles (which don't rotate at all) reducing the force of attraction on the former.

13 Inside the mine, the net force on the object is the sum of gravitational forces coming from all around it.

14 Accurate measurements of gravity are used to detect the presence of mineral resources.

15 Friction is another concept whose intuitive meaning we assume grasped by the reader.

tance is the same and they are interacting with the same celestial body), i.e. the body with the **greater** *gravitational mass* will be attracted down with the **greater** *force*.

On the other hand, the *acceleration* gained by each body will be the **greater**, the **greater** the *force* and the **smaller** its *inertial mass*. From this, we conclude that if the *gravitational* and *inertial masses* of those two objects (in fact any object) were not equal, the body with **greater** *gravitational mass* would not reach the floor necessarily at the **same** *time* as the body with a **lesser** *gravitational mass* -- because if the *inertial mass* of the first were sufficiently **big** and the *inertial mass* of the second sufficiently **small**, the **bigger** *force* of attraction on the first body could be compensated and surpassed by the effect of its **greater** inertia, reaching the floor after the body with **less** *gravitational mass* (**less** *force* of attraction) but also **less** *inertial mass* (**greater** *acceleration* for a given *force*).

In more compact terms: the *force* acting on a body in free-fall is, on the one hand, proportional to the product of its *gravitational mass* with the *gravitational mass* of our planet and divided by the square of the *distance* between them (Newton's Law of Gravitation) and, on the other hand, it is proportional to the product between its *inertial mass* and its falling *acceleration* (Newton's Second Law). Equating both expressions for the *force*, we conclude that if the *gravitational* and *inertial masses* are equal, they cancel each other out and the *acceleration* of the body in free-fall is the **same** regardless of the actual value of its *mass*. Finally, if the falling *acceleration* for two bodies of **different** *masses* is the **same** throughout their trajectories when **simultaneously** released from the same elevation, the two objects reach the floor at the **same** *time*.

The legend goes that Galileo (1564-1642) let two objects with very **different** *mass* fall from the Leaning Tower of Pisa, showing that they reached the floor at the **same** *time*. Reality is that Galileo inferred this experimental fact letting small polished spheres fall on inclined planes[16] -also thoroughly polished to minimize friction- and measuring the *time* taken by them to reach the floor. Newton himself, in 1680, measured the oscillation *periods* of various pendulums with the **same** *length* but **different** *masses* and **different** *materials* (wood, gold, silver, etc.) finding that all their *accelerations* were **equal** within 0.1%, thereby elevating this experimental equality to the category of *fundamental principle* in his master piece [7].

Let me reiterate: the only way that any two bodies of any *material* and any *mass* may fall with the **same** *acceleration* so as to reach the floor at the **same** *time* (assuming again friction is nil) is if, for all of them, their *gravitational* and *inertial masses* are **identical** so that the difference in *acceleration* produced by the different *mass* (due to the different *force* of attraction) is exactly compensated by the difference in *acceleration* produced by the different *inertia*. More pithily: a *gravitational field* is equivalent to a *field of accelerations*, because **all** bodies of **all** *materials* and **all** *masses* have the **same** *acceleration* in the same place. This *acceleration* on our Earth and at sea-level is approximately 9.81 meters per second every second (9.81 m/s^2). For example: if we let the object fall from repose (zero initial velocity), after the first second its

16 He used inclined planes to prolong the *falling time* so as to measure it with greater accuracy.

speed will be 9.81 m/s regardless of its *mass*; after two seconds it will be 19.62 m/s, and so forth. Given then that the *acceleration* is independent of its *mass* and *composition*, its trajectory is determined exclusively by its initial *position* and *velocity*.

Curiously enough, we use the verb ‘to plummet’ in English, and ‘desplomarse’ in Spanish, both coming from the Latin word *plumbum* alluding to the heavy chemical element *lead*, to refer to a fast or noisy collapse of something, giving the wrong impression that the heavier an object is, the faster it falls -- as per Aristotle’s teachings. It is not uncommon that linguistic expressions which convey an idea clearly and efficiently (that of a quickly falling body) propagate for centuries or millennia, even though the expression in itself simultaneously evokes a totally false statement.

After this exhaustive discussion on the reasons why two objects with **different** *masses*, absent of friction, reach the floor at the **same** *time*, I cannot but mention the simplicity and elegance of an argument Galileo gave in his *Discourses and Mathematical Demonstrations relating to Two New Sciences*. In essence, the idea was that, if we split a body in two equal parts, we get two bodies with **equal** *mass* but **half** that of the original. Assuming Aristotle was right, then each half with identical *mass* falling side by side would reach the floor at the **same** *time* while the original body (with **double** *mass*) would take **less** *time* to reach the floor than its half parts. But... what is the difference between the two parts falling side by side in unison and the original single-piece object? How elegantly and with how much grace a titan of science can demolish the philosophical edifice of another giant that had been in place and intact for more than 1,500 years!

During the Apollo 15 mission to the Moon in 1971, Commander David Scott let a falcon feather (0.030 kg-mass) and a geological hammer (1.32 kg-mass) fall at the **same** *time* from a height of 1.6 m and, given that there is no atmosphere on the Moon, as Galileo had shown 350 years before, both objects touched the lunar floor at the **same** *time*, corroborating once again the **equality** between *gravitational* and *inertial masses*.

It is worth mentioning that, most probably, even Einstein in 1905 still accepted (though not without reservations) the opinion that the numerical equality between *gravitational* and *inertial masses* was merely a coincidence -- however curious and interesting this might appear to be. It required precisely the genius of Einstein, while conceiving his General Theory of Relativity during the 1905-1915 period, to understand that such a ‘coincidence’ was hiding something essential and intrinsic to our millenary notions of *space* and *time*.

Here, I cannot resist pausing and exclaiming: What a mysterious and universal *force* gravity is! Its effect (the *acceleration*) is exactly the same for all types of *materials* whatever their *masses*! How different it is to, e.g. *temperature*: if the effect of heat on glass were not different from that on mercury, a thermometer would not work, because both *materials* would then equally expand for a given *temperature* change and mercury would not ‘rise’ with our fever. More over, we cannot block the effect of a *gravitational field*! There are multiple ways of having a bad reception in our cellular phone (i.e. to block the *electromagnetic field*), but there is no way of blocking the gravitational effect of our planet on our natural satellite. It is interesting to realize that it was precisely the **universality** of the phenomenon of *gravitation* which prompted Einstein to suspect that what we call the *force of gravitation* does not exist as such! But let us refrain from getting ahead of the heart and soul of this book, and continue instead our gradual process of deeply understanding the needed basic concepts.

We agree,thus, from now on that the term *mass* will be employed with the univocal scientific meaning of *inertial/gravitational mass* and **not** with the old meaning of *quantity of matter*. *Matter* is **tangible**, but *mass* is **intangible**, like *energy*. The *mass* of an object is a measure of its *inertia*; period. Einstein could not have been more precise in the title of his famous publication in 1905.[17]

Once again the curious and now probably impatient reader must be wondering if, with this semantic agreement on the word *mass* -which makes it as **intangible** as *energy*- the folkloric expression 'the atomic bomb **converted** *matter* into *energy*' could be turned into objective truth by simply replacing the word *matter* with the word *mass* obtaining 'the atomic bomb **converted** *mass* into *energy*'. But **NO**; not even this latter proposition is correct! Let us go little by little and see now what Einstein had in mind when using the term *energetic content* referring to an object.

2. The Term 'Energy' and its Semantics.

Fortunately, the term *energy* does not have the confusing historical baggage we saw the term *mass* has. Given that this book is written for the non-scientist who has a genuine interest in understanding Nature, we go back again to the Merriam-Webster dictionary and find, for the word *energy*, in its acceptation 3 for Physics: "A fundamental entity of Nature that is transferred between parts of a system in the production of physical change within the system and usually regarded as the capacity for doing work". The first part of this definition ("fundamental entity") gives the erroneous impression that *energy* can stand by itself when, in fact, as we already stated, it is like *mass*, a **property** of a physical system, namely: its capacity for doing *work*. Without a physical system, i.e. without an object doing *work* or capable of doing *work*, it makes no sense to talk about *energy* -- even though many scientists do so.

As usual, for maximum comprehension, we now look for the word *work* and find acceptation 2c which reads: "The transference of *energy* that is produced by the motion of the point of application of a force and is measured by multiplying the *force* and the *displacement* of its point of application in the line of action". Again, we need to discard the first part (transference of *energy*) because, if not, we would be defining *work* in terms of *energy* and *energy* in terms of *work*! Given that we have accepted the term *force* for now with its intuitive anthropomorphic meaning of *pushing* and *pulling*, we can define the *work* done by a displacing *force* as the product between that *force* and the *distance* along which its point of application acts. A *force* has magnitude as well as direction[18]; for example, if we want to lift **vertically** a body against gravity by applying a **horizontal** *force* on it, it will not happen because our *work* on

17 A. Einstein, *Does the Inertia of a Body depend on its Energetic Content?* Annalen der Physik, 18:639-641, September 27, 1905.

18 Mathematicians say a *force* is a *vector* univocally characterized by its *magnitude* and *direction*.

the **vertical** direction will be **zero** and the body will instead slide **horizontally** against friction on the table. Instead, if we apply a **vertical** *force* to the body, i.e. right against gravity, we will lift the object and will do **zero** *work* on the **horizontal** direction. For any other intermediate direction of *motion*, the direction of the *force* is to be projected on the direction of the trajectory to calculate the actual *work* done by the *force*.[19]

Now we ask: does this product between *force* and *distance* roughly correspond to our intuitive concept of *work*? In another way: does the numerical behavior of this product reflect our natural experience in that the more we *pull* or *push* and the longer the *path* over which we do the pulling or pushing the more *work* we do (the more tired we are)? Yes, that is precisely the algebraic behavior of the product of two numbers, which increases or decreases as those numbers increase or decrease.

The capacity to do *work* appears to us in three basic ways: a) *Kinetic Energy*; b) *Potential Energy*; and c) *Radiant Energy*.

2.1. Kinetic Energy

Kinetic energy (from the Greek *kinesis*, meaning *motion*) is that part of an object's *energy* exclusively due to its state of *motion*. Every car driver knows the catastrophic consequences of colliding at high *speed* and intuits that destruction (*energy* displayed through *work*) did not occur merely because the car was moving with respect to the road but specifically because it was moving relative to the car with which it collided. Pithily: the destruction is not produced by the *speed* as listed on the police report, but by the *relative speed* between cars at the moment of the clash.

Being a little more technical (in preparation for future chapters), when driving a car, our *kinetic energy* varies with our *frame of reference*: if we use our own car as the *reference frame*, then our *kinetic energy* will be zero along the whole trip (we cannot collide with ourselves!); if our *reference* is the road, then our *kinetic energy* at the moment of the accident corresponds to the *speed* indicated in the police report. As explained above, none of those different *kinetic energies* is the direct responsible for the material destruction.

We also intuit that the destruction not only depends on the ***relative*** *speed* between the colliding vehicles but also upon their *masses* (if we hit a bicycle we are more worried about its driver than about us inside the car). The product of the *mass* with the *velocity* is what physicists call *quantity of motion* or *linear momentum* and corresponds to the everyday meaning of the word *momentum* in our lives: it is hard to get something going (whatever it is) but then it takes much less effort to keep it going and a lot to stop it. Because *velocity* is a *vector* (i.e. it has magnitude and direction), and *mass* is what mathematicians call a scalar (it only has magnitude), the *linear momentum* is also a *vector* which takes its *direction* from that of the *velocity*, and its *magnitude* from the product of the *mass* by the *magnitude* of the *velocity* (i.e. the *speed*).

There exists another associated concept which also varies with the *frame of reference*, and about which I need to comment here, if only briefly: If our *ref-*

19 Notice that *work* in Physics does not entirely correspond to our intuitive meaning. For instance, thinking constitutes a large part of my daily *work* but I do it without moving (as a whole) and ergo my *work* is zero.

erence frame is our own vehicle, then I am authorized to say that my *position* has not changed during the whole journey[20], even though when I got into the car I was in front of the *Mormon Temple* in Salt Lake City and, when getting out, I was at the main entrance of the *Caesar's Palace* in Las Vegas. So far, I do not think the reader doubts the correctness of what I said, but it would have been a pill much harder to swallow had I said that the *distance* I travelled during my trip was **zero**. The reason why this latter assertion sounds *prima facie* incorrect is because there exists a tacit agreement in considering *terra firma* as our natural *frame of reference* and, if we do so, of course that my travelled *distance* was about 672 km.

Which is the **real** *distance*? Our tendency would be to say 672 km instead of zero but, in order to eliminate any doubts, I will remind the reader that the *distance* I travelled with respect to our Sun was about 865,000 km![21] Which one of the now three *distances* is the one that I **really** travelled? All of them and none of them are, because the concepts of *trajectory* and its associated *distance* are only univocal if they are defined with respect to a given *frame of reference*. For two *reference frames* in **relative rest**, the *distance* travelled by me will be **the same**; for two *reference frames* in **relative motion**, the *distance* will be **different**. During my trip, all three *frames of reference* (vehicle, *terra firma*, and Sun) were in **relative** *motion* and that is why all three travelled *distances* were **different**. None of those *distances* is more or less **real** than the others.

Returning to the concept of *energy*, we can define *kinetic energy* of an object with a certain *mass* and a certain *velocity* as the *work* necessary to *accelerate* it from the state of *repose* to that *velocity* (in the chosen *frame of reference*). The curious reader is probably wondering why the object's future capacity of doing *work* at the moment of collision (my *kinetic energy*) is defined as the *work* that my vehicle had to do in the past since I left home until the collision. The answer is that the future *work* to be done in order to *decelerate* my car from its collision *speed* to *rest* is identical to the *work* necessary to *accelerate* it from *rest* till that *velocity*. That is why we say that the *kinetic energy* is an **attribute** of the object that only changes with its *mass* and *velocity* and not with its *trajectory* (in the chosen *frame of reference*).

When I return home and park the car in my garage (fortunately the accident was imaginary), my *kinetic energy* is identical to the one I had right before starting my trip (independently of where and how I could have gone, and of which my *frame of reference* might have been). In summary: the *kinetic energy* of an object is a property of the object and of its *state of motion*; it is transported by the object through *space*, and depends upon the *frame of reference*.

2.2. Potential Energy

Potential energy is defined in the dictionary as: "The energy that a piece of *matter* has because of its position or nature or because of the arrangement of parts". This definition clearly confirms our assertion that *energy* **is** a **property**, **not** an entity by itself, and that this type of *energy* is due to either the object *position* or its *internal structure*. However, the only way the mere *position*

20 We ignore the multiple stops to go to the bathroom and having coffee.

21 The Earth travels around the Sun with a speed of approximately 30 km/s.

could confer *energy* to an object is if there was an external *field of force* interacting with it. Consequently, the *potential energy* of an object can be a joint property of both the object itself and the *field* in which is immersed. Thus, depending upon the nature of the *field* (gravitational, electrical, magnetic, nuclear, chemical, etc.) there are different types of *potential energy*.

An everyday example is the immediate *space* in which we live in. Everyone knows that if we lift an object and release it, the *field of force* due to *gravity* will increase the object's *speed* until it returns to earth. We say that the body on its initial *position* (when about to be released) has *gravitational potential energy* which is gradually being converted into *kinetic energy* as the body goes down. The *gravitational field* does *work* on the object as it falls. Choosing the floor as the *frame of reference*, when the object is released, its *potential energy* is a **maximum** and its *kinetic energy* is a **minimum** (zero, as it is *at rest* in the frame). While it is falling, the body has at every *instant* a *velocity* (an increasing *kinetic energy*) and a new *position* (a decreasing *potential energy*). At the *instant* previous to impact, the body's *potential energy* is a **minimum** (zero, because it is at the origin of the *frame of reference*), and its *kinetic energy* is a **maximum** (the greater, the higher the original *position* was).

If we ignore the air friction that the object has to counteract while falling, the **decrement** of *potential energy* is equal to the **increment** of *kinetic energy* throughout the free-fall. If we consider the friction of the air, immediately before the object hits the floor, its *kinetic energy* will be a little lower than the initial *potential energy*, with their difference being the amount of heat (thermal energy) generated by friction. It is clear then that this *potential energy* due to the body's *position* is, like its *kinetic energy*, an **external property** of the object. Choosing an appropriate *frame of reference*, this **external** *kinetic and potential energies* can be made **zero** and, ergo they are **extrinsic properties** of the object, i.e. they **vary** with the *frame of reference*.

Frictionless free-fall is a patent example of **conversion** of *energy* of one class (gravitational) into *energy* of another class (kinetic), so that the sum of both is constant during the process, obeying the *Law of Conservation of Energy*. Similar processes give birth to the concepts of *electromagnetic potential energy, chemical potential energy, nuclear potential energy*, etc.

To summarize, the *Law of Conservation of Energy* states that when we consider all possible types of *energy* during a physical process, their summation is always **constant**[22]. This summation stays constant through the **conversion** of one form of *energy* into another. It is paramount thus to understand that in order for a **conversion** of one type of *energy* into another to take place, the quantity of the first has to **decrease** exactly what the quantity of the second **increases**.

There exists also the *Law of Conservation of the Linear Momentum*, which correspondingly says that the total *linear momentum* of a closed system (i.e. when all interactions are considered) is always **constant**. Regardless of whether the collision between two objects is almost perfectly elastic (two billiard balls) or almost perfectly inelastic (two snow balls), or any other intermediate case, the *total momentum* (the sum of the momenta of both objects) before

22 Here, the word 'conservation' does not mean 'saving' as in Economy, Political, or Environmental sciences.

the collision is to be **equal** to the *total momentum* after the collision.

The other part of the definition for *potential energy* is the one due to the object *internal structure*. How does *internal structure* produce *energy*? And what do we mean by *internal structure*? Science says that a body is composed of molecules, and molecules are comprised of atoms, and atoms are made of subatomic particles[23] (protons, neutrons, and electrons), and protons are... stop please! For now, we only need to know that an atom comprises a nucleus made of protons and neutrons plus one or more distant shells of electrons. Ernest Rutherford (1871-1937) demonstrated that in spite of the apparent solidity of the atom, it is almost empty! This means that the volume occupied by the nucleus and electrons is insignificant relative to the volume of the atom (the nucleus' diameter is about 100,000 times smaller than the atom's diameter).

These constituent blocks of *matter* (notice I said *matter*, not *mass*) have *potential energies*: protons and electrons have their own *gravitational* and *electrical* fields due to their *gravitational mass* and *electrical charges* respectively, and neutrons (without electrical charge) have a *gravitational field* due to their *mass*. In addition, they are immersed in the combined *field* generated by the rest of the particles thereby having external *potential energies*. Even more, all of them are *moving* in the chosen *frame of reference* (in which the body as a whole is *at rest*[24]) and therefore they have individual non-zero *kinetic energy*. This relative *motion* between electrically charged particles and the relative *positions* between the nucleus and the electrons, provide them with both *electromagnetic potential* and *gravitational energies*, i.e. a combined **property** of the *particles* and the *field of force* existent in the immediate microscopic *space*. Similarly, the protons inside the nucleus electrically repel each other but, even so, the nucleus maintains its integrity due to attractive *forces* called (for obvious reasons) *nuclear*, that counteract and exceed the electrical repulsion, establishing once again a *nuclear potential energy*.

Summarizing: even when its *external kinetic and potential energies* are nil (by choosing adequately the *frame of reference*), a body has an **intrinsic** (internal) *energetic content* that can be used to do *work*. This *energetic content* is the sum of all the **intrinsic** contents of its components, plus their *kinetic energies*, plus their *potential energies* due to their *spatial* distribution.

The summation of both **intrinsic** and **extrinsic** *energetic contents* constitutes what Einstein called *energetic content* of a body in the title of his publication[25]. Once more: the **intrinsic** *energetic content* is obtained when we choose a *frame of reference* such that the body **extrinsic** *kinetic* and *potential energies* (due to its *position* and *velocity* as a whole) are zero.

2.3. Radiant Energy: The Classical Distinction between Particle and Wave (Field)

The concept of *radiant energy* also requires a little history in order to avoid the existing semantic confusion, because we use it when referring both to

23 The discovery of subatomic particles negates the etymology of the word *atom* which means 'indivisible', but its use was already solidly established when the discovery took place.

24 As we already intuit, and it will be discussed in detail later, stating that an object is *at rest* has no meaning unless we choose a *frame of reference* such that there is no relative *motion* between the object and the frame.

25 A. Einstein, *Does the Inertia of a Body depend on its Energetic Content?* Annalen der Physik, 18:639-641, September 27, 1905.

energy **transmitted** by *electromagnetic waves* (radio, TV, light, etc.) and to *energy* **transported** by subatomic particles (*matter*). Notice I use the verbs 'to transmit' and 'to transport' for *waves* and *matter* respectively. In fact, the Merriam-Webster's says for *radiation*: "The process of emitting radiant energy in the form of waves or particles".

Let us start then with the concept of *undulatory radiation* and, in particular, of *electromagnetic waves*. Once again, I assume the reader intuits what a wave -as opposed to a particle- is and, for instance, understands that our voice (sound) is transmitted throughout the air from our mouth to the interlocutor's ear by means of a **local** modification of the mechanical pressure in the air (alternative compression and rarefaction), which is initiated by the orator and propagates until it reaches the listener ***without even a single minuscule portion of air being physically transported from one point to the other***. What propagates (or is transmitted) is the local perturbation of the air pressure (or density), and not the air as pieces of *matter* and, as a result, the *energy* generated by the transmitter reaches the receiver doing *work* by mechanically stimulating our auditory system. Again, the *undulatory energy* is transmitted through *space* without any net transport of *matter*. In the case of *sound*, given that the local perturbation is a physical state of the air, we say that air is the *propagating medium*, being thus obvious that *sound* cannot be transmitted in vacuum (where there is no pressure/density to be perturbed and propagated).

Newton thought that *light* was composed of "corpuscles", while his contemporary Christian Huygens (1629-1695) proposed that *light* was a *wave* and that its different colors were its components with different *wavelengths*. The *wavelength* of a *wave* is the *distance* it travels in the *time* needed for the oscillating magnitude to go through a full cycle (i.e. to return to its original value). Human beings can perceive *light* with *wavelengths* between 380 nanometers (nm)[26] which we see as violet, and 750 nm which we see as red. This range is called the visible spectrum.

When, in 1803, Thomas Young (1773-1829) experimentally demonstrated the *diffraction* of *light*, the ideas of Huygens were finally accepted. *Diffraction* is the phenomenon by which, when the *wave* encounters a body of a size comparable with or smaller than its *wavelength*, instead of producing a sharp shadow behind the object, surrounds it in the same way a sea wave encloses a small boat while continuing its journey. The Newtonian theory which assumed *light* to be *matter* could not explain such a phenomenon. The so-called *wave interference* (a phenomenon associated with *diffraction*) was, instead, the paradigm of the *undulatory* behavior, certainly not obviously shared (at least not in those days!) by *moving* material *particles*.

In 1800, William Herschel (1738-1822) put a thermometer on the luminous spectrum obtained by a crystal prism with the objective of measuring the heat emitted by each color, and discovered that the temperature was always higher close to the last color (red) in the spectrum -- where there was no *light*! In this manner, he demonstrated that *heat* (thermal *energy*) can be radiated by way of an invisible form of 'light'. Herschel named this invisible *radiation* 'caloric rays'; today we use the term *infrared radiation*, whose *wavelengths* are be-

26 A nanometer is the billionth of a meter.

tween 750 nm and 0.1 mm.

In 1801, Johann Wilhelm Ritter (1776-1810) discovered that *radiation* with *wavelengths* shorter than that of the color violet darkened a piece of paper impregnated with salts of silver. He called this phenomenon 'deoxidant rays' to emphasize its chemical activity; today we use the term *ultraviolet radiation* whose *wavelengths* are between 10 nm and 380 nm.

In 1865, James C. Maxwell (1831-1879) demonstrated that the then known laws for the phenomena of electricity and magnetism, together with the interaction between those phenomena, predicted the phenomenon of *electromagnetic propagation*: if a physical system named transmitter produced an electrical field variable in *time*, a magnetic field would appear in its contiguous *space*, not only variable in *time* but also in *space*, and this magnetic field, in a sort of inert symbiosis, would reinforce the electrical field which, in turn, would reinforce the magnetic field restarting the process *ad infinitum*, with the result that the combined field (electromagnetic), once generated, would have a *spatial* and *temporal* existence independently of its source (in the same way our words, once emitted, will reach the interlocutor no matter how much we might repent for having open our mouth), and would propagate through *space* with a fixed *speed* that Maxwell was able to calculate from the referred laws and a pair of already-existing experimental measurements.[27]

It is thus clear that a *wave* of any nature propagating in *space* is equivalent to a *field* variable in *time* and in *space*. In this way, physical **reality** in the 19th century was conceived as either showing itself as *particles* (*matter*) or as *waves* (radiation). This vision of the Universe proved to be extremely useful as we shall see in subsequent chapters. However, starting the 20th century scientists realized that something was not right with this dichotomic vision of **reality** -- giving birth to the Theory of Quantum Mechanics. But...what is **reality**? Well... **reality** is the subject of my next book. For now, all we can do is to extend our intuitive notion of physical **object** to include not only *particles* (*matter*) but also *waves* (*fields*). Then, from now on we shall use the term **object** to generically refer to either *particles* or *waves*. Please, do not forget this.

As we said, in the case of sound, the mentioned *field* is one of mechanical local pressures and, therefore, the existence of a *material medium* (air, water, etc.) is essential for its propagation. In the case of and electromagnetic wave, the field is a combination of electrical and magnetic *fields* and the necessity or not of a *material medium* of propagation was the subject of a passionate controversy for some time -- until Einstein resolved the apparent paradox in 1905.

When Maxwell calculated the *speed* of his electromagnetic waves, he was astonished to find out that it was equal to the value known for the *speed of light* at the time and, at a stroke, he concluded that it would have been too much of a coincidence had light not been also an electromagnetic wave -- unifying the two major scientific fields (until then entirely independent) of *electromagnetism* and *optics*. By analogy with sonic waves, it was immediately postulated that there had to exist a *material medium* for the propagation of *electromagnetic waves* (a crass error indeed, as we shall see in Chapter 5), and

27 These measurements were the electrical force between two electrical charges at rest, and the magnetic force between two electrical currents (electrical charges in motion).

this hypothetical substance was called *ether*.

Almost a decade after Maxwell's death, Heinrich Hertz (1857-1894) experimentally confirmed Maxwell's theory by generating and receiving in his laboratory what we today call radio waves. Ironically, while conducting this grand corroboration of the ***wave*** character of *electromagnetic radiation* in general (including light), he stumbled against what is currently known as the 'photoelectric effect' which, in 1905, would prompt Einstein to postulate the ***discrete*** character of *light* (as opposed to the ***continuous*** nature of a *wave*).

In December 1895, Wilhelm Conrad Röntgen (1845-1923) wrote a report on a very powerful "new class of radiation" that he called "X-rays", a discovery for which he, in 1901, received the first Nobel Prize of history. Later on, it was determined these rays were also *electromagnetic waves* with wavelengths in the range 0.01 nm to 10 nm.

In 1896, the semantics associated with the word *radiation* became convoluted. Antoine Henry Becquerel (1852-1908) serendipitously discovered, while working with salts of uranium, a phenomenon he interpreted as an "emission of rays" capable of going through a piece of paper and, because of that, it was believed these rays were similar to the X-rays. In 1898 Marie and Pierre Curie discovered, working with uranium ores, two new chemical elements to which they called Radium and Polonium, and which had strong radioactivity -- a term which was created to refer to their spontaneous emission. In 1899, Becquerel demonstrated that his 'rays' were subatomic particles (*matter*) charged with electricity (because they were affected by an electromagnetic field). In the same year, Ernest Rutherford called those particles 'alpha rays' and also discovered other more penetrating rays, to which he gave the name of 'beta rays'. Subsequently, an even more penetrating kind of *radiation* was discovered and naturally named as 'gamma rays'.

The origin of these three types of 'rays' is the spontaneous disintegration of complex chemical elements, which gives birth to simpler elements while emitting the mentioned *radiation*. In 1903 Becquerel shared the Physics Nobel Prize with Pierre and Marie Curie "for their discovery of the spontaneous radioactivity". Further research concluded that the 'alpha rays' are nucleuses of the chemical element Helium (two protons and two neutrons), the 'beta rays' are electrons -- i.e. the energy transfer is attained in both cases by the transport of *matter* in space. As for the 'gamma rays', it was instead concluded that they were *electromagnetic waves* with extremely short wavelength (under 10 picometers[28]), i.e. the energy is transmitted without transport of *matter*.

3. Finally the Equation $E = m.c^2$

Describing the same phenomenon from two different *frames of reference*, and applying the results of his previous publication, Einstein proved[29] that every change in the energetic content (E) of an object is accompanied with a change in the same direction of its *mass* (inertia!) and that this change of *mass* (m) can be numerically determined by dividing the change in *energy* by the

28 A picometer is a millionth of a millionth of a meter.

29 A. Einstein, *Does the Inertia of a Body depend on its Energetic Content?* Annalen der Physik, 18:639-641, September 27, 1905.

square of the *speed of light in vacuum* (c^2)[30]. Because of the tremendously high value of the speed of light (and because it still has to be multiplied by itself!) the change of *mass* associated with a modest change of *energy* is infinitely more modest.

From this relation between the *changes* of *mass* and *energy* of an object, it is easy to prove that the same relation is valid between *mass* and *energy* (not just their changes) and, consequently, the ***more** energy* a physical system possesses, the ***more** inertia* it has ($E = m.c^2$). In this sense, the *mass* of an object is a measure of its *energy*. Notice that, if the expression 'a measure of' were eliminated, there would be important semantic consequences.

Let us cogitate a little: if the *energy* ***increases***, the *mass* (inertia) ***increases***, and if the former ***decreases***, the latter ***decreases***. According to the most famous equation in history, *mass* and *energy* go hand in hand and, therefore, they do not meet the necessary condition for one to be converted into the other, namely that what one ***decreases*** is what the other one ***increases*** (like, as we already discussed, *potential* and *kinetic energies* of a body in free-fall without friction).

Because *energy* is conserved, when the *energy* of an object ***decreases/increases*** (together with its *mass*), the ***deficit/excess*** of *energy* has to go ***out/come*** into/from the external world, and this *energy* that ***goes/comes to/from*** the exterior can be numerically calculated if we know the *change of mass* experienced by the object. Einstein responded affirmatively to his own question "Does the inertia of a body depend on its energetic content?" by saying: Yes, the *mass* of an object depends upon its *energetic content* and vice versa; ergo, the *mass* is a measure of its *energy*, and vice versa. The rest is folklore.

Nonetheless, Einstein himself contributed inadvertently to the folklore when casually referring to 'the Principle of Equivalence between Mass and Energy' in 1907 and 'the Law of Equivalence between Mass and Energy' in 1946 [1] -- as if it was something with an unequivocal meaning. Even the eminent Bertrand Russell (1872-1970) fell in the trap by literally interpreting Einstein's expression and asserting in his book *ABC of Relativity* that *mass* is the same thing as *energy* [3].

In order to understand what Einstein really meant by the term 'equivalence', we again check the dictionary and find: "Equal in value, measure, force, effect, significance, etc". Are *energy* and *mass* equal in value? NO! But they are related by a universal constant (the speed of light) which, employing a suitable physical unit for velocity, e.g. light-years/year, would render the famous equation to a plain equality between *energy* and *mass*! ($E = m.1^2 = m$). Is numerical equality (forced by the use of specific physical units) the same as conceptual equality? Of course NOT! (despite what some renowned physicists may say). To be fair to Einstein, he simply referred to the validity of the equation which relates two different concepts, by choosing an unfortunate expression, first because it is not a principle and second because of the polysemic character of the word 'equivalence'.

30 As will be discussed profusely throughout the book, an essential postulate of Special Relativity is that *light speed in vacuum* is always the same, regardless of the state of *motion* of the source and observer. General Relativity shows, in turn, that *light speed in vacuum* can vary from place to place.

Since throughout this book we will deal with another 'Principle of Equivalence' enunciated by Einstein, we can improve the meaning of the term 'equivalent' by stating that, in order for two things to be *equivalent* in some sense, there has to exist at least one property of them which is indistinguishable from some point of view. Only in the case when all properties of them are equal from every conceivable point of view, can we affirm that one thing is the same as the other. Being *equivalent* or not is thus relative to an explicit or implicit perspective, and when the latter is ignored, the concept is misused and abused, leading to erroneous scientific folklore.

3.1. Chemical Reaction vs. Nuclear Reaction

Combustion is a chemical reaction because the atomic nucleuses of the reactants remain unaltered while the molecules are the ones which undergo transformation (carbon and oxygen combine to produce carbon dioxide); dynamite's destructive power is based on chemical reactions which release the potential **electromagnetic** energy of atoms and molecules. The atomic bomb and the process that takes place in a nuclear power plant, instead, are the result of a nuclear reaction, precisely because it is the nuclei of the atoms which undergo transformation, liberating potential **nuclear** and *electromagnetic* energy. In both cases, the *energetic* contents of the reactants change but, of course, the summation of all -intrinsic and extrinsic- *energies* of the reactants and of the products remains **constant** throughout the reaction.

From our discussions on potential *energy*, we learnt that the *energetic content* of an object can change without changing its *quantity of matter* -- it would be enough to spatially rearrange the electrons around the nucleus[31] to change its *electromagnetic* potential in the same way as we change our gravitational potential when climbing a mountain. Consequently, the *mass* of a physical system can change while its *quantity of matter* remains unaltered, forcing us to conclude that the *Law of Conservation of Mass* proposed by Lavoisier (1743-1794) in the eighteenth century **may** be false and that the reason why, in a chemical reaction, the sum of all *masses* (reactants and products) appeared and continues to appear as **constant**, may be simply because its **change** is ridiculously small and, ergo, out of reach at Lavoisier's time and even today. This change of *mass* is, contrariwise, perfectly measurable in a **nuclear** reaction -- simply because its magnitude is ***one hundred million times bigger***. Is the inability to detect something proof of its non-existence? Of course not! But then, was Lavoisier right or not? Hold on.

In summary: the equation $E = m.c^2$ is as valid for a chemical reaction as it is for a nuclear reaction but, for the time being, the equation has no practical value for the former type of reaction. This conclusion explains -but does not justify- the second part of the folklore denounced at the beginning of this chapter, demystifying the idea that the equation is the foundation, essence, and archetype of nuclear energy. The equation is valid for every conceivable physical process; period.

Another conclusion, as ineludible as apparently bizarre, is that the *mass* of an object varies with its *velocity*: if we chose a *frame of reference* with re-

31 Heating a body is a way of spatially rearranging the electrons inside its atoms.

spect to which the object has a non-zero *speed*, from what we have learned, the body has a non-zero *kinetic energy* and, ergo its *energetic content* has ***increased*** with respect to the one it has when in repose but... according to the most famous equation in history, this ***increase*** in *energy* has to be accompanied with an ***increase*** in *mass* so... in our *frame of reference*, the **faster** the object *moves*, the **greater** its *mass* will be!

Again, for this conclusion not to appear as pure fantasy, let us remember that *mass* simply means *inertia* (**not** *quantity of matter*), so what we are plainly saying is that the **faster** an object *moves*, the more difficult it is to accelerate it, namely, to attain, in a given *period* of time, a given additional ***increment*** in its *speed* (that is what *inertia* means) -- and this certainly does not appear to be as unbelievable as it did earlier in this chapter[32]. When the moment is ripe, we shall see that this difficulty of increasing the *speed* of an object becomes harder to overcome the closer it is to the *speed of light in vacuum* (mathematicians say the body's *mass* tends to infinity when its *speed* tends to the *speed of light in vacuum*).

The increment of *mass* with *velocity* had been observed, without explanation, while working with beta-rays (high speed electrons) emitted by radioactive material, before Einstein predicted it with his equation in 1905. The electron trajectory had been registered in the so-called Cloud or Wilson Chamber so its *masses* at different *velocities* could be calculated, with the conclusion that they were different from one another and from the known *mass* of the electron at lower speeds. This incomprehensible increase of the electron *mass* (against Newton's mechanics) had been erroneously explained by saying that "the ether inertia was being added to the known electron's inertia producing an apparent increase of the electron mass".

The above considerations suggest the creation of the concept of *rest* or ***proper*** *mass*, which is analogous to the already-discussed concept of *internal* or ***intrinsic*** *energy*. Reiterating, the latter is the *energy* a body has in a *frame of reference* in which it has no ***extrinsic*** *energy* (neither kinetic nor potential) and therefore, by definition, the ***internal*** energy is independent of the object's *position* or *state of motion* with respect to other *frames of reference*. Similarly then, the ***proper*** or ***rest*** *mass* of an object is the one it has when measured in a *reference frame* within which it is *at rest*. The term 'rest mass' is not a happy choice because it may give the wrong impression that an object can be *at rest* without choosing a *frame of reference*. I prefer and will always use the term 'proper mass'.

Summarizing: the total *energy* of an object is, after choosing a *frame of reference*, the summation of its ***intrinsic*** or ***proper*** (reference-independent) *energy*, plus its ***extrinsic*** (reference dependent) *energy*. *Mutatis mutandis*, the object *mass* is the sum of its ***proper*** and ***extrinsic*** *masses*.

Now, there are still more surprises: even the ***proper*** *mass* is not univocal: the fact that it is independent of the *reference frame* does not mean it is always the same for a given object. For instance, when heating a body, its *internal energy* ***increases*** and, ergo, its *proper mass* ***increases*** a minuscule amount...

32 I hope this proves the practical relevance of our previous semantic discussions. Once we free ourselves of the semantic confusion, apparently unintelligible and bizarre concepts become easier to assimilate by our intellect.

but ***increases***! And when a body emits *light*, following the same argument, its ***proper*** *mass* ***decreases*** a minute amount... but ***decreases***! The ***proper*** *mass* is thus *intrinsic* to the body because it does not depend on its *state of motion*, but is variable as a function of the physical interactions the body may have with the external world and, ergo, it is not inherent to the object.

Another erroneous folkloric assertion says that what the equation $E = m.c^2$ really means is that neither *energy* nor *mass* are conserved but that instead it is mass-energy that is conserved -- whatever that means; or even worse: that the "sum of mass plus energy will always remain constant"! [4]. The closest Einstein was to such statements is "mass is energy and energy has mass; the two conservation laws of mass and energy are combined by the relativity theory into one: the conservation law of mass-energy" [6]. We are now finally very close to understanding the essence of what Einstein really wanted to say when he joined the words *mass* and *energy* with a hyphen.

Reality is that, in the light of the famous equation, *mass* -with its classical meaning only associated with material objects-, does not conserve, while *energy* does conserve precisely because of the relation between *mass* and *energy* discovered by Einstein. As we already suggested, Lavoisier's *mass* conservation law may have been just the result of our inability to detect tiny changes in the *mass* of reactants and products in a chemical reaction. Hmm... it does not sound like the kind of Universe Einstein's God (Spinoza's) would create!

Let us meditate a little: if I said that *energy* is the one conserved precisely because of the equation $E = m.c^2$, why can we not reformulate the equation as $m = E / c^2$ and say that what is conserved is *mass* and not *energy*? Of course, from the mathematical point of view both interpretations are equally valid and therefore not one but both are to be equally conserved. And, what about from the physical point of view? Well... the problem is that a physical process can also have -besides *matter*- *undulatory radiation* as one of its products and, so far, we know that *radiation* carries *energy* but we never thought of assigning *mass* to something immaterial! However, we already understood that *mass* is **not** *quantity of matter* but *inertia* so... what is the problem? Let us be courageous (easy, for Einstein took the plunge first!) and accept that *radiation* also has *mass*, with its value being its *energy* divided by c^2, as the most famous equation in history suggests.

This is why Einstein said that his equation implied that *undulatory radiation* (immaterial) also has *inertia* (mass!). This *mass* associated with a *wave* must not be confused with the *mass* of the material *medium* some *waves* require, e.g. the air in which our voice is transmitted or the water in which sound and sea waves propagate.

We are finally now prepared to understand the *energy* and *mass* balances taking place in the uncontrolled nuclear fission that occurred on August 6th 1945, as well as in the mundane combustion discovered 1.5 million years ago. The next two sections may be a little too technical for some readers; if so, the detailed discussions on *energy* and *mass* balances for combustion and nuclear fission can be simply browsed or even bypassed and, even so, the reader will not miss the gist of the message I am trying to convey.

4. Energy and Mass Balances in Combustion and Nuclear Fission

Once again, to put the most celebrated equation in history in the appropriate perspective, let us remember that humanity learnt how to control fire and Alfred Nobel (1833-1896) invented dynamite much before Einstein was born, but that this does not mean his equation is less valid for chemical reactions than for nuclear ones. The following analysis will allow us to understand, for combustion as well as for the fission of uranium, why the denounced folklore associates Einstein's equation only with nuclear fission.

Let us start with the fission of uranium, which occurs in an atomic bomb and that, in essence, is a chain reaction during which the absorption of a neutron by uranium atoms is repeated in an uncontrolled fashion. We can symbolically represent this absorption as follows:

Uranium nucleus + slow neutron ⟶ nuclear fragments + fast neutrons + gamma radiation

The terms on the left are the *reactants*, while those on the right are the *products* of the reaction. When the uranium nucleus incorporates a neutron, the inner balance between electrical and nuclear forces is altered and the nucleus splits (fissions) into two fragments (e.g. nucleus of Barium and Krypton), a few fast neutrons, and *electromagnetic radiation*. These neutrons are, in turn, absorbed by other uranium atoms and then the number of fissioned atoms quickly multiplies. Upon fission, the electrical repulsive force between the fragments (both with positive electrical charge) imparts a great *kinetic energy*[33] to them. The totality of the *energy* before and after the fission reaction is conserved, and the total number of particles **does not** change either (though it could, as will become apparent later in this chapter). Protons, neutrons, and electrons are simply spatially rearranged, producing different amounts of *potential electromagnetic* and *nuclear energy* (and its corresponding *mass*) contained in the fragments.

It is the fragments *kinetic energy* and *radiation energy* which produce the explosion in an atomic bomb, or convert into electrical energy in a nuclear power plant where the reaction is under control. The *energy* produced by the fission of a single atom is minuscule, but the Hiroshima bomb contained about one hundred million trillion atoms of which approximately 1% exploded. Let us apply the *Law of Conservation of Energy*, using the convention E(something) to indicate the energy of that something:

E(Uranium nucleus)+E(neutron)=E(fragments)+E(neutrons)+E(gamma radiation) [I]

Applying what we learned in this chapter, we know that the *energy* of an object (*matter* or *radiation*) can be expressed as its ***intrinsic*** *energy* (*at rest* with our *reference frame*) plus its *kinetic energy*, and that the former is calculated multiplying the object ***proper*** *mass* by the square of the *speed of light in vacuum*. Given that for an atom of uranium to absorb a neutron, the latter has to be slow[34], we choose a *frame of reference* in which both the uranium atom and the neutron can be considered *at rest* (i.e. we neglect the *kinetic energy of the slow neutron*). Adopting the convention mp(something) to indicate the

33 The fragments are expulsed with speeds in the order of 4 million km/h.

34 This unexpected fact was found experimentally by Enrico Fermi in 1934 and we will talk again about it in my next book on Quantum Physics and folklore.

proper *mass* of that something, we obtain:

mp(Uranium+neutron). c^2 =mp(fragments+neutrons). c^2 +kinetic and radiant energies [II]

And, by rearranging terms:

{mp(Uranium+neutron)-mp(fragments+neutrons)}. c^2 = kinetic and radiant energies [III]

The small positive difference between the ***proper*** masses of the reactants (Uranium+neutron) and those of the products (fragments+neutrons) multiplied by the square of the speed of light is equal then to the **total** energy released by the fission reaction. Given that the ***proper*** *masses* have been measured and tabulated for a long time, and that its difference is measurable with accuracy, Einstein's equation allows us to calculate the **total** *energy* liberated by the atomic bomb by multiplying that difference in ***proper*** *mass* between the reactants and products by the square of the *speed of light in vacuum*. In order to have a practical order of magnitude, this difference of ***proper*** *mass* is:

0.000000000000000000000000000311 kg mass[35]

If now we carry out the same *energetic* balance for combustion (remembering that the molecule of carbon dioxide has one atom of carbon and two atoms of oxygen) we obtain:

{mp(Carbon)+2.mp(Oxygen)-mp(Carbon Dioxide)}. c^2 = kinetic and radiant energies [IV]

Here, the term on the right (*kinetic* and *radiant energies*) is ***one hundred million times smaller*** ($10^{(-8)}$) than the corresponding term in a nuclear reaction (Equation III) and, ergo, the difference of ***proper*** mass between reactants and products in a chemical reaction is one hundred million times smaller than in a nuclear reaction (the factor c^2 is the same in both types of reaction). To compare, this difference in ***proper*** *mass* would be:

0.00000000000000000000000000000000000311 kg-mass

This minute difference in *mass* cannot be currently measured with accuracy, which explains why the most famous equation in history has no practical value to predict the *energy* released in a chemical reaction. Another more vivid way of understanding this fact is to realize that a conventional power plant needs more than 100 railroad cars full of coal per week, while a nuclear power plant needs only a truck of uranium per year. But, once again, the equation is as valid for a chemical reaction as for a nuclear one.

We might be tempted to believe and say that because the products *proper mass* is lesser than the reactants *proper mass*, this 'missing' *mass* has been converted into *energy*. But we know by now that this ***deficit*** of *mass* is not a ***deficit*** of *matter* but simply that the **total** *proper inertia* of the ***products*** is **lesser** than the **total** *proper inertia* of the ***reactants*** because part of the *potential energy* of the ***reactants***, by their spatial rearrangement, has been converted (here the word 'conversion' is correctly employed) into *kinetic energy* of the products plus *radiant energy*. In other words, and in order to debunk the

35 A useful comparison is that this difference in *mass* is only a fifth of the *mass* of a proton.

folklore once and for all:

The intrinsic energy of the products is smaller than the intrinsic energy of the reactants, and this energy deficit reappears as the kinetic plus the radiant energies of the products -- completing the total energy of the products which has to be equal to the total energy of the reactants.

Notice that the word mass is not mentioned at all in order to explain the energetic balance. Finally, let us rearrange Equation III by dividing both sides by c^2 to obtain:

mp(Uranium+neutron)=mp(fragments+neutrons)+kinetic energy/ c^2 + radiant energy/ c^2 [V]

The *kinetic energy* of the fragments and neutrons divided by c^2 is their ***extrinsic*** *mass* (*inertia*) because fragments and neutrons are moving (in our *frame of reference*) at high *speeds*. The summation of the two first terms on the right is then the **total** *mass* (***intrinsic+extrinsic***) of the fragments and neutrons. If now we agree (as we already did heuristically) in interpreting the term 'radiant energy/ c^2 ' as the *inertia* (*mass*) of the *radiation* emitted by the reaction, and note that the combined ***extrinsic*** *mass* of the uranium atom and the slow neutron is ***zero*** because they are *at rest* in our *frame of reference*, we obtain the following equation (using the symbol mt(something) for the **total** *mass* (***intrinsic+extrinsic***) of that something):

mt(Uranium + neutron) = mt(fragments + neutrons) + mt(radiation) [VI]

Another way, thus, of effectively putting the last nail in the folklore's coffin is:

The intrinsic mass of the products is smaller than the intrinsic mass of the reactants, and this deficit of intrinsic mass reappears as extrinsic mass of the products -- completing the total mass of the products, which has to be equal to the total mass of the reactants.

We should remember that the ***intrinsic*** *mass* of the products is **smaller** because their ***intrinsic*** *energy* is **smaller**, and this is because there is conversion of *potential energy* into *kinetic energy* as well as into *gamma radiation*. ***Notice that the word 'energy' has not been mentioned at all in order to explain the mass balance.***

The surprising feature of this reformulation is that, having extended the concept of *mass* (*inertia*) from being applied only to *matter* to legitimately assigning it also to *radiation*, we have vindicated Lavoisier because now the *mass* is also conserved! In plain words and pithily: employing the generalized concept of *mass*, the equation expressing the ***balance of energy*** can be considered as expressing also the ***balance of mass***, and this is exactly what Einstein meant (without much success!) when he used the expression "Equivalence between Mass and Energy" or "Conservation of the Mass-Energy". Certainly, he did **not** mean that *mass* is the same thing as *energy*.

Undoubtedly, once again we confirm that Truth is not as pompous and romantic as myth... but it has the immeasurable value of being the Truth.

5. Is it really possible to transform Matter into Energy?

By a curious coincidence, precisely while I was writing this section for the Spanish version of this book in 2008, I had hired a company to measure the concentration of radon[36] in the basement of my new house where we were building the bedrooms for my children and my laboratory. Besides its high toxicity, this gas is associated with a rare and perplexing phenomenon.

In 1930, the existence of an elementary particle identical to the electron but with opposed electrical charge (positive instead of negative) was experimentally confirmed and called positron (or anti-electron). In 1955, using a particle accelerator, an anti-proton was artificially created and detected. In this manner, the concept of antimatter was accepted by conceiving -in principle- the existence of anti-atoms (made of anti-protons, anti-neutrons, and anti-electrons) for each chemical element we know here on Earth, i.e. the anti-version of all the *matter* in the Universe. In fact, the particle accelerator at CERN (Geneva) in 1995, and soon after the one at Fermi Lab (outside Chicago) have created tiny amounts of anti-hydrogen [10].

A puzzling phenomenon is the one that takes place when an electron and a positron collide. In principle, the same phenomenon would occur when an atom collides with an anti-atom. In this elementary particle encounter, what we consider *matter* (electron and positron) disappears and, in its place, we only see *gamma radiation* (immaterial electromagnetic waves!). Given that, as we saw, an *electromagnetic wave* transmits *energy* without transporting *matter*, we are tempted to admit that in this case, as opposed to nuclear fission, *matter* disappears by 'converting' into *energy*.

Is this language legitimate? First, let us agree that we are not saying that something is transforming into nothing, because *radiation* is something! Besides, as the reader must recall, we emphasized that *energy* is not an object, but an ***attribute*** of an object, and that both *matter* and *radiation* are physical objects so... if *matter* is converting into something, this something is properly named as *radiation* -- not as *energy*!. What perplexes us is that *matter* and *radiation* are such basic and ergo indefinable concepts that when we look into the subatomic world, our intuition and common-sense fail us miserably, because we cannot liken the tangibility of *matter* with the intangibility of *radiation* (let alone when we mistakenly refer to it as *energy* instead of as *radiation*). It is our philosophical notion of **Reality** what is at stake here, and my next book, which is entitled *What is Reality? - Einstein, Quantum Physics, and Folklore*, will tackle this subject in depth.

Let us, again, help our thinking with symbols. The annihilation taking place when an electron and a positron meet can be described as:

$$\text{Electron} + \text{Positron} \longrightarrow \text{Gamma radiation} \qquad [VII]$$

The reverse reaction is also possible, i.e. when *gamma rays* impinge on a plate of lead, this latter element acts as a catalyzer for the transformation of *undulatory radiation* into two particles: an electron and a positron. In symbols:

36 Chemical element which is colorless, odorless, tasteless, very heavy, radiotoxic, and may indicate seismic activity.

$$\text{Gamma radiation} + \text{Catalyzer} \rightarrow \text{Electron} + \text{Positron} \quad [\text{VIII}]$$

Is this perplexing complete transformation of *matter* into *radiation* and *radiation* into *matter* (which we now know does not happen in the nuclear fission of Uranium) also governed by the most well-known equation in history? Of course it is! It is simply a special case where the ***proper*** *mass* of the *products/reactants* is ***zero*** *after/before* the reaction, respectively.

To improve our understanding of the phenomenon, let us write down the energetic balance as we did with the uranium nuclear reaction and with combustion. Let us choose a *frame of reference* in which the electron is *at rest* and use the same conventions E(), mp(), mt() for *energy*, ***proper*** *mass*, and ***total*** *mass* respectively, obtaining for the annihilation of an electron and a positron:

$$\text{mt(Electron)}.c^2 + \text{mt(Positron)}.c^2 = \text{E(Gamma radiation)} \quad [\text{IX}]$$

And, given that the ***proper*** *mass* for the electron and the positron are identical while the *kinetic energy* for the electron is ***zero*** (with respect to itself), the equation transforms to:

$$2.\text{mp(Electron or Positron)}.c^2 + \text{Kinetic energy(Positron)} = \text{E(Gamma radiation)} \quad [\text{X}]$$

In this case the ***intrinsic*** *energy* of the reactants disappears completely with the only product being *gamma radiation*. In other words, the **deficit** of ***intrinsic*** *energy* is equal to the **total *intrinsic*** *energy* of the reactants which, together with the *kinetic energy* of the positron, reappear in the *energy* of the *gamma radiation*. It is paramount for us to realize that *mass* is still as conserved as *energy* is! This is easily confirmed by dividing both terms of Equation IX by c^2, obtaining:

$$\text{mt(Electron)} + \text{mt(Positron)} = \text{E(Gamma radiation)}/c^2 = \text{mt(Gamma radiation)} \quad [\text{XI}]$$

As before, the **deficit** in ***intrinsic*** *mass* is the sum of the ***intrinsic*** *masses* of electron and positron, and this **deficit** together with the ***extrinsic*** *mass* of the positron (due to its *speed*), reappears as the *mass* (*inertia*) of the electromagnetic *gamma radiation*. Succinctly: there are **no** *particles* among the products but there is *mass*, because *radiation* also has *inertia*.

It is clear from the above considerations that both *energy* and *mass* are conserved, and that the latter does **not** convert into the former. The same considerations can be made for the reverse phenomenon, namely the 'miraculous' generation of an electron and a positron from *gamma radiation*.

It is the shocking fact that there are **no** *particles* and only *radiation* as a product of the reaction -while there are particles as reactants (*mutatis mutandis* for the reverse reaction)- that allows non-rigorous scientists and sloppy writers to pompously proclaim that "matter is converted into energy" and, in doing so, to confuse *radiation* with *energy* (which is only one of its properties).

Despite the consistency of the *energy* and *mass* balances, the idea that two *particles* can disappear from this world, leaving only *radiation* in their place, is mysterious -if not mystic- and hard to assimilate. The reason is that we tend to expect elementary *particles* to behave as we perceive *material particles* in our macroscopic world do -- for instance when two billiard balls collide. After some cogitation, it is difficult not to agree with Herman Bondi (1919-2005) who in his book *Relativity and Common Sense* [8] says:

> *... The surprising thing, surely, is that molecules in a gas behave so much as billiard balls, not that electrons behave so little like billiard balls.*

As the astounding success of Quantum Physics attests, the subatomic world is governed by probabilistic causal relations, instead of the deterministic causal relations of the Theory of Relativity. Again, Quantum Physics and its folklore will be the subject of my next book.

In summary, the conversion between *matter* and *radiation* (**not** *energy*) is possible but, fortunately, extremely unlikely because, otherwise, our bodies could mysteriously disappear in a burst of *electromagnetic radiation*. Nonetheless, *matter* annihilation occurs daily (in the microscopic world) in the basement of houses where there is a slight concentration of radon. This gas is highly radioactive and, in its process of disintegration produces positrons. When these positrons get in contact with the electrons in the air or in our bodies, annihilation between elementary particles and antiparticles takes place. Once again, this phenomenon is not the one that produces the tremendous amount of energy released by the atomic bomb.

6. Einstein and the Atomic Bomb

Another folkloric belief, as unfortunate from the human point of view as technically incorrect, is that the equation $E = m.c^2$ was responsible for the human catastrophe caused by the atomic bomb, prompting Einstein to be on the defensive until his death in 1955. As we said, both scientists here in the USA and those in Germany did not need to know the existence and validity of the most celebrated equation in history to achieve technological success or failure in the development of the atomic bomb -- in the same manner that combustion was exploited to its ultimate consequences much before Einstein was born. Physicist Robert Serber (1909-1997), who participated in the 'Manhattan Project'[37], expressed this fact in the following way:

> *Relativity Theory is not necessary to discuss atomic fission. The theory of atomic fission is not relativistic in the sense that the relativistic effects are too small to affect significantly the dynamics of the process.*

Einstein was neither invited to participate in the Manhattan Project nor ever officially informed about its progress[38]. In a few words: despite profuse amounts of folklore, the invention and development of the atomic bomb did not depend on the discovery of the relation between *energy* and *mass*. Nevertheless, even the eminent Stephen Hawking, in his bestseller A Brief History of Time (Chapter 1), says that "Relativity has given us nuclear energy".

Tired of distortions due to ignorance and pseudopolitical reasons, Einstein wrote a letter to a Japanese newspaper in 1952 stating: "My participation in the development of the atomic bomb consisted of a single act: I signed a letter[39] to President Roosevelt". In a letter to a French historian written in 1955

37 'Manhattan' was the code name for the development of the atomic bomb.

38 Niels Bohr, Einstein's intellectual rival in Quantum Physics interpretation, did participate in the project.

39 Einstein refers here to the letter with a fragment of which this chapter begins.

(the year of his death), Einstein was much more specific: "You suggest that I should have, in 1905, foreseen the possible development of the atomic bomb. That is practically impossible because the feasibility of a nuclear chain reaction depended of the availability of empirical information impossible of being foreseen in 1905... Even if that information had been available in 1905, it would have been ridiculous trying to hide the particular conclusion[40] resulting from the Special Theory of Relativity. Once the theory existed, the conclusion also existed." [5]. Clearly, Einstein's moral and ethical values compelled him to defend something that the historian should have understood with minimal effort.

Einstein once commented to his secretary: "Had I known the Germans would not succeed in producing the atomic bomb, I would have never moved a finger. Not a single one!" Of course! But Einstein's decision of using his scientific stature and fame to persuade the USA government to build the bomb was probably the correct one -- when we consider that neither Einstein nor anybody else can predict the future, and accept the premise that it is obviously much better not to go to war though... once the abhorrent decision has been made, its purpose can only be to win or to die.

If, as is my fervent desire, this first chapter was for the reader as interesting and instructive as startling, let us tightly fasten our safety belts because our demystifying adventure is just beginning.

Additional Recommended Reading

[1] Einstein, Albert, "$E = m.c^2$". New York: Science Illustrated, April, 1946.

[2] Baierlein, Ralph, *Newton to Einstein - The trail of light*. Cambridge: University Press, 1992.

[3] Russell, Bertrand, *ABC of Relativity*. London: Routledge, Taylor and Francis Group, 1997.

[4] Bodanis, David, , *A Biography of the World's Most Famous Equation*. New York: Berkley Books, 2001.

[5] Nathan, Otto; Norden, Heinz, *Einstein on Peace*. New York: Simon & Schuster, 1960.

[6] Einstein, Albert; Infeld, Leopold, *The Evolution of Physics*. New York: Simon&Schuster, 1966.

[7] Sir Isaac Newton, *Principios Matemáticos de la Filosofía Natural*. Traducción parcial de *Philosophiae Naturalis Principia Mathemática* (1686) en *Grandes Obras del Pensamiento, Albert Einstein y Otros*. Barcelona: Altaya, 1993.

[8] Bondi, Hermann, *Relativity and Common Sense - A New Approach to Einstein*. New York: Dover Publications, Inc., 1964.

[9] Goldsmith, Donald, *The Ultimate EINSTEIN*. New York: Byron Press Multimedia Books, 1997.

[10] Kaku, Michio, *Physics of the Impossible*. New York: Anchor Books, 2009.

40 Einstein refers here to the equation $E = m.c^2$.

Chapter 2

The Perception of Time...

When a man sits with a pretty girl for an hour, it seems like a minute. But let him sit on a hot stove for a minute and it's longer than an hour. That's relativity[1]. Trivial assertion attributed to Einstein *ad nauseam*, even though it constitutes the antithesis of his Relativity Theory.

and its Measurement

The power of appearance leads us astray and throws us into confusion... whereas the art of measurement...would have caused to live in peace and quiet abiding in the truth. Plato.[2]

In Chapter 1, I assumed that all of us shared the intuitive and classic notions of *time* and *space* using the concepts of *instant*, *duration* (interval of time), *simultaneity*, *position*, *distance* (interval of space), plus those magnitudes -derived from their combination- which are needed to describe *motion*: *velocity* and *acceleration*. In this chapter, we leave the notion of *space* again as intuitive and shared by all of us, while we attempt to understand the complex and elusive concept of *time*.

To begin with, given that in our colloquial discourse we tend to confuse them, I would like to spell out the difference in meaning for the adjectives *subjective* and *relative*. The Merriam-Webster's says in its acceptation 3a for *subjective*: "characteristic of or belonging to reality as perceived rather than as independent of mind". It is clear then that a statement is *subjective* when it contains elements of pure ***perception***, i.e. which do not belong to the ***exterior*** *objective* world but that depend on the person who asserts them. In brief: the antonym of *subjective* is *objective*.

Given that Physics is the discipline which studies our ***external*** world (the *objective* world), we can say that the desideratum for this science is that the only *subjective* part of it must be its practitioner (the scientist). Consequently, when we talk about the 'observer', we do not mean necessarily a human being. The 'observer' can be just a human being (but conscientious of his *subjective* baggage); this human being can, in turn, be assisted with the needed instrumentation to *objectively* observe a particular phenomenon; or the 'observer'

1 Abstract of a fictitious publication: A. Einstein, *On the Effects of External Sensory Input on Time Dilation*, Journal of Exothermic Science and Technology (JEST, Vol. 1, No. 9; 1938).
2 As cited in *The Discovery of Time* (Mccready, Stuart, Editor, Illinois: Sourcebooks, Inc., 2001).

can be exclusively a piece of technology to register the phenomenon by collecting experimental data which will eventually be analyzed (it does not need to!) by a scientist and shared within the scientific community.

On the other hand, the Webster's thesaurus says for the word 'relative': "being such only when compared to something else". We learnt in Chapter 1 that both the *mass* and the *energy* of an object are ***relative*** to the *frame of reference*, so that only by specifying the latter we can know the former. However, it must be understood that this dependence of those properties upon the *frame of reference* has absolutely nothing to do with the *subjectivity* of the scientist. On the contrary: this dependence is and ***objective*** fact resulting from the meaning of those properties, the meaning of a *frame of reference*, and from how the ***external*** world actually is. It is, thus, concluded that *relativity* has nothing to do with *subjectivity* and, ergo, can be ***objectively*** established. Summarizing: ***subjectivity*** is strange to science, while ***relativity*** is a conspicuous part of it -- if, and only if, it can be ***objectively*** affirmed.

One of the culprits of the confusion that remains regarding the essence of the Theory of Relativity is precisely the abuse of the term 'observer', without the previous semantic elucidation and, therefore, conveying to the reader the mistaken impression that the *relativity* the theory refers to, is due to the human being who *observes* **reality**, and not to **reality** itself. In other cases, as in the book *ABC of Relativity*, Bertrand Russell specifically and correctly states the meaning of the word 'observer', though he employs repeatedly the term 'subjective' to indicate something which is 'relative' and, ergo, inadvertently promoting the same erroneous interpretation of the theory.

Finally, objectively asserting that something is *relative* is not indicative of any lack of knowledge but, contrariwise, it is a sign of being closer to the *objective* truth. As José Ortega y Gasset (1883-1955) masterfully says in *The historical meaning of the Theory of Relativity*: "Relativism here does not oppose to absolutism; on the contrary, it fuses with the latter and, far from suggesting a defect in our knowledge, it confers absolute validity to it."[3]

Let me finish this semantic discussion citing Einstein himself: "The meaning of the Theory of Relativity has been misinterpreted by a large majority: philosophers play with the word relativity like a kid plays with a toy... Relativity does not mean that everything is relative in life." I would pithily say that it was precisely the obsessive search for the **absolute**, what made Einstein to conceive the Theory of Relativity.

1. What this Book is all about

This is not a thriller where an impatient reader has to go straight to the end so he can discover the mystery -- cheating on the author's ability to create the suspense so necessary for a book to be interesting and engaging. To prove it, here is the conclusion of the next four chapters:

*Let us make it clear right away: in classical Physics (Newton's and Galileo's), **time** reigns in an **absolute** fashion and independently of **space** which, in turn, reigns in an **absolute** manner and independently of **time**.*

3 Translated from *Complete Works of José Ortega y Gasset*, Revista de Occidente, Madrid 1947, Vol. III, pp. 231-242.

*The existence of any one of the two is not a sine qua non condition for the existence of the other. In Relativistic Physics, instead, both **time** and **space** are **relative** and **interdependent**.*

My objective is for the reader to understand how and why *time* and *space* are relative and interdependent without falling into the folkloric trap that 'time is the fourth dimension of space'[4]. *Time* and *space* were intrinsically different notions before Poincaré and Einstein, and they are still different today, more than a hundred years later. Even though I am revealing upfront the whole plot, I still expect to achieve the suspense to engage the reader, because I am convinced that neither you nor me are sure of understanding the previous paragraph that I, ironically, started with "Let us make it clear right away:" There are no more prime and fundamental notions than those of *time* and *space* and each time humanity believes to have understood them, they slip away from our intellectual grasp to the light of new discoveries.

Newton says in his *Principia*[5]:

... Absolute, true, and mathematical time, of itself, and of its own nature flows equably without regard to anything external, and by another name is called duration: relative, apparent, and common time, is some sensible and external (whether accurate or unequable) measure of duration by the means of motion, which is commonly used instead of true time; such as an hour, a day, a month, a year.[6]

It is, thus, evident that for Newton, it is the ***measure*** of *time* "by the means of motion" that is ***relative*** and ***apparent***, and used in place of the 'true' time, which is ***absolute***. This conception is no less and no more than the one all of us subconsciously feel in our guts: we all know that 'before and after Christ' only defines a completely arbitrary point of reference in order to measure a *time* that we wrongly believe is *absolute*. In short: we accept the arbitrariness of our 'year zero', and a few are aware of the degree of arbitrariness of the *unit of time*, but we are not prepared to easily concede that the 2010 years that have past since our 'zero' do not constitute an ***absolute*** *interval of time*, i.e. common for all of us here on Earth as well as in the Andromeda galaxy. As it is apparent by now, despite this chapter being dedicated to *time*, it is impossible for me to talk about it without involving the notion of *space*, and vice versa in the next chapter -- reinforcing the suspicion that our intuitive and everyday understanding of both, à la Newton, may not be correct.

A natural tendency, somehow pernicious for the progress of scientific knowledge, is to issue ontological statements about the Universe based on incorrect philosophical interpretations of correct mathematical descriptions. When Newton developed his laws of motion, which are valid with great exactitude (except for the refinements of Relativity Theory), he interpreted his almost correct equations in terms of ***absolute*** *force*, *space*, and *time*, putting science on the wrong philosophical track for more than two centuries, until Einstein showed that *space* and *time* are ***relative*** and also that the concept of *force* is not only relative but can disappear by a change of *frame of reference* (don't

4 Philosophers of the caliber of Henri Bergson erroneously sustained that physicists did not understand the concept of *time* and had "spatialized" it by treating it as another dimension of *space*.

5 Sir Isaac Newton, *Mathematical Principles of Natural Philosophy*, July 5th, 1687 (known as the 'Principia').

6 As cited in *Great Experiments in Physics*; New York: Dover Publications, 1959.

despair: I'm just setting the stage for suspense).[7]

Semantics, Epistemology, and Ontology again? Of course! What else does understanding mean? Tempted to jump to the end, evading what promises to be pure and boring philosophy? Why? As far as *time* and *space* are concerned, you already know the end of this thriller and you could close this book for ever but... then, you would miss the opportunity of comprehending *space* and *time* while we travel together through the former during and throughout the latter.

2. What is Time? - The Ontology of Time

In the history of natural philosophy, the notion of *time*, due to its direct relation with our physical deterioration and its culmination in death, has always carried a heavy emotional load -- sometimes bordering with the irrational. The term *secular* (from Latin seculāris) which originally referred to the passing of a hundred years (*century*), evolved into an epithet referring to the terrestrial world governed by the 'tyranny' of *time*, as opposed to a supposed 'superior and divine reality exempt from the flux of time'. Different versions of this notion of a superior reality were elaborated by Parmenides (530-515 BC) and Plato (circa 427-347 BC). As Hans Reichenbach (1891-1953) [3] says, it is curious that this 'superior reality' free of the flux of *time* was labeled by the philosophers as 'eternal', a term which naturally evokes the notion of *time* and hides a profound desire of immortality -- irrationally ignoring the objective interest these prominent philosophers undoubtedly had in understanding and accepting the **real** world.

Heraclitus (544-484 BC) had a more objective and optimistic vision. He is known by the famous aphorism saying that we cannot enter the same river twice because we are, and we are not the same or, a better-known account of it (spread by Plato) stating that we cannot enter twice the same river. The essential meaning of these two statements is that both the person and the river continuously change and, ergo, they are not the same between encounters. And this change is not simply that the person changed her political party, religion, or shirt, or the river was colder the second time because the first one occurred in summer: neither the molecules of the water in the river, nor the totality of the atoms in our bodies are physically the same between any two occurrences.[8]

Notwithstanding, despite of their incessant change, both person and river maintain an easily-reckoned identity (within certain limits). But those changes are of such a nature and magnitude that, instead of talking about different things/people, we say that the same person or the same river evolves through different states ***in time***, until the changes are of such a magnitude that the notion of identity cannot be upheld any longer. Tersely: beings and things are persistent patterns (structure); not a dull collection of particles. This is the line of thought which sustains that our notion and experience of *time* emanate

7 Jesús Zamora Bonilla wittily points out to me: "And how can you know whether it is wrong or not? In general, it is only discovered after long discussion and experimentation, perhaps taking even centuries. Besides, some times the incorrect ontological interpretation promotes great scientific progress... even when it will eventually lead to a dead-end". This is precisely the historical relation between the ontological interpretations of Newton and Einstein.

8 Every atom in one's body is replaced every seven years or so (from *Geometry, Relativity, and the Fourth Dimension*. New York: Dover Publications, 1977).

exclusively from our perception of *change* in things around us and the relation this *change* has with the idea of *identity*. Stretching this philosophical position to its ultimate consequences, in an absolutely stagnant (**no** *changes*) world, or in one where the *changes* were so drastic and erratic that we could not conceive the notion of *identity*, there would be no need for creating the concept of *time*.

For George Berkeley (1685-1753) *time* is "the succession of ideas which flow uniformly and of which all beings take part"; for David Hume (1711-1776) *time* is "a sequence of indivisible moments". Immanuel Kant (1724-1804) formulated ancient philosophy in logical terms, and his persistent reference to immortality explain his efforts -in a much more elaborate fashion than Parmenides- to 'prove' that the *flow of time* was merely ***subjective***. But Kant lived in a *time* completely different from that of the ancient philosophers; Newton's mechanics had proven amazingly exact and effective in predicting *motion* for terrestrial and celestial objects. Pierre Laplace[9] (1749-1827) had issued his celebrated parable about a superior intelligence who, knowing all the forces of Nature and all pertinent circumstances, and having the cyclopean capacity of carrying out all the necessary mathematical calculations, could predict the future for both the macro and the micro-cosmos.

Kant only needed to extend the validity of Newton's physics to the world of the atoms of which we are made of, to conclude that the determinism of Newton was opposed to our venerated free-will -- given that our future would be, thus, as determined as the past, turning our faculty to act by reflection and election into a mere illusion[10]. Kant, desperately though laudably, wanted -at any cost- to save our freedom to choose, as well as its associated notion of morality, but he could not ignore the reality of Newton's determinism success. Kant then stated that Science only discovers appearances, without revealing the *objective* **reality**, a **reality** which, according to him, is not governed by *causal* laws. With the commendable objective of saving our freedom and morality, he asserted that both *time* and *causality* are mere creations of our mind and not intrinsic properties of the universe we live in.

Once again, Hans Reichenbach [3] sharply points out that Kant's philosophy about ***subjective*** *time* and ***determinism*** "is a form of escapism", because it does not resolve the antagonism between ***determinism*** and ***free-will***, and it does not clarify our irrefutable experience of the *flux of time*. I would personally add that, it is precisely the objective existence of ***causal*** relations -which establishes a *temporal order* in this, our mysterious world- that allows us to plan and work hard to modify our future, as well as to establish the difference between good and evil. As Rudolf Carnap [12] said:

> *Without causality in the world, there would be no point in educating people, in making any sort of moral or political appeal. Such activities make sense only if a certain amount of causal regularity in the world is presupposed.*

Newton[11] and Gottfried Leibniz (1646-1716) maintained a fierce dialectics about the notion of *time*; the former proposing an ***absolute*** and ***universal***

9 His masterpiece entitled *Mécanique Céleste* (5 volumes) was published between 1799 and 1825.

10 As we shall see in my next book, Quantum Physics negates determinism both à la Newton and à la Einstein.

11 In reality, it was Samuel Clarke who (on behalf of Newton) corresponded with Leibniz.

time, the latter reasoning more in accord with Heraclitus. Newton won the diatribe, prevailing for about two centuries, but his reign began to collapse when the Austrian physicist and philosopher Ernst Mach (1838-1916) (in his book *The Science of Mechanics*, originally published in German in 1883) shrewdly opposed the Newtonian notions of *time* and *space*. Mach's writings influenced Einstein who, resuming Leibniz's ideas, would later on resurrect the concept of ***relative*** *time* without taking away its *objective reality*[12]. Mach said about Newton's notion of *time*:

> *... but we must not forget that all things in the World are connected to and depend upon one another, and that we ourselves and all of our thoughts are part of Nature...*
>
> *...Time is an abstraction to which we arrive through the change in things, and that abstraction is made because we are not restricted to any particular measure, all things being interrelated as they are...*
>
> *...This absolute time cannot be measured by comparison with any motion; therefore, it is deprived of both practical and scientific value, and nobody has reasons to say he knows something about it. It is a useless metaphysical concept...*[13]

For Jorge Luis Borges (1899-1986) time is easily refutable by our senses but not so much so by our intellect, and "its essence seems to be inseparable from the concept of succession". Borges wrote an essay (*A New Refutation of Time*, 1946) to demonstrate that *time* is an illusion. He starts noting the contradiction existing in the title when using the adjective 'new' and, as the great lover of ambiguity he was, ends up the essay stating: "And even so, denying temporal succession, denying the self, denying the astronomic universe, are apparent desperate measures and secret consolations... The world, unfortunately, is real; I, unfortunately, am Borges."

Thus, what is time? St. Augustine asked himself in his book *Confessions*, to end up confessing: "if nobody asks me, I know what it is; if I have to explain it to somebody, I do not know". I think everyone of us feels as St. Augustine, regardless of our religious tradition -- though I know that if I do not attempt to explain it (even though I may not be able to convince you), nobody will believe that I understand it -- or, better, that I am trying hard to understand it!. Hold on... here it goes.

2.1. The Classical Notions of Time and Space

Before we attempt to understand the modern notion of *time*, let us agree in what the classical meanings for *space* and *time* are (as they were used in Chapter 1). We choose for *space* the Merriam-Webster's acceptation '4a': "a boundless three-dimensional extent in which objects and events occur and have relative position and direction".

Remembering that, in Chapter 1, we extended the meaning of the word *object* to include *radiation*, if follows that we think of *space* as an immaterial

12 Einstein referred to Mach as the "precursor of Relativity Theory". To Einstein's dismay, Mach rejected the theory.

13 Translated into English from *La teoría de la relatividad: sus orígenes e impacto sobre el pensamiento moderno*, Barcelona: Altaya S.A., 1993.

container of *matter* and *radiation* (fields) where *events* take place. How powerful our imagination is! We are talking about a 'box' made of an immaterial material! Besides, for an instant, this definition sent a shiver down my spine, because I felt the word 'extent' was simply used to avoid saying, again, 'space'! This corroborates how primitive and basic the notion of *space* is.

The dictionary also states that our *space* is three-dimensional, i.e. we can change our *position* along three independent directions (length, width, and height), but any other change in *position* is a combination of changes along those three directions. This experimental fact can also be expressed by saying that a solid body cannot have more than three edges meeting at one point, and being mutually perpendicular. A mathematician would say that in order to localize a sufficiently small object in *space*, only three numbers are enough. For instance, the *position* of an airplane (referred to our planet) is known when we know its latitude (which parallel), its longitude (which meridian), and its altitude.

In summary, for a given *frame of reference*, any *point* in *space* has one and only one ordered set of three numbers associated with it and, vice versa, every trio of numbers identifies one and only one *point* in *space*. It is also worth-noting that, by its very classical definition, *space* exists independently of its content, giving birth to the notion of *vacuum*, i.e. if we imagine that there exist neither *matter* nor *radiation*, we still insist -perhaps irrationally- that there is *space*, and we call it empty space or simply *vacuum*.

What about *time*? Well... by now I do not expect the reader to be surprised, if I say that the dictionary would get once again the first prize in a **circularity** contest! Acceptation '1a' reads: "the measured or measurable ***period*** during which an action, process, or condition exists or ***continues***"; and acceptation '1b' says: "a non-spatial continuum that is measured in terms of events which ***succeed*** one another from ***past*** through ***present*** to ***future***". Certainly, the words 'period', 'during', 'continues', 'succeed', 'past', 'present', and 'future' are all so inextricably intertwined that none of them can be used to effectively define any of the others. So much for understanding what *time* is! Nevertheless, and in agreement with the dictionary expression 'non-spatial continuum', we feel in our bones that to quantify (measure, know) *time*, it is sufficient to specify just one number and, ergo, *time* is one-dimensional.

The everyday notion of *event* (occurrence) is the combination of the three numbers needed to identify the place in *space* (where it occurred) and the fourth number needed to specify the *time* (when it happened). A mathematician could not resist the temptation of combining the first trio with the fourth number to form a quartet and say: in a given *frame of reference*, every *event* can be specified with one and only one quartet of numbers and, vice versa, every quartet identifies one and only one *event*. With the characteristic concision of a mathematical mind, s/he would assert: *space-time* has four dimensions. In this manner, we realize that the concept of *space-time*, as a mathematical tetra-dimensional construct, is not something intrinsic to the Theory of Relativity, but an abstraction as natural as perfectly legitimate in classical Newtonian physics as well. The utility of this abstraction, though, is minimal in classical physics and maximal in modern physics.[14]

To wrap up the classical notion of *space*, we all subconsciously feel *space* (and its relation with *time*) as a tridimensional scenario -as impalpable as immutable- where *matter* and *radiation* dwell in and events occurred, are occurring, and will occur. We say this conception describes *space* as ***absolute*** and ***independent of time***: *position* is relative to the *spatial frame of reference*, but the *distance* between any two *points* in *space* -at a common instant- is ***absolute***, i.e. the **same** for the whole Universe, independently of the *reference frame*. *Past, present*, and *future* are *spatially* ***universal***, i.e. the **same** for us here on Earth, as for any inhabitant of a recondite planet in a remote solar system in a lost galaxy running away from us at thousands of kilometers per second.

To wrap up the classical notion of *time* (and its relation to classical *space*), we all sense it as something -as impalpable as inevitable- that *flows* in an irreversible direction, allowing us to order the *events* we perceive occurring in *space* by means of the dynamic intuitive notions of *past*, *present*, and *future*. We say this conception describes *time* as ***absolute*** and ***independent of space***: an *instant* is **relative** to the *temporal frame of reference*, but the *duration* (time interval) between the two *instants* at which two given *events* occur is for us ***absolute***, i.e. the **same** for the whole Universe, independently of the *frame of reference*. The notion of *simultaneity* seems extremely natural to us, and nobody would doubt that saying that two *events* -one here and another one happening outside our galaxy- occurred at the **same** *time*, is a fully rational and logically acceptable statement; *position* and *distance* are ***universal*** with respect to *time*, i.e. they were the same 10,000,000 years ago as they are at the very moment I am writing this paragraph.

Let us see then how far we can go in our adventure of understanding the modern concept of *time*, while concurrently maintaining the intuitive and classical notion of *space* (until next chapter).

2.2. The Modern Ontology and Epistemology of Time

A not so well-known quote from Einstein, though one of the most damaging to our mental sanity, is found in a letter to the family of his friend Michele Besso (1873-1955) after he died. Einstein said: "People like us, who believe in physics, know that the distinction between past, present, and future is only a stubbornly persistent illusion". It is to be clear that Einstein does not state that *time* is an illusion à la Kant or à la Borges, but that, instead, the chimera consists of our subdivision of it into *past*, *present*, and *future*. And, why is such a view troubling for our mental sanity? Because such a pithy assertion can only mean that *time* does **not** *pass* (flow) but, instead, that *time* (including what we call *future*) **is**... in the same manner we say *space* **is** and, consequently, the notion of *time* as something *flowing* is ***subjective*** -- where now the ***subject*** is humanity as a whole!

Such a collective delusion, if it had actually occurred and continues occur-

14 In fact, it was not Einstein who introduced the term space-time but his Professor Hermann Minkowski (1864-1909) after reading Einstein's publication of 1905. To be fair with Poincaré, it has rarely been mentioned that this eminent French philosopher published a paper at almost the same time as the 1905 Einstein's paper, and before Minkowski, where he already treated space and time as a tetra-dimensional mathematical entity.

ring, can only have its causes explained by the neurophysiology, psychology, and linguistics of the human species (I may venture, of any living species). To be honest, I still do not have a solid and sharply defined opinion on this matter, though I tend not to agree with Einstein. Perhaps, during the process of explaining my ideas and feelings with the greatest possible clarity along this book, I may reach a more clear-cut judgment on this profound philosophical issue. We shall see!

Let us start by *understanding* what is that we *understand* when we naturally say that *time* 'flows' or 'flies' or 'passes'. I respectfully challenge the reader to explain what 'to flow' means without using the idea of *time*. To me, the idea of 'flowing' is inseparable of the idea of *time*, so... if what we mean is that *time* changes with or 'flows' *in time*, it has little sense, has it not? What sense could it make to say that time 'passes' at a rate of one second ***per second***? Or, worse, that time 'flies' at one hour ***per second***! What sense does it make to talk about the *pace* of *time*? Once again this damned circularity that produces only tautologies![15]

Very succinctly: saying that *time* passes ***rapidly*** or ***slowly*** is absurd, and saying that *time* flows at its own *pace* is a tautology. Ironically, those three idioms are understood by all of us without ambiguity, and the idea behind them is the favorite theme during birthday parties.[16]

But we measure *time* with *clocks*, and *clocks* are physical processes to which the idea of *pace* (speed) can be applied without circularity, so they can fall behind or be ahead with respect to another *clock* (process). Would it not be possible then that we are abusing our language and incorrectly likening what is ***measured*** (*time*) with what ***measures*** (*clock*)? Grammatical structures of human languages evolved by choosing a scaffold resting on the millenary concepts of *past*, *present*, and *future* and, if these latter notions were to be fallacious, it is inevitable to infer that a human language is not the most appropriate tool to discern the ontology of *time*. Paraphrasing Borges: "The idea of succession is in the essence of every language; languages are not suitable to ponder the eternal or the atemporal" [6].

But, this grammatical structure is all we have for now! Let us, thus, use our rich and powerful English the best possible, with the certainty that this structure will continue evolving in the light of the inexorable progress of our understanding of the Universe and, as a result, what we today call Metaphysics and Philosophy, tomorrow shall be knowledge -- solidly and effectively expressed through the new structure of our language.

2.3. Subjective Time (Perception)

I must have read the following famous quote attributed to Einstein a dozen times: "When a man sits with a pretty girl for an hour, it seems like a minute. But let him sit on a hot stove for a minute and it's longer than any hour. That's relativity". In all and every one of those attempts, it was impossible for me to understand what he meant. Why? Because the situation at hand corresponds to

15 A tautology in Logic is a proposition which is always true or always false and, ergo, without information content. In colloquial language, it refers to a useless and vicious repetition.

16 For a kid, time goes slowly; for an adult, time passes rapidly; for the ostentatious of possessing 'deep' knowledge, time flows at its own pace. None of these three asseverations has any logical sense.

perception -not measurement- of *time* and, therefore, even though ***perception*** is the natural subject-matter of study within the sciences of Psychology and Neurophysiology, it could never be considered the subject-matter of Physics. The latter is the science of *objective* **reality**, and the way to define the *objective* world is by developing reproducible techniques for ***measurement*** and ***discernment***, so we can attain conclusions which are **independent** of the ***subject*** who formulates them.

Given that Einstein's theory establishes the limited relativity of the *objective* world -not the complex unlimited relativity of the *subjective* realm- I initially thought his quote was simply a superficial attempt by Einstein to convey the idea of relativity to the layperson. In fact, it is believed that this funny account of his theory was given to his secretary Helen Dukas so she could deal with non-scientific people and media reporters[17]. Moreover, this piece of folklore affirms that the quote "is one more proof of Einstein's genius to explain the theory to the layperson". What a bunch of bull! It is doubtful whether this phrase was actually used by Einstein or not but, if it was, he did it surely with the only intention of being funny and, maybe, of getting rid of the reporters -- otherwise he would have been conveying the antithesis of his theory and of the very essence of Science.

As a second thought, I considered that, since Einstein never liked the name "Theory of Relativity"[18], his incursion into the realm of *perception*, was only to emphasize the eminently *subjective* and *relative* nature of being human, as opposed to the search for the ***objective*** and ***absolute*** which characterizes any scientific theory and, in particular, the Theory of Relativity. Though, most surely, I may be splitting hairs because it is impossible to know what he meant -- if he really ever said so.

2.3.1. Mrs. Chaplin

Adding wood to the folkloric fire, Steve Mirsky, a humorist of science, wrote in *Scientific American* (September 2002, Vol. 287, 3, Page 102) a satiric piece taking advantage of Einstein's well-known weakness for the opposite sex, and the funny side of his supposed 'pretty girl' quote. In this article, Mirsky suggests that Einstein did more than just utter the famous quote: in order to prove its validity, he invited Charles Chaplin's wife, Paulette Goddard, to meet at a bar in New York so they could chat for a *period of time* which seemed to him (due to Mrs. Chaplin's physical[19] beauty and intelligence) like a minute though, after checking his clock, almost an hour had really elapsed. After this pleasant experience, Einstein, in the best spirit of science, sat on a hot stove and, when he estimated that an hour had elapsed, he confirmed that only a minute had really passed.

From the above 'meticulous' experimental work, the master of the scientific arts supposedly concluded that "The state of mind of the observer plays a crucial role in the perception of time". An Argentinean could not help but utter a popular expression which offers you -sarcastically- 'chocolate as a prize for

17 James B. Simpson, Best Quotes of '54, '55, '56, (1957); Alice Calaprice, *Expandable Quotable Einstein*, (2005).
18 Einstein never used the word relativity in his original (1905) publication. Max Planck and others suggested the name a year later. Einstein liked the name Theory of Invariants better, inspired by Minkowski who used the expression "Einstein's postulates of invariance".
19 Here 'physical' has very little to do with our Physical Science!

your profound and unexpected discovery'. The comparative study between the *time* that really elapsed and the one *perceived* by a person when s/he is having fun or suffering, is worth carrying it out in Psychology and Neurophysiology but, once again, it has no objective value beyond understanding human beings as intelligent organic entities and, as such, they are not part of Physics. I am sure that the reader, like me, realized very early in his/her life how unreliable our ability to judge the passage of *time* is.

Up to now, everything sounds as funny as ludicrous; and the readers may be asking themselves what a comic article is doing in this book. Well... Steve Mirsky not only invented the story but, in its making, he also made up a scientific publication named "Journal of Exothermic Science and Technology"[20] in Vol. 1, No. 9 (1938) of which Einstein published a paper entitled "On the Effects of External Sensory Input on Time Dilation", where he scrupulously describes the 'scientific method' used to arrive to his silly results. With careful mastery of the English language and the satiric-comic genre, and perhaps with a subconscious malice associated with his knowledge of human nature, Mirsky limited himself to comment that the scientific magazine was "defunct" in 2002 -- leaving its real existence in 1938 to interpretation and, by association, the reality of Einstein's ridiculous publication.

That is how history is written! Mirsky's article not only authenticated the dubious quote but gave birth to a scientific magazine that never existed[21], and a trivial paper 'authored' by Einstein which, of course, he never wrote. Very few realized the fictitious character of this charade; the majority started citing the fake publication as true; another few, not sure of its existence, cited the quote through citing Mirsky's paper. Be as it may, very few resisted the temptation of taking advantage of the halo of wisdom and authority accompanying everything that Einstein said -- particularly when the glamorous and mysterious expression *space-time* is involved.

A patent and recent example of this phenomenon is the scientific article entitled *Space-Time Relativity in Self-Motion Reproduction* published in the *Journal of Neurophysiology*, 97:451-461, 2007 where Einstein's alleged conclusion is cited several times as real and extended from being applied to *time* to being applied to *space*, namely: "Moreover, the conclusion posited by Einstein (1938) that "the state of mind of the observer plays a crucial role in the perception of time" can be extended by the perception of space during motion".

Psychology and Neurophysiology are the scientific disciplines which *objectively* study the *subjective*. Given that, naturally, the notions of *space* and *time* these disciplines refer to are specifically those as *perceived* by the ***subject***, the conclusion affirming that the state of the mind influences the perception of *time* and *space* is obvious, and a Spaniard could not help but exclaim "¡Menuda perogrullada! (What a truism!)" ... And we already know that an Argentinean would offer you chocolate as a reward for the platitude. ***Physics, and specifically the science of making physical measurements, were developed precisely to stay away from the multitude of different perceptions***

20 Its acronym is 'JEST' which, transformed into a word, means 'joke'. This fact alone should have alerted people, but it didn't!

21 I am grateful to Mr. Robert Behra (librarian of the University of Utah) for his assistance in documenting the fictitious character of the JEST magazine, consulting WorldCat, EBSCOhost, Science Citation Index, and several other specific documents on Einstein's writings.

experienced by different individuals.

I do not want to be misunderstood: the work mentioned above is interesting and valuable, as far as the analysis and quantification of a well-known effect are concerned; what I am talking about here is the irresistible temptation of using Einstein's authority to assign more credibility to the authors' results. If, on top of it, we know that the referred Einstein's paper never existed, it is not hard to understand why there is so much folklore and confusion around the Theory of Relativity.

The indiscriminate (and many times incompetent) use of the prestige and success attained by the science of Modern Physics is still more evident in the realm of Spirituality, the interaction between mind and body, and its indisputable effects on our health, happiness, and personal success. An extremely successful writer is Deepak Chopra who, in his book *Ageless Body, Timeless Mind - The Quantum Alternative to Growing Old* (chapter on "The Metabolism of Time"), assumes the famous 'pretty girl' quote was seriously meant by Einstein and, ergo, an assertion about ***subjective** time* receives an unnecessary support and prestige from what Modern Physics has to say about ***objective** time* (which, as we shall see, is surprisingly far from trivial). Chopra received in 1998 the Physics 'Ig Nobel'[22] (satiric) Prize for "his unique interpretation of quantum physics as it applies to life, liberty, and the pursuit of economic happiness". This parody of the Nobel Prize consists in praising those who "first make people laugh, and then make people think". There is an awful lot to think about in what Chopra says -- without the need to take refuge in the respected objectivity of Relativity and Quantum Physics.

2.3.2. Mrs. Albizuri

It was *vox populi* that the ***clock*** on the big wall in the assembly hall of the elementary school[23] was ***running fast***. With just six years of age, immanently timid and with a volume of voice genetically very reduced, there I was by the little staircase ready to step up to the stage, so as to recite a verse which I instinctively perceived to be as ridiculous as uninspiring. At such an early age, and not very innovatively indeed, I had fallen in love with my first-grade teacher. Mrs. Albizuri had the face of an angel, the body of Aphrodite, and she always emanated an intoxicating feminine fragrance. Very little had more significance in my life than what she could think of me.

Space was not the problem: the embarrassment of being on stage before hundreds of merciless kids and parents, trying to declaim a silly poem, could be simply avoided if I quickly figured out how to ***stop** time*. Simultaneously, I knew I had to go ahead because crying and running away were certainly no solution (humiliation would be even greater) so... to the high sign from the master of ceremonies, I would begin my trajectory in *space-time* towards the dais center.

To grab some courage from nowhere, I thought all I had to do was to initially not try to reach the center of the podium where my stellar performance would take place, but simply to shoot for traveling ***half*** that *distance*, because it seemed to me a target more easily to achieve and tolerate and, once there, I

22 The term 'Ig Nobel' was purposely chosen because it sounds like ignoble (plebeian).
23 Elementary School 'General Belgrano', Necochea, Buenos Aires, Argentina, Earth, Milky Way, Universe, 1954.

would think again. Well... not much to think! I would challenge myself again to go the next ***half***... and so forth. Finally, I did get to the final *spatial position*; I did perform (miserably, as I recall), and time went on inexorably till this moment, in which I am remembering that journey of mine through *space-time* that occurred more than 50 years ago (*time interval*) and at a place more than 10,000 km afar (*spatial interval*). As any sensible person would have expected, I could not ***stop*** *time*.

But, just a *moment*! *Time* did not stop, but the process of setting my mind to achieve mini-objectives so as to accomplish a much bigger one, altered my *perception of time*, ***dilating*** it, making me *perceive* a *time interval* ***shorter*** than the **real** one and, ergo, transforming a painful experience into another much more tolerable and transient. Had I not thought of that instinctive stratagem, my judgment of how much *time* had elapsed would have been much greater -- with a much more prolonged suffering on my part. Even so, my *perception* of the elapsed *time* was much ***greater*** than the *interval of time* displayed by the *clock* on the wall (even though it was *running* ***fast***!), and surely different from the *duration* perceived by the bored audience with a multitude of worries and personal aspirations.

Dilating it? I feel compelled to clarify the meaning of the expression ***dilation*** of *time* because it will be used profusely from now on. When I employed the verbal mode 'dilating it' referring to *time*, what I meant was that the ticktack of my 'internal clock' was altered giving me the illusion that *time* was running more ***slowly*** (dilated) and, therefore, the *duration* perceived by me was ***shorter***. In sum: if my *time* is ***dilated***, the *pace* of my 'internal clock' is ***slower*** so it lags ***behind*** producing ***shorter*** *time intervals*; if my *time* is ***contracted***, the *pace* of my 'internal clock' is ***faster*** so it hastens ***ahead*** producing ***longer*** *durations*. Mrs. Chaplin dilated Einstein's *time*; his hot stove ***contracted*** it. Our linguistic license, when using the idiom 'dilation/contraction of time', consists in referring to *time* when we are actually talking about the behavior of our 'internal clock' with respect to the *clock* on the wall! Likewise, when I say the clock on the wall was *running* ***fast***, we are comparing its ticktack with that of another 'standard clock' supposedly more 'exact' (hold on, please).

Once my suffering ended, the angel I had as teacher hugged and kissed me. I wished that *moment* of happiness would had *lasted* for hours; what probably *lasted* just one ***second*** seemed to me as a ***microsecond***. Mrs. Albizuri ***dilated*** my *time*. When my sensible mother wanted to know the *time* to go back home to cook, she did not take me to the psychoanalyst's couch to ascertain the relation between my *perception of time* and the *time* that had 'really' elapsed, or to find out why a six-year old kid falls in love with and adult beautiful woman (that's easy!). My Mom, with little more than elementary school education, knew the difference between ***subjective*** and ***objective*** *time*. I grasped that disparity during my brief excursion through stardom.

Neither Einstein's 'internal clock' nor mine are used as instruments to measure time in Physics.

2.3.3. The End of Innocence... and the inevitable Emergence of the most Profound Questions

About a decade after, being already in college[24], when I started thinking about the process of assigning numbers to physical magnitudes (*measuring*[25]) and when I learned mathematicians say that between any two numbers (no matter how close to each other), there is always another number, it occurred to me that, perhaps, my infant instinctive subterfuge of subdividing my trajectory in halves 'to kill time' could have effectively *made my day* (avoided my embarrassment) -- because no matter how close I could have been to the platform center, the remaining *distance* could have been halved and there would I have been, setting my mind again to walk the next **half** *ad infinitum*, without never ever having to confront the audience and deliver the detestable verse.

Looking at it that way, my *spatial* trajectory was divided into an ever-increasing number of small trajectories, the *distance* of which was shorter and shorter (without ever becoming zero![26]), the closer I was getting to the target position (without ever reaching it!). As a result, if both the total *distance* and total travel *time* had to be the summation of those tiny partial *distances* and *times*, my happiness was assured -- given that those sums involved an ***infinite*** number of terms and, ergo, an ***infinite*** number of decisions and actions naturally impossible of being done by our miserable human condition. Preposterous!

Let us keep on being ridiculous: in Chapter 1, I used phrases like "the instant previous to impact", or "the instant previous to starting the car", or "at every instant" and most probably you, the reader, accepted them as full of logical sense. Well... let us now think of the *instant* ***previous*** to starting my odious walk to the center of the stage and let us see once again how I could have avoided my discomfiture. Our intuitive idea of an *instant* is an extremely short period of time[27] but... how short? How about one second? Or, maybe one millisecond? And, why not choose a microsecond? Let us select any one of them, say a ***second***, and ask ourselves in which *instant* -after the ***previous*** *instant*- I began to walk. Well, of course, one ***second after***! However, employing the same mental scheme I did before, to subdivide one second into as many intervals as I please and thinking of that interval as the summation of an ***infinite*** number of ***infinitesimal*** *intervals* at the end of which I had to decide whether to start walking or not, it is fallaciously concluded that I would have never started moving my legs -- another silly trick to avoid embarrassment! But, then... how could I have had a position in *space* 'at every instant' along my trajectory in *space-time* towards the center of the stage if I never started to *move*?

These apparent absurdities related to *motion* are known as Zeno's paradoxes, among which the most famous is that of the race between Aquilles and a tortoise[28] whereby, if the latter -zoological archetype of sluggishness- had a head-start of just a ***microsecond*** (or any other ridiculous non-zero period of

---

24 Universidad Nacional del Sur, Bahía Blanca, Buenos Aires, Argentina, Earth, Milky Way, Universe, 1966.

25 Measuring *time* and *space* consists in assigning to them what mathematicians call 'real' numbers. A most important feature is that they form a 'continuum', so that between any two numbers there is always another one.

26 Infinitely small but non-zero numbers are called "infinitesimals" by mathematicians.

27 The Merriam-Webster reads for the meaning of 'instant': "an infinitesimal space of time".

28 This paradox can be seen also as the result of the ill-fated "Principle of the excluded middle" considered valid from Aristotle through Descartes and Leibniz, and responsible for the so-called 'Ontological argument for the existence of God', finally refuted by Kant.

time), the 'swift-footed" Aquilles, following the previously described fallacious line of argument, could not ever catch up with the turtle. Zeno utilized these astute arguments to 'prove' the impossibility of *motion* and, consequently, the truth of the doctrine of the monist[29] Parmenides who believed that *time* and *motion* were illusions. So much for these great philosophers not having ever seen a turtle winning a race -- not even against a baby in diapers!

Back in those days of intellectual virginity, it was very difficult for me to grasp the difference between a mathematical ***model*** or ***theory*** and the Universe itself (***objective reality***) and even harder to understand that such a ***model*** could be very effective and exact but, at the same time, miserably fail when we asked too much from it, or used it inappropriately. But... is it really too much to ask that if we are to represent *space* and *time* with numbers (determined through *measurement*), those numbers should allow us to describe something so simple and mundane as stepping up to the stage? After all, I do not have a scintilla of doubt that my embarrassing experience happened in a **finite** non-zero *interval of time* while I traversed a **finite** non-zero *distance*.

Of course that it is not too much to ask: to my intellectual delight, I also soon learnt that the summation of an **infinite** number of summands does **not** have to be **infinite** but that, in the case of my puerile ruse, this summation is **finite** and equal to the *distance* between the center of the dais and its rim, or to the *time* elapsed to travel that *distance* (according to whether the summands represent the *spatial* or the *temporal* intervals respectively). The mathematical tool I am talking about is called *differential and integral calculi*[30] and it is surprisingly effective to describe and predict physical **reality**. My mistake resided in imagining that my real trajectory involved an infinite series of decisions and corresponding actions that, naturally, a human being cannot physically carry out. Most probably, what I did that theatrical day was to complete -at most- two or three of the half-distance planned stages, when I already found myself trying to sloppily regurgitate the damned poem.

Likewise, with respect to the word *instant*, I learned that it is only a term to refer to an idea whose practical quantification changes with the circumstances, but that its theoretical meaning when referring to *time* is the same as for the word *point*[31] (*position*) when we refer to *space*: useful geometrical abstractions[32] which are impossible to be realized in the real world.

The pseudoparadox emerging from realizing that every finite change involves an infinite number of infinitesimal changes, with the detected contradictions, could be avoided by denying the ***continuous*** character of *time* and *space* and postulating that there is a ***minimum*** *interval of time* -let us call it a quantum of *time* or chrono- and a ***minimum*** *interval of space* -let us call it a quantum of *space*- both of which cannot be subdivided any longer. But how do we choose those quanta? Is there any physical basis for such a choice? Would *Integral Calculus* -modified accordingly- still accurately correspond to and predict physical **reality**?

Despite all that, I have to say that, for now, there would be no practical

29 Monism is the philosophical view that reality is one unitary organic whole with no independent parts.
30 *Calculus* was invented independently by Newton and Leibniz.
31 This is why a person who gets to a meeting at precisely the agreed time is called 'punctual'.
32 Next chapter will cover in detail the difference between mathematical geometry and physical geometry.

benefit for us in doing so because, as we mentioned, *Calculus* -based on the *continuum* hypothesis- is sufficiently powerful to provide the correct results (wait though for my next book to be disappointed!). All we have to do is not to fall into the trap of literally interpreting the abstract notions of *instant* (punctual *time*) and *position* (punctual *space*), and let the mathematics do its job without philosophizing unnecessarily. As we shall see in this book and the next one, Science had to confront real inconsistencies -much more substantial than these inoffensive paradoxes- to be compelled to develop Relativity and Quantum Physics during the twentieth century.

2.4. Objective Time (Measurement)

It is, thus, beyond doubt that human *perception* of *time* is little more than merely ***qualitative***: during the fictitious encounter between Einstein and Paulette Goddard, the former could have erroneously estimated his period of pleasure as 15 minutes or even 30 minutes instead of 1 minute, and still all of us would feel empathically identified with him... even though we would still distrust his ability to remain objective during romantic, entertaining, or intellectually-engaging situations.

The same can be said of when Mrs. Albizuri hugged and kissed me even though, because a kiss of a beautiful teacher to a pupil naturally only lasts for -at most- a second (what a pity!), I purposely said "it seemed a microsecond" to me. A *millionth of a second* is well below our sensorial threshold and, ergo, inaccessible to our ability to discern two events (start and end of the kiss) that close in *time*. Due to the briefness of the kiss, my possibilities to describe the *subjective* character of my estimate were more limited than those of Einstein with Mrs. Chaplin, so I turned to dramatization so as to assure the message came across to the reader. As in the case with Einstein, no sensible person would trust my chronometric skills under those circumstances.

Now, even if we assume for a moment that Einstein and I lived in a 'neuro-physiologic superior reality' allowing us to discern ***infinitely short*** *time intervals* (my 'microsecond' would now turn from *impossible* into *subjective*), in spite of the great flexibility we had to estimate the *time interval* and still stay within the limits of the *subjective*, there is something that neither he nor I could have said without sensible people labeling us as demented or simply liars: *that the sensation of pleasure* ***preceded*** *the encounter*[33] But why? This distinction between the *subjective* and *madness* is normally expressed by saying that our *psychological* experience of *time*, even though highly unreliable, 'flows' in the ***same*** *direction* as physical *time*. However, the latter phrase may look witty and pithy but tells us very little, until we deepen into the two *topological* (qualitative) properties of *time*: ***order*** and ***direction***.

For now, let us note that the sensation of pleasure was the ***effect***[34] of the interaction between both people; the *event* of the kiss occurred in a *place* and at a *moment* constituting the ***cause*** of my satisfaction, which was another *event* occurring in a *place* and at a *moment*. We say that the pleasure ***succeeded*** the kiss or, equivalently, that the latter ***preceded*** the former, this judgment being totally independent of our physiological and psychological limitations.

33 I discard, of course, the pleasure of thinking what is going to happen!

34 I am sure that the reader has a clear -though perhaps only intuitive- idea of the concepts of *cause* and *effect*.

But... let us be a little more inquisitive: the folkloric aphorism says that "the *cause* ***precedes*** the *effect*" giving the impression that the notion of 'preceding' (i.e. the notion of *time*) is more basic (fundamental) than the concepts of *cause* and *effect* and, consequently, the adage should be asserting something subtle and essential about the notion of *causality*. Quite the contrary! As we shall discover right away, *causality* -in our macroscopic world[35]- is a basic fact observable and characterizable without any *a priori* reference to the notion of *time*, and from which our notion of *temporal* ***order*** emanates -- one of the two *topological* properties of *time*. In order for us to understand all this, we need to leave the realm of *perception* (***subjective***) and enter that of physical *measurement* (***objective***) so we can discuss the epistemology of *objective time*.

2.4.1.Topological Properties of Objective Time: Order, Topological Simultaneity, and Direction

Time is completely characterized by its qualitative (*topological*) and quantitative (*metric*) attributes. ***Order*** and ***direction*** are *topological* attributes that convey the essential nature of *time* and sharply differentiate it from *space*. Brainstorming these two qualitative properties will allow us to conceive how to define the needed *metric* attributes so we can ***measure*** *time*.

Leibniz was the first one who suggested that our notion of *temporal* ***order*** was associated with the more basic idea of *causal* ***order***. To begin with, let me convince the reader that the *cause-effect relation* is in fact more basic than, and independent from, the notion of *time*. After that, I will demonstrate that the *temporal* ***order*** between two *events* occurring in our macroscopic objective world can be established from our ***perception*** and ***discernment*** of the *cause-effect relation* between them. When we finally reach Chapter 6, where we shall discuss Special Relativity Theory, we will corroborate the logical consistency and objective validity of this conclusion. Let's go for it!

Given two events A and B, we say that A **is** a *cause* of B when a small variation of A **is** associated with a small variation of B, but a small variation of B is **not** associated with any variation of A. Correspondingly, we say B **is** an *effect* of A. As an example, imagine we turn a lantern on (event A) and point it towards a screen which becomes illuminated (event B) and stays that way while the flashlight is on. All of us understand what we mean when we say that the illuminated zone on the screen is *caused* by the lantern. How can we prove it by using our definition of *cause-effect*? If we interpose a blue glass between the lantern and the screen, close by the former (variation of A), the bright zone on the screen turns blue (variation of B). We only need to insert and take out the glass a few times to convince ourselves of this correlation between the changes in A and the changes in B. However, there is **no** conceivable variation of B that will reflect on a variation of A (e.g. no matter where you insert the glass and which color you manage to turn the illuminated screen into, the light produced by the lantern will be always white). We say then that the bright zone on the screen is *caused* by the light propagated from the lamp, i.e. B **is** an *effect* of A and, moreover, B **cannot** be a *cause* of A.

35 As we shall see in my next book about Quantum Physics and folklore, the *cause-effect* relation between two events in the microscopic world is a relation not between the events, but between their probabilities of occurrence.

From the above considerations, we see that a fundamental property of the *cause-effect relation* is its ***asymmetry***, i.e. if A **is** a *cause* of B then B **is not** a *cause* of A. Besides, if A **is** a *cause* of B and B **is** a *cause* of C then A **is** also a *cause* of C because we only need to understand that A being a *cause* of B, a variation of A will imply a variation of B, but because B **is** a *cause* of C, the variation of B, in turn, implies a variation of C, so A **is** a *cause* of C. We say that the *cause-effect relation* is then ***transitive*** and that the *events* A, B, and C constitute a *causal chain*.

Asymmetry, in turn, implies that the *cause-effect relation* is ***irreflexive***, i.e. any event A cannot be its own *cause* because a variation of A will of course reflect on itself (A influences B=A) but, just because of that, it violates the second part of the definition (B=A cannot influence A!). Finally, the *cause-effect relation* is ***incomplete***, which means that given any two *events* A and B, it is possible that neither A is a *cause* of B nor B is a *cause* of A. Colloquially we say that neither A can affect B nor B can affect A.

It can be proved that these four properties (***asymmetric***, ***irreflexive***, ***transitive***, and ***incomplete***) of the *causality relation* can be used to establish a ***partial order*** between *events* in our Universe. Let us graphically express this *partial order* as follows: *events* are represented by small circles and, when an *event* is the *cause* of another, we draw an oriented line from the *cause* towards the *effect*.

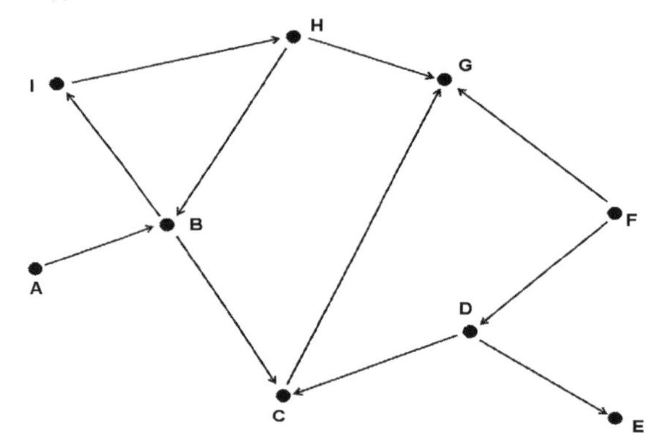

Figure 1. Causal Network for the Events A-I

From the *causal* graph in Figure 1, the following are examples of *causal chains*: (A,B,C,G), (I,H,G), (B,C,G), (H,B,C,G), (F,D,E), etc. Given the definition and properties of the notion of *causality*, it is not difficult to understand that there are two impossible entities in a *causal* graph or net: a) single lines which emanate from and die on a given ***single*** *event*; b) cycles, i.e. a *causal chain* that starts and ends on the ***same*** *event*. In both cases, that particular *event* would be the direct or indirect *cause* of itself, a situation prohibited by the definition of the *cause-effect relation*.

2.4.1.1. Temporal Order and its Absence (Topological Simultaneity)

It is paramount to realize that the ***order*** of *events* defined by the *cause-effect relation* exists exclusively in virtue of the observable phenomenon of *causality* and its properties, and that we have not made a single reference to the notion of *time*. It is indisputable then (despite the existing folklore) that the notion of *causality* is more fundamental than the notion of *time* or that, at least, the former does not require the latter for its formulation.

Now I beg the reader to take great pains in getting rid of her/his preconceptions regarding the notion of *simultaneity* (exclusively relying on our everyday local experiences) and try hard to digest how intuitive and natural the following definition of ***before***, ***after***, and ***simultaneous*** is:

> *Given two events A and B, if A is a direct or indirect **cause** of B, we say that A occurs **before** B or, equivalently, if B is a direct or indirect **effect** of A, we say that B occurs **after** A. If, instead, neither A is a cause of B nor B is a cause of A, we say that A and B are **indeterminate** in their temporal order and we will call them, by definition, **topologically simultaneous**.*

With this definition of *temporal* ***order***, the network of *causal* ***order*** in Figure 1 becomes now a network of *temporal* ***order*** and we easily conclude, for instance, that A ***precedes*** G (through B and C or through B, I, and H), and that A is ***simultaneous*** with D (because there are **no** *causal chains* joining A and D). The *cause-effect relation* entitles us to ***partially order*** any set of *events* from ***before*** to ***after***: all those *events* that could be *effects* of a ***present*** *event* constitute the ***future***, and all those *events* that could have been the *cause* of a ***present*** event constitute the ***past***. Please notice that I am talking about the ***past*** and the ***future*** with respect to a single *event* that I call the ***present***. I cannot but emphasize that this definition gives us a criterion to, given any two events A and B, establish which one occurred first and that, when this rule fails (because A cannot be either the ***cause*** or the ***effect*** of B), then we are forced to declare them as *topologically* (qualitatively) ***simultaneous***.

This definition of *simultaneity* has the virtue of reflecting the essence of our intuition which tells us that two *events* are ***simultaneous*** when it is impossible for us to *order* them and, at the same time, it has the not-less-valuable virtue of avoiding the folkloric definition (it would be circular at this moment!) which reads "two events are simultaneous when they occur at the same time". And why circular? Because we do not know yet how to ***measure*** *time* and, ergo, we do not know what it means to say that the *times* of occurrence for two *events* are **equal** (i.e. that they happen at the **same** *time*).

What is not so good news is that our definition of *topological simultaneity* establishes a relation between *events* which, contrary to our intuition, is **not** *transitive*: we see in Figure 1 that A is ***simultaneous*** with D and D is ***simultaneous*** with H; but alas, because there exists the *causal chain* (A, B, I, H), A and H are **not** *simultaneous*! Here is the crux of the matter and the genesis of the ***relativity*** of *time* but, once again, let us not haste; we need to learn how to ***measure*** *time* and then we will be able to transform our *topological* (qualitative) *simultaneity* into ***metric*** (quantitative) *simultaneity*, and this latter shall be ***transitive*** -- as our intuition and common sense demand.

It is interesting to realize that the oriented line joining a *cause* with an *effect* in a *causal* graph exceeds its symbolic meaning as it also suggests almost explicitly the idea of a physical signal propagating from the *cause event* to the *effect event*. In fact, in our macroscopic physical world, if two *events* are *causally* related, there must exist a physical signal, e.g. a light beam, which can propagate from one *event* to the other. Otherwise, how could one be affected by the other? From this, we conclude that if we can prove that there is no physical signal in the Universe that can connect two given *events*, then they cannot be *causally* related. This is known as the *Principle of Locality*, of which I will have an awful lot to say in my next book.

As I already emphasized, we still do not know how to objectively ***measure time***, so when in the following discussions I use, e.g. the symbol ‘ t_A ’, I am referring to that something called ‘time’ that, for now, can be ordered, but not to a mathematical *number* that ***measures*** it.

Any two places in *space*, say our Earth and Mars, can be connected through a signal -e.g. a radio signal- which by virtue of propagating, defines a *causal chain* beginning on the former and ending on the latter (Figure 2). Notice I said ‘places’, not ‘events’, which allowed me to employ the indeterminate adjective ‘any’. Let us call A the *event* occurring on Earth at time t_A (departure) and B the event on Mars at time t_B (arrival). Our Martian neighbors return the signal immediately as a token of friendship, defining another event C on Earth at time t_C. From our previous arguments, it is clear that B occurs after A, and C after B and, of course, after A.

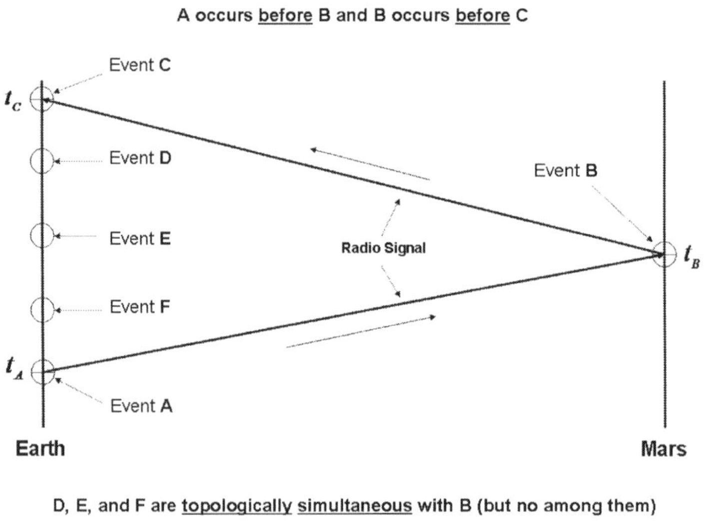

A occurs before B and B occurs before C

Event C
t_C
Event D
Event B
Event E
Radio Signal
t_B
Event F
t_A
Event A
Earth
Mars

D, E, and F are topologically simultaneous with B (but no among them)

Figure 2. Topological Simultaneity between Earth and Mars

Now we ask: what *temporal order* exists between event B (arrival and departure on Mars) and all the happiness and misery that occur on Earth with times between t_A and t_C (e.g. events D, E, and F)? The answer is that, in order for us to be able to establish a *temporal order* between any one of those terrestrial events and the event B on Mars, there must exist a physical signal that can connect them on Earth with the event on Mars. But we already saw that any signal departing from Earth after t_A has to arrive after t_B, and any signal departing from Mars at t_B cannot arrive on Earth before t_C. In sum: all events on Earth with *times* between t_A and t_C are *causally* ***isolated*** with the event B on Mars and, therefore, they are ***indeterminate*** in their *temporal* ***order*** with respect to the event B on Mars. According to our definition of *topological simultaneity*, based on the *causality* of our Universe, we arrive at the following surprising result:

> *All events on Earth with times between those of events A and C -independently of their temporal order between themselves- are* ***topologically simultaneous*** *with the event B on Mars.*

Something interesting to note is that we did not consider at all the relative *motion* between our planet and Mars and, consequently, our conclusion has nothing to do, as a matter of principle, with the relativity of *motion* and, even less to do -of course- with our *subjective time*. The *interval of time* elapsed between t_A and t_C (which we still do not know how to measure) is the *time* that the signal takes to travel from Earth to Mars and back to Earth so, the ***faster*** the signals propagates, the ***shorter*** that *interval* will be until, at an ***infinite*** *speed* we obtain $t_A = t_B = t_C$, and both planets would share a common time -- the ***absolute*** *time* of Newton. The same conclusion can be attained for a signal with a ***finite*** *speed* of propagation, but assuming that the two places in *space* are closer and closer, and that is why in our daily life (involving very ***small*** *distances*) ***absolute*** *time* and the ***uniqueness*** of *simultaneity* appear to us so viscerally logical and natural.

However, it is an established experimental fact that any physical signal takes a ***finite non-zero*** *time* to travel a ***finite non-zero*** *distance* and, even more, that the *speed of light in vacuum* is the limit for any other physical process in the Universe. In short: there is **no** *causal chain* that propagates faster than *light in vacuum*. We can therefore affirm that:

> *Every event (B) occurring at a place in the Universe is* ***topologically simultaneous*** *with a set of events (with times between those of A and C) at another place (A) of the Universe, this set corresponding to the interval of time defined at the second place (A) when light goes to the first place and comes back (A,B,C).*

Once we learn how to ***measure*** *time* (combining its *topological* and *metrical* properties), we shall return to the concept of *simultaneity*. Before that, and to finish this section, let us briefly present the second *topological* property of *time*: its ***direction***.

2.4.1.2. The Arrow of Time

The ***direction*** or ***arrow*** of *time* is a feature commonly expressed as: "time flows from past towards future". Does this statement have a meaning different from the one that is implicit in the definitions of the terms 'past' and 'future'? Could our language exchange the meanings of those two words without obliterating our whole Science and vision of the world achieved after millennia? Tersely: is this contention a tautology or not? The science of Thermodynamics was the first to tackle this problem.

Arrow (direction) and *order* are frequently confused. We know that the points of a straight line can be *ordered* using, for instance, the relation 'to the left of', because this relation is ***asymmetric*** and ***transitive***; but this *order* does not give the straight line an inherent *direction*. Why? Because there is no structural difference between that straight line so ordered with the one we would have obtained had we ordered it with the inverse relation 'to the right of'. We just need to turn the straight line 180° to see that nothing has changed in its structure to enable us to talk about a *direction* -- without reference to the definitions of ***left*** and ***right*** and their dependence upon the orientation of the observer. This clearly indicates that *order* and *direction* are different notions, i.e. we already learnt that *time* has an ***order*** but we cannot deduce from this fact that *time* has a ***direction***.

Imagine we throw a ball into the air and film its trajectory until it hits the floor. We then comfortably sit down and watch two movies; the first one consists in seeing the sequence of images in the same *temporal* ***order*** they were taken (from ***before*** to ***after***), and the second movie shows us the reverse sequence (from ***after*** to ***before***). We know the first movie shows us the **real** process: the ball went from a place A through the air up to a place B. Now we ask, based on the mere observation of the two movies (without having pre-marked which place is A and which is B): is there a way of knowing which one of the two movies is the **real** (forward) one?

The second (reverse) movie could be the result of having thrown the ball from B to A and then watching this film in reverse would be indistinguishable from the first one we watched as the forward version of throwing the ball from A to B. There is no characteristic whatsoever in both versions of the same phenomenon which could allow us to determine which one of the movies is the one played backwards: ***both versions can occur in the real world***. Processes as these are called ***reversible***.

Now imagine we film a person walking normally and, again, watch the movie forward and backward. We now see in the second movie a person walking backwards and causes us some surprise but... can we say the first movie corresponds to the real process and the second does not occur in the real world? Of course not! We walk the way we walk because it is much easier to control our movements using our eyes, but it does not exist any physical law impeding us to walk backwards and, if we are skillful and the movie does not last long, the observer will be incapable of discerning whether s/he is watching what **really** was filmed or just the movie in reverse! We have ***order*** again, but without an intrinsic ***direction***. Another ***reversible*** process.

To conclude, I sit down with my wife to watch *Casablanca* and focus on a brief scene where Humphrey Bogart refers naturally to Ingrid Bergman as 'kid' while he is, of course, smoking. We then run the scene in reverse and observe carefully what happens. We already know from the previous discussion that some movements will call our attention because they do not correspond to the normal way human beings move -- but they are practically imperceptible because 'Rick' and 'Ilsa' are steady, talking.

But there is something much more interesting: the smoke, instead of going ***away*** from the cigarette seems to get ***into*** it, and the butt instead of ***short-***

ening while the tube of ash ***lengthens***, seems to grow ***longer*** while the ash tubing ***shortens***. Watching the movie backwards produces an ***intrinsic*** change in the ***order*** of the images: overwhelmed with intellectual enthusiasm, any observer could discriminate between the two movies and pinpoint with absolute certainty which one corresponds to **reality** (i.e. between the forward and backward versions). Why? The observer would explain that one of the movies does not correspond to a process in the **real** world: no mortal has ever seen the smoke going into the cigarette while its unburned body keeps growing until the smoke disappears and the cigarette emerges intact. Clearly, the movie showing this **unreal** process has to be simply the backward version of the one which was **actually** filmed. We say that *combustion* is an ***irreversible*** process, i.e. its inverse process does **not** occur in Nature.

Does Science have a quantitative way of discriminating between *reversible* and *irreversible* processes? Yes; and it does it through the concept of *entropy* of a physical process, which distinguishes between *energy* that can be reverted (transferred back and forth) and *energy* which has only one ***direction*** for its transfer. In reality, the concept of *reversibility* is an idealization reached as a limit of an *irreversible* process in which the part of the *energy* transfer which cannot be reverted is negligible. The reader may remember my insistence in Chapter 1 with respect to the ***inexistence*** of *friction* on the Moon during the free-fall of an object, so as to be entitled to neglect the *irreversible* conversion of *mechanical energy* into *heat* and, consequently, be able to assure the complete *reversible* conversion between *kinetic* and *gravitational energy*.

Once this concept of *entropy* has been defined, besides the *Law of Conservation of Energy* we saw in Chapter 1, we can formulate the so-called *Second Law of Thermodynamics*[36] asserting that the entropy always ***increases*** in a closed system (i.e. when all interactions are included). The existence of a **real** physical process corresponding to the *backward* version of *Casablanca* would imply that its *entropy* ***decreased***, and that is forbidden by the *Second Law*, explaining why no mortal has ever seen something like that (except in a movie in reverse).

It seems, thus, as if the science of thermodynamics had found the proof that *time* does have an inherent ***direction***, because saying that 'time flows from past to future' is not just due to the very definitions of *past* and *future*, but to the fact that there is a ***direction*** permitted, and another prohibited by Nature, thereby defining the so-called ***arrow of time***. As it usually occurs, this solution was too simple and direct to be correct: Ludwig Boltzman (1844-1906) discovered in 1872 that the *Second Law of Thermodynamics* was not a strict or ***deterministic*** law but a ***probabilistic*** one, which means that the law, instead of being expressed as 'entropy always increases', must be formulated as 'it is highly probable that entropy will increase'. Summing up, the observer of *Casablanca* could, with justified wisdom, bet all his fortune that the movie showing the smoke going into the cigarette is **not** the one which was ***actually*** filmed but, instead, its ***backward*** version... though... he could also end up the rest of his life under a bridge!

In this fashion we end our treatment of the *arrow of time*. The reason why

36 The *First Law of Thermodynamics* is simply the *Law of Conservation of Energy*.

we do not deepen further into this important problem is because it cannot be resolved within relativistic physics, as it requires the notion of probabilistic causal relations, instead of the deterministic ones assumed by Relativity. We will resume the subject in my next book entitled *What is Reality? - Einstein, Quantum Physics, and Folklore*.

2.4.2. Metrical Properties of Objective Time: Unity, Uniformity, and Metric Simultaneity

It is well-known that in order to *measure* a physical magnitude we need a definition for what is called a ***unit*** of measurement. For instance, the *standard meter* still zealously kept in Paris, was the *definition* for the *unit of length* in the international metric system for a long time. The process of ***measuring*** consists in comparing the object/process being measured with the *standard unit* so, to undertake such a comparison in the case of *time*, we need another *definition* which will allow us to determine when two *intervals of time* are **equal**. These *metric definitions*, as such, have a high degree of ***arbitrariness***, so it is paramount we learn how to distinguish between what is ***conventional*** and what is reflecting physical **reality**.

Humanity developed the notion of *time* as 'something' that flows by means of the observation of *recurrent* natural phenomena: day and night, phases of the Moon, *motion* of the Sun on the sky, *motion* of the stars on the celestial sphere, etc. In the same way the measurement of *length* requires that the concept of *unit* be associated with a physical object, humanity started associating the concept of *unit of time* with a ***cyclic*** natural process, defining the *unit of time* as that 'something' elapsed between two ***consecutive*** observed *occurrences* of the chosen cyclical phenomenon. In this way, three natural *units* were born: day, month, and year respectively associated with the rotation of the Earth around its imaginary axis, the rotation of the Moon around our planet, and the Earth rotation around the Sun.

Next step was to define when two *intervals of time* were **equal.** In the case of measuring *distance*, the standard *definition* of **equality** between two *lengths* is one of *congruence* or physical superposition: two bodies have the **same** *length* if, when put together side by side, their extremes **coincide.** As simple as that! In the case of *time*, though, we cannot take two *intervals of time* and put them one by the other to check whether their ends coincide. What could be done about it, then?

2.4.2.1. Direct Measurement of Objective Time - Uniformity

The most primitive way of ***measuring*** *time*, once a natural cyclic process was selected, consisted in counting the number of its cycles that occurred from an arbitrary event taken as 'zero'. But the immediate concern is: how do we know that the 'something' elapsed between two consecutive occurrences is always the same? If it was not, saying that '45 moons have elapsed' seems not to carry much information because only God knows how much of that 'something' we are measuring the Moon took to change its phases for the cycle, say 33, and how much it took for the cycle, say, 21. As we saw, it is impossible to take two cycles, put them side by side to see if their 'ends' coincide and, obviously, our ambassador of humanity thousands of years before Christ[37] could not pick up a chronometer from his pocket and measure each one of the cycles -- because he still had not defined when two *intervals of time* were supposed to be **equal**! (neglecting of course the fact that chronometers did not exist in those days).

In another way: it is only when the selected cyclic process shows ***uniformity*** in its cycles that we can choose the ***common*** cycle as a *unit of time* and, then, simply counting the number of cycles delivers a ***measure*** of the *time* elapsed between the beginning of the 'first' cycle and the end of the 'last'. But... how do we know that all cycles 'last' the same? The answer to this epistemological dilemma will surely astonish the reader, but it is the only one available: There is no way of knowing it! It is not a matter of **knowing**; it is a matter of **defining**!

The solution to the problem of ***measuring*** *time* demands another decree-style *definition*: once we have selected the cyclic process, its cycles are considered, ***by definition***, as having the **same** *duration*; full stop! The standard process is the paragon that rules, by the divine grace of a ***convention***, the *uniform flow of time*, and any one of its cycles (all of equal duration) constitutes the **unit** adopted for ***measuring*** any other physical process in the Universe. In this fashion, we can only measure integer multiples of the *standard cycle* but, choosing wisely the *standard process*, resolution and accuracy for the ***measurement*** of *time* could be gradually improved.

Philosophically, the only criterion to identify physical processes which are appropriate to be used as *standards of time* is that they should approximate the ideal of a closed (isolated) system, i.e. that they are not subject to external influences. This is because, according to the *Principle of Causality*, a perfectly isolated system which reaches a physical state identical to one where it was before, being free of any imaginable external influence, inevitably will repeat itself over and over. In the practical world, it is the posterior usage of the ***measurements*** and the simplicity and consistency of the ***physical laws*** obtained with those ***measurements***, what determines the appropriateness of the selected *standard process*. Accordingly, the process used as the *standard of time* has evolved through history, the ***atomic clock*** being the one used today because of its supreme *resolution*, *stability*, and *accuracy*.

A superficial reading of some physical laws leads us to believe that they 'predict' facts as, for example, the equality of the cycles of a pendulum. The truth is that those laws carry an *intrinsic* content which predicts the *objective* **reality**, as well as an *extrinsic* component which is a mere consequence of the ***arbitrariness*** of the ***definitions*** (conventions) needed to implement a ***measurement*** technique and, ergo, they say nothing inherent to the ***objective*** world. When we measure the period of a pendulum to prove the validity of *Newton's Law of Gravitation*, we use clocks which were calibrated in accordance with the assumed equality of its cycles, so it should not surprise us that, naturally, the physical law so obtained would 'predict' the constancy of the pendulum period. Discovering these things makes us all feel a little betrayed (at least I do) by the great majority of the Physics textbooks.

37 The first observations of the angular displacement of celestial bodies occurred in the mesopotamia (today Iraq) between 5,000 and 6,000 years before Christ. For this reason, given that the Babylonian civilization used the number 60 as the base for its numerical system, the divisions of the hour (minute and second) are still sexagesimal today, and have the same names as the divisions for geometrical angles.

2.4.2.2. Indirect Measurement of Objective Time - Uniformity

The direct ***measurement*** of *time* using the cycle of a recurrent standard process does not allow us to ***measure*** *durations* ***shorter*** that the *standard cycle*, unless we **change** the process. Another way of ***measuring*** *time* without changing the *standard process* is indirect, i.e. through the ***measurement*** of *distances, angles, weights*, etc. For example, if our *standard process* is the rotation of our planet around its imaginary axis, we chose as the *standard cycle* what we call a 'day' and declared that all 'days' have the **same** *duration* (otherwise we could not use it as a standard!). If we now introduce another *definition* through which we decree that our planet covers **equal** *angles* in **equal** *times*, then we can subdivide the 'day' in **shorter** *intervals of time* which will be considered **equal** if they correspond to **equal** *angles* of rotation. If the number of angles (meridians) is 24, we have defined the *unit of duration* called 'hour', and we accept the **equality** of each one of the 24 hours the day has, because we accept the assumption that the Earth rotation is **uniform**, i.e. it covers **equal** *angles* in **equal** *times*.

For thousands of years the shortest duration was the 'hour' which, due to the use of sundials, was initially defined as the twelfth of the day or of the night, so its *duration* differed between day and night, from place to place, and from season to season -- an inconvenience finally fixed with the advent of mechanical clocks. The sixtieth of an hour was called *minuta primam* (Latin for 'first small part') from which comes the word 'minute'. By the end of the sixteen century, the sixtieth of a minute was in use and called *minuta secundam* (Latin for 'second small part') from which it comes the word 'second'.

It is difficult to resist the temptation of asking: why did they not corroborate that Earth rotation was indeed ***uniform*** by *measuring* instead of assuming it? Well... because it was precisely the ***hypothesis*** of *uniformity* that permitted them to *define* the ***fractions of time*** necessary to be able to *measure* it with a resolution better than a 'day'! And, naturally, if we attempt to *measure* the process chosen as *standard*, we will obtain the *hypothesis* we adopted in the first place! Only the selection of a more *appropriate process* as a *standard* can break this vicious circle.

Now, the more we used the astronomical *units of time* (day, month, and year), the more complications and contradictions we found[38] and, little by little, after innumerable millenary adjustments, by the beginning of the second half of the twentieth century, we decided to leave the macrocosm and search for a better *unit of time* in the microcosm. It was only when we decided to base the **unit** on another more appropriate physical process, i.e. when we shifted our *definition* of ***uniformity*** to a new *standard process*, that we could *measure* the ***uniformity*** of the Earth rotation revealing the causes of contradiction, and finding for example that its angular velocity is decreasing very slowly so that the solar day is increasing at a pace of about 1.5 milliseconds per century! Or that its imaginary axis of rotation wobbles in an easily measured though unpredictable way. What initially appeared to our senses as ***cyclic*** and highly

38 Julius Caesar called the year 46 BC "the end of confusion" when he established the so-called Julian Calendar, defining the 'year' with 365 days and adding one more day to February every four years. Because Easter is biblically related with the Moon phases and the spring equinox, with the consequent undesired shifting of the religious festivities, Pope Gregory XIII instituted, in 1582, the so-called Gregorian calendar (the one actually in use). In those days, politics, religion, and science were inseparable.

uniform, because our techniques of physical ***measurement*** became more and more elaborate, revealed itself as ***irregular*** and even ***chaotic***.

2.4.2.3. Galileo's Water Clock - Uniformity

As another example of ***indirect*** *measurement* of *time*, illuminating for me, we saw in Chapter 1 that Galileo proved experimentally (letting little polished spheres fall down polished inclined planes) that, when there is **no** *friction*, all bodies fall with the **same** *acceleration* independently of their *composition* and *mass*. At that moment we explained, as a footnote, that he employed inclined planes to ***increase*** the *duration* of each experiment so as to *measure time* more accurately.

And, how did Galileo measure *time?* Here is an excerpt from the master [9]:

For the measurement of time, we employed a large vessel of water placed in an elevated position; to the bottom of this vessel was soldered a pipe of small diameter giving a thin jet of water, which we collected in a small glass during the time of each descent, whether for the whole length of the channel or for a part of its length; the water thus collected was weighed, after each descent, on a very accurate balance; the differences and ratios of these weights gave us the differences and ratios of the times, and this with such accuracy that although the operation was repeated many, many times, there was no appreciable discrepancy in the results.[39]

Clearly, Galileo *measured time* ***indirectly*** through the *measurement* of the ***weight*** of water collected during each experiment, and assumed that the ***weight*** of accumulated water was *proportional* to the *time* elapsed between opening and closing the valve and, ergo, that the *difference of weights* was *proportional* to the *difference of times* and the *ratio of weights* were *equal* to the *ratio of times*. The natural process he selected as *standard* was that of water[40] flowing out of a tank, and the *unit of time* was defined by that 'time' which a given ***weight*** of water took to discharge. The idea is exactly the same behind a sand clock. And, what was Galileo's rationale to decree that water ***weight*** and *time* were *proportional* to each other? Allow me to modestly put myself into Galileo's shoes, imagining how his thought process might have been.

First, most surely it was very easy for him to realize that, repeating the same experiment, say letting the little ball fall down the whole channel length, the ***weight*** of accumulated water in the glass from aperture to closing of the valve ***changed*** with the level of water in the tank and, ergo, the *time* he wanted to *measure* (supposedly the same for the same experiment) would be different for different loads of the tank. In those circumstances, he could **not** use ***weight*** to measure *time*. Nonetheless, after a couple of experiments, he noticed that the bigger the vessel was, the more ***insignificant*** the differences detected by the balance were when ***weighing*** the different amounts of water discharged in the glass and, in this fashion, he determined the ***minimum*** volume of the tank he refers to as "a large vessel". If he then was careful enough to execute all the experiments with the vessel about full, one and only one ***weight*** of accumulated water would correspond to any *duration*, and vice versa. This rela-

39 As cited in *Great Experiments in Physics*; New York: Dover Publications, 1959.

40 The water clock (clepsydra) is of Egyptian origin and was used mainly during night when sundials did not work.

tion -called by mathematicians 'bi-univocal' or 'one-to-one'- was essential to legitimately consider the ***weight*** of accumulated water in each experiment as a ***measure*** of its *duration*.

Second, he also immediately realized that if the *weight* of the collected water in the glass was excessively ***small***, then human reaction to open and close the valve would be ***significant*** with respect to the ***duration*** he was trying to ***measure***, adulterating the results. In this manner, Galileo determined the ***minimum weight*** of water he had to collect in the glass for this ***weight*** to be a reliable *measure* of the actual *time* the little ball would take going down the plane.

And here is when the genius of Galileo takes perhaps subconsciously -but nonetheless brilliant- the epistemological decision of assuming the ***uniformity*** of the relation between ***weight*** of water and *time*: he arbitrarily defined that the ***weight*** of accumulated water was *proportional* to the *time* he wanted to ***measure*** and, ergo, once the ***unit of time*** was *defined*, the *differences in weight* would be *proportional* to the *differences in time* and the *ratios of the weights* would be *equal* to the *ratios of the times* -- as he describes it when explaining his experiment and its results. Galileo could have assumed any other more complicated relation and then his law of free-fall would have been different due to a different *definition of time*, even though the new law would still be describing the same *objective* **reality**!

Flabbergasted? I am. Besides, when I consider these epistemological matters, I cannot help but feel a little bitter for not having been taught these fundamental concepts in College. However, to be fair, most probably I was not intellectually prepared to understand them by then... though a mere warning about the existence of arbitrary elements hidden behind what they were teaching me as sacrosanct, would have certainly assisted me in accelerating my slow intellectual maturation.

2.4.2.4. From Topological to Metrical Simultaneity

The *definition* of ***uniformity*** solves then the problem of discerning when two ***successive*** *intervals of time* in the **same** *place* are **equal**. However, there are situations in which we need to compare two *time intervals* which are 'parallel' and that, in general, occur in **different** *places*, and this takes us back to the problem of *simultaneity*. The situation in which these two ***concurrent*** *intervals* occur in the **same** *place* is a special case of the general one; the important thing here is that the two *intervals* are ***parallel***, as opposed to being ***successive*** as with the cyclic processes we already discussed.

Our question is: given the *definitions* of ***unit*** and ***uniformity***, are we entitled to objectively determine the ***metric*** *simultaneity* of two *events* or do we still need another *definition*? Until Einstein's paper in 1905, scientists believed the notion of *simultaneity* was ***absolute*** and did not need any *convention* beyond those for ***unit*** and ***uniformity***. According to Newton's ***absolute*** *time*, given any two events in the Universe, no matter how far away or how fast they were moving with respect to each other, it was always possible to assert without ambiguity and with universal validity that one of them occurred ***before***, or **after** the other, or they were ***simultaneous***. Well, get ready! Because we are by now suspecting that this is not *a priori* true but... let us first go back to our admired Galileo.

For Galileo, during his celebrated experiments, ***ordering*** the *durations* for his different trials consisted simply in ***ordering*** the *weights* of collected water. A ***greater*** *weight* implied an experiment with ***greater*** *duration*; a ***smaller*** *weight* implied a ***shorter*** one. Two trials, during which the *weights* of collected water were the ***same***, had the ***same*** *duration*. How could then he prove that any two bodies of different *materials* and *masses* launched *simultaneously* would get to the end line *simultaneously*?

One way would be to set up two identical inclined planes with identical water clocks to be synchronized (valves open at the same time) and then we should determine whether the closing of the valves were *simultaneous* or not (whether when one operator closed the valve, the ball in the other plane had arrived or was still running). Even though this description seems natural and simple, it would be really cumbersome to implement and highly inaccurate in its results, because it involves human ***perception*** of *simultaneity* instead of its ***objective*** *measurement*. Both the supposed public demonstration in the Leaning Tower of Pisa and the one on the Moon by Commander Scott were dramatically conceived to achieve maximum popular impact by using the human perception of the *simultaneous* launch and the *simultaneous* arrival. A much more accurate and simple way is to forget about human perception of *simultaneity* and, instead, infer *simultaneity* from the *objective measurement* of the *durations* of different experiments with different *materials* and *masses*.

Proving that different balls *simultaneously* launched reach the end line *simultaneously* is equivalent to proving that the *durations* of different experiments carried out with each one of the different balls are all the **same**. And all these experiments can be done one after the other on the same inclined plane. All Galileo had to do was to conduct the same experiment twice, one with a ball with a **large** *mass* and one with a **small** *mass*, *weigh* the corresponding glasses of collected water and, if these ***weights*** were **equal**, then the *durations* were **equal** and, therefore, had the two balls been launched *simultaneously*, they would have arrived *simultaneously*. Galileo concluded that, within the precision of his experimental technique, the ***weights*** of water were always the **same** independently of the *masses*, thereby establishing the **equality** between ***gravitational*** and ***inertial*** *masses* -- as we discussed at length in Chapter 1.

How could Galileo imagine that three centuries later, Einstein -in a period of intense inspiration and intellectual effort prolonged for 10 years- would deduce from this **equality** of *gravitational* and *inertial masses* that "space-time is curved"? Impossible! Both things seem to be as related to each other as "velocity is to bacon" (as my father would say in Spanish) and, notwithstanding, I aspire for the reader to finish this book with the feeling that it is possible to achieve an acceptable understanding of Relativity Theory, based on the *objective* **truth**, and free of folklore (popular as well as scientific).

2.4.2.5. Perception vs. Measurement of Simultaneity - Lighting vs. Thunder

But... what is the problem with the ***perception*** of *simultaneity*? For ***local*** *events*, i.e. when the *distance* between them is comparable with the human dimension, and as long as we accept as practically *simultaneous* those events whose degree of non-simultaneity is under our physiological threshold of discernment, there is no problem but... let us think about the well-known *temporal* relation between thunder and lightning involving *distances* quite greater than our human size.

About 5 seconds after having seen the lightning, I hear the thunder while my watch says 15:20. We all know that this *time* is indicative of the moment the thunder reached my ear and not of when it was produced in a place we suppose considerably afar, and we also know that had that place been much farther, or much closer, the *time* displayed by my watch would have been very different. Observers located in ***different*** *places perceive* and *measure* ***different*** *times* -- even when they are in relative *repose*. Determining the *time* at which the thunder occurred, besides our *perception* and **local** *watch*, requires the intervention of our *intellect*: if we know the *distance* between the *place* of the electrical discharge and our *position*, and we know the *speed* with which sound propagates in air, we can calculate the *interval of time* that the thunder's ***sonic*** wave took to reach my ear and, subtracting it from 15:20, we calculate the *time* at which thunder happened, say at 15:19:55. We can say the same regarding the ***light*** wave triggered by lightning though, due to its much greater *speed* of propagation (as compared to the *speed* of sound), we can assume that the *time* displayed by my watch is practically the same as the *time* of discharge, i.e. 15:19:55. In other words, had I been at the *place* where lightning and thunder occurred, our senses and our ***local*** *clock* would have indicated that both *events* were *simultaneous*.

Summarizing, two *events* occurring both in the same *place* far away from us, appear to our senses (audition and vision) as clearly ***separated*** in *time*. Moreover, even though both *events* were practically *simultaneous*, our **local** *clock* (an instrument for ***objective*** *measurement*) registers them as ***separated*** in *time* by 5 seconds. The only way for us not to be deceived by the apparent temporal priority of lightning, so as to discover the ***objective*** world, is to combine our senses and **local** *clock* with our *intellect*.

We have learned so far that determining the *interval of time* between two *events* ***separated*** in *space* comprises two steps: 1) a **local** comparison between our *clock* and the *arrival* of a signal that propagated from the faraway *place* to our ear/eye; and 2) an ***inference*** based on additional knowledge. Our *clock* plus the ***inference*** led us to conclude that the *time* of thunder was 15:19:55 in our *place*, far away from where the electrical discharge took place. Can we say that a *clock* located at the *place* of the discharge would have also indicated 15:19:55 when it happened? Or, rewording, is *simultaneity* between ***distant*** *events* an ***objective*** and ***univocal*** fact?

As we saw, the alluded *inference* requires our knowledge of the *distance* between *events* and the *velocity* of propagation of the signal connecting them. Suppose the *distance* was ***measured*** and try now to ***measure*** the signal *velocity*. This latter ***measurement***, knowing the *distance*, requires the ***measurement*** of two *times* (departure and arrival). The *distance* divided by the difference between those two *times* (elapsed *interval*) will give us (by definition) the signal *speed*. Now, those two *times* correspond to *events* occurring in very ***distant*** *places* and, if their values are to have any meaning for our purpose, the **local** *clocks* employed to determine them are to be *a priori* ***synchronized.*** By 'synchronized', we mean that they (departure and arrival clocks) displayed the **same** *time* when the signal departed; what compels us to accept that in order to ***measure*** *speed* we need to know when two ***distant*** *events* are *simultaneous*. But... damn it! Then...

*To determine **simultaneity** between two **distant** events, I need the **speed** of a signal propagating between them, and to determine that **speed**, I need to establish **simultaneity** between those events!*

Got it! So I do not have to synchronize two distant *clocks*, I simply choose ***light*** as the messenger, position a mirror at the ***distant place*** to reflect the signal and... presto! With only **one** *clock*, comfortably relaxed at my *place*, I determine the *time interval* between the *light beam* departure and its round-trip arrival -- with the only caveat that now the *distance* used in my calculation is, of course, ***twice*** as big as the *distance* between the ***distant*** *places*. Clearly, I have opened the vicious circle and it seems that, after all, determining *simultaneity* of ***distant*** *events* is not that big of a deal. Not so fast!

How do we know that light or any other signal propagates with the **same** *velocity* in both directions if we avoid measuring it? Epistemologically, the only way to know it, is to measure the *time of arrival* at the remote place before it returns to my cozy place and, for that, we need again to *synchronize* two *clocks* (one at each *place*), i.e. ***we need to know what we want to determine!*** Taken aback again? You betcha! How in the world an everyday notion so darned elemental as *velocity* can be logically vitiated?

I do not give up. How about *synchronizing* both *clocks* at my *place* where I can assure they display the **same** *time* and keep running with the **same** *pace* long enough for me to feel confident; then taking one of them aboard my private Concorde[41] to where the electrical discharge is about to occur, leaving it with my assistant so s/he can read the *time* on this ***transported*** *clock* when lightning strikes, and coming back to my place to hear it happen. Given that I did **not** need to *synchronize* ***distant*** *clocks*, it looks like now victory is in sight.

Sorry buddy, says Einstein to me: how do you know that two *clocks synchronized* at one *place*, stay *synchronized* when they are ***transported*** to different *places*? - Oh, Come on! You love splitting hairs! After the fireworks, I go back and put both *clocks* side by side confirming they still run *in unison*. - Astute! But not enough: all you would have proven is that both *clocks* are *synchronized* when they are together, but not that they were so while afar. Too much for me (and for the reader)! Let us continue the gradual path to comprehension, waiting for Chapter 6 where full enlightenment for this conundrum will unfold[42]. To wrap up, even with this last smart trick of ***transporting*** already *synchronized clocks*, we have to conclude that another *definition* (convention) is needed if we want to talk ***objectively*** and ***univocally*** about ***metric*** *simultaneity*.

Here we are face-to-face with what Einstein called the 'conventionality of simultaneity': in order to objectively and univocally ***measure*** and compare

41 I bought it as a bargain in 2003 when they retired!

42 In fact, the Theory of Relativity sustains that clocks lose their synchronization when transported.

time within **large** regions of *space*, we need to humbly accept the necessity of a third *convention* regarding the notion of *simultaneity*. And let us make it clear at the outset (debunking another piece of folklore): this ***necessity*** of another *definition* to establish *metric simultaneity* has absolutely nothing to do with the *relativity of motion*. In fact, we only assumed that the two *events* were ***distant***, not that they were in relative *motion*. However, most of the popular science books fallaciously affirm -and some of them even prove!- that the ***relativity*** of *simultaneity* is a consequence of the ***relativity*** of *motion*. Don't despair! Humanity believed *time* was ***absolute*** for millennia so... what are a few more chapters?

As young as sixteen years of age, Einstein imagined he was galloping with a *light wave*, i.e. he was riding on the *wave* and therefore traveling at the **same** *speed*. In such a hypothetical situation, 'horse' and 'rider' would be in relative *repose* and, consequently, Einstein could observe the *spatial* variation of the *electromagnetic field*, but **not** its *temporal* changes, i.e. for him the *wave* would be ***stationary***[43]. But Einstein was not a regular adolescent as we were: he immediately realized that Maxwell equations, which governed the propagation of the *wave* when it was observed from Earth, were invalid when observed from a *reference frame* traveling *in unison* with the *wave*. To a mind like Einstein's, it seemed instinctively inconceivable that the laws of Nature could be valid in one *frame of reference* and not in another, and dedicated his life to prove he was right. Ten years later, his introduction of the ***relativity of simultaneity***, i.e. the abolition of ***absolute time***, constituted the key to resolve the paradox that tormented his adolescence.

In order to transform the inherent ***multiplicity*** of *topological simultaneity* (Figure 2) into the indispensable ***uniqueness*** that *metric simultaneity* is to have, Einstein wisely assumed the simplest: that, to reach the remote place, the *light signal* took ***half the time*** it took for the round-trip, i.e. that *light* propagates with the **same** *speed* in both directions. This simple hypothesis was essential for his Special Theory of Relativity[44], but it is **not** epistemologically necessary. As with any *definition*, there is a degree of inherent arbitrariness which, inevitably, will inconspicuously creep up to our description of the **real** world and, if we are not careful thinkers, we may end up erroneously believing it is telling us about **reality** more than it actually is. With the same latitude Galileo had to choose his definition of ***uniformity***, Einstein could choose from a multitude of possibilities, all of them epistemologically tantamount -- though producing more complex descriptions of the same physical **reality**. A very important advantage of using Einstein's convention is that *metric simultaneity* becomes a relation which is ***symmetric***[45] and ***transitive***[46], as it was when considered absolute. Remember that *topological simultaneity* is **not *transitive*** (Figure 1).

Interested in knowing which Einstein's options were? More precisely: which was the range of values that the immeasurable arrival-to-the-remote-place *time* could have ***by convention*** without logical contradictions and without vio-

43 This becomes gradually clearer, the closer we get to Chapter 6 entitled *Galloping with Light - The Special Theory of Relativity*.

44 Einstein, *On the Electrodynamics of Moving Bodies*, Annalen der Physik, 17:891-921, June 30, 1905.

45 As a reminder: if A is simultaneous with B, then B is simultaneous with A.

46 As a reminder: if A is simultaneous with B and B is simultaneous with C, then A is simultaneous with C.

lating any basic laws of the Universe? Again, we have to resort to our sacrosanct *Principle of Causality*.

Figure 3 is a *space-time* diagram similar to the one in Figure 2 but adapted to the situation at hand. Let us remember that if thunder took -as we assumed- only 5 seconds to reach my ears, then light arrived on my eyes virtually in ***zero** time*. Nevertheless, so we can handle simple numbers, let us assume the *distance* between where the future electrical discharge will take place and my *place* is sufficiently long for *light round-trip* to take 2 minutes. Let us further assume that the *light beam* leaves my place at 16:58 and returns at 17:00. Einstein decided, for the sake of simplicity and convenience, that the *light beam* arrived at the remote place at 16:59, i.e. light took one minute to reach the place of the future electrical discharge, and another minute to get back to me.

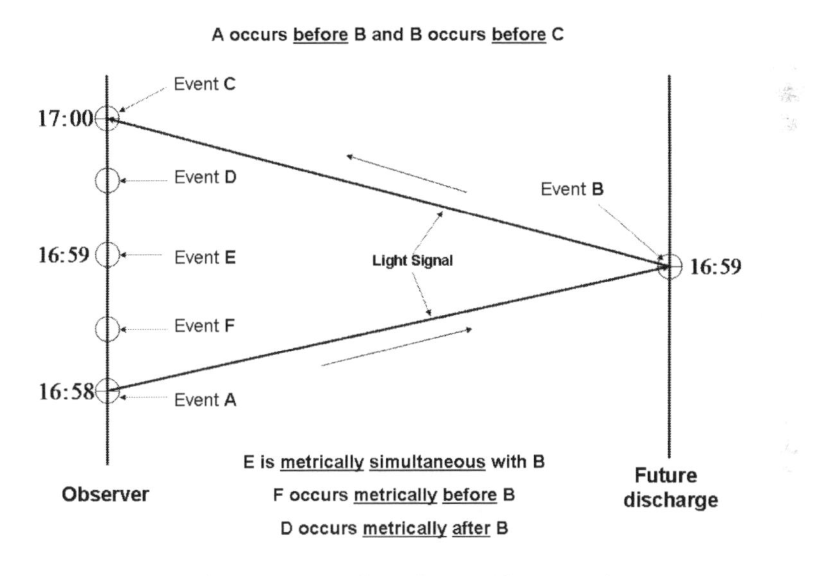

Figure 3. Einstein's convention of simultaneity between distant events.

Given the demonstrated logical impossibility of ***measuring*** the *simultaneity* of ***distant** events* (without a definition), the *arrival time* at the remote *place* could be defined arbitrarily as long as it is ***greater*** than 16:58 and ***smaller*** than 17:00. Why? Because if it was ***smaller*** or ***equal*** than 16:58, the light beam would have arrived at the remote place either ***before*** departure or in ***zero** time* (departure is the *cause* of arrival through the process of wave propagation) and, if it was ***greater*** or ***equal*** to 17:00, it could have not arrived back to me at 17:00, unless it took ***zero** time* to come back -- both cases against the archetype of a *causal chain* that *wave propagation* is. Any other number inside the open interval (16:58,17:00) can never produce any logical or physical contradiction.

We are thus entitled, with this degree of arbitrariness, to select one of the innumerable *time* values in (16:58,17:00) -all of them corresponding to *events topologically simultaneous* with the *arrival event* (B)- as the value which will be, *by convention*, ***metrically** simultaneous* with B. Einstein decided to choose the middle point (16:59) and, in this manner established, *by convention*, that the *speed of light* is the same in both directions. Also, given this *convention*, it follows that all *events* with *times **below*** 16:59 occur ***before*** the event B, and all events with *times **above*** 16:59 occur ***after*** the event B.

How do we then *synchronize* the ***distant** clocks* using Einstein's convention? Conceptually, the procedure -now finally free of circularity- is the following: (Figure 3): a) we send the *light signal* annotating the *time* given by our clock (16:58); b) we receive the signal back taking note of its *arrival time* on our *clock* (17:00); c) we subtract both numbers to come up with the *round-trip time* (2 min); we decree that the *time* elapsed till the signal arrived at the remote place is ***half*** the *round-trip time* (1 min); e) we now send again the signal, this time with crucial information: the summation of the *departure time* as indicated by **our** clock with the *time* that, by decree, we know the signal/information will take to reach the other *place*; and f) when the person/computer in charge of the ***distant** clock* receives the signal plus the needed information, s/he/it adjusts the clock to display a *time* which is that sum of our *departure time* plus the *propagation time*. From then on, both ***distant** clocks* will run *in unison*, and in accord with the adopted *definition of simultaneity*.

Conclusion: independently of the *definition of simultaneity* we use to *synchronize* the *clocks*, one at my *place* and the other where the thunder will occur, both will indicate the **same** *time* when the lightning bolt occurs and both will indicate the **same** *time* when the thunder reaches my ear. Articulated in this way, it does not seem we have learned a lot. Yes, we have! We have learnt that this conclusion, apparently so logical and natural, is the partial result of a *definition*, and not of an inherent property of the ***objective*** world. *Simultaneity* (topological) was **not** unique, but we made it unique (metrical) by means of a *convention* and, besides, using Einstein's convention, we made it (as we shall understand in Chapter 6) ***relative*** to the *frame of reference*. Quite an awful lot, indeed!

Wrapping up, it is to be noticed that the uncertainty (degree of arbitrariness) associated with the *definition of simultaneity* is much more reduced for *light* than for *sound*; and it would be much less for any signal that hypothetically traveled in *space **faster*** than *light*. If Newton's action at a *distance* (***infinite** speed* of propagation) were possible, then there would not be any arbitrariness, and *simultaneity* for ***distant** events* would be univocally determined -- with no need for any *convention*. Instead, as we said, it is a physical fact that every *casual chain* propagates in *space* with a **finite** *speed*, i.e. the effect requires a non-zero *time interval* to manifest itself at a non-zero *distance* from where the *cause* takes place.[47]

As I anticipated in Chapter 1 and repeat now, we shall see that the *speed of light in vacuum* is the physical limit for the propagation of any *causal chain* in the Universe. Thus, by selecting *light* as the physical process to define *simultaneity*, we obtain the ***minimum*** uncertainty possible, and by choosing Einstein's convention, we attain the ***maximum*** descriptive simplicity. Besides, had we used, for example, the process of *sound propagation* for defining *simultaneity*, then it is not difficult to prove that the *Principle of Causality* would be

47 As we shall see in Chapter 7, the gravitational field also propagates with a speed equal to that of light in vacuum.

violated, because a signal **faster** than sound would get to destination **before** having departed!

2.5. The Current Unit for Time - The Atomic Time and the Theory of Relativity

The basic ***unit of time*** in the current international system is the *second*, and we saw it started as an astronomical unit by being defined as the sixtieth of the sixtieth of the twenty-fourth of that 'time' which takes our planet to complete a rotation around its imaginary axis. We also learned that the cyclic processes in the macrocosm are plagued with irregularities that led -together with human struggle for political, religious, and scientific supremacy along history- to a complex evolutionary process which, little by little ended redirecting our focus towards the microcosm, and thereby achieving accuracies not surpassed by any other physical magnitude in the history of Science.

The industrialized nations of the world gathered in Washington DC for the 'International Meridian Conference' in 1884, establishing the *Mean Time* (1847, GMT) as the international *standard* which divided our planet in the current *time zones*, and made possible for the first time to know the *time* anywhere in the world. The US Naval Observatory began broadcasting the *time* in 1905. Later on, multiple scales of *time* were defined. The so-called scales of *Universal Time* are based on the ***daily*** rotation of the Earth (1928, UT0, UT1, UT2, UT2R, UTS) and were replaced in 1956 with the *Ephemeris Time*, based upon the ***annual*** rotation of our planet around the Sun.

From the sundial through the clocks of water, candle, pendulum, balance wheel, etc., we got to a clock controlled by a quartz crystal with substantial improvements for stability, resolution, and reproducibility, to the point of detecting the Earth rotation irregularities. Nevertheless, all these clocks are subject to variations in their fabrication as well as aging. The secret to stability, resolution, precision, and reproducibility was in the atom itself: any two atoms of the same chemical element are ***identical*** and they emit *undulatory radiation*[48] of a ***unique*** *frequency*, when properly excited. This uniqueness, characteristic of a given chemical element, makes the atomic emission of radiation an ideal cyclic process to be used as *standard* for *time*.

The first atomic clock was built in 1949 at the National Bureau of Standards (NBS, USA, today NIST). In 1957, Louis Essen of the National Physical Laboratory (United Kingdom) improved the precision for the measurement of Cesium 133 resonance *frequency* to about one part in 10,000 million (0.00000001%). Finally, in 1967, the '13th General Conference on Weights and Measures' discarded any other astronomic unit and decreed that the unit *second* is defined as "the duration of 9,192,631,770 cycles of the radiation corresponding to the transition between the two hyperfine levels of the ground state of the Cesium 133 atom at a temperature of 0°K".[49]

The *Atomic Time* (TAI)[50] became the supreme *standard* for *time*. As we ar-

48 When throwing some salt into a flame, the latter will turn yellow because the sodium atoms in the grains of salt (Sodium Chloride) emit radiation with a frequency our eyes/brain interprets as 'yellow'.

49 After years of studying astronomical data, it was determined that the *second* in Ephemeris Time corresponded to 9,192,631,770 cycles of Cesium 133 radiation.

50 From the French 'Temps Atomique International'.

gued exhaustively, we selected a cyclic natural process (this time in the microcosm), we assumed uniformity (i.e. all its cycles are of equal duration), and we defined Cesium's cycle as the 9,192,631,770th part of the unit of time. Another scale named *Coordinated Universal Time* (UTC[51]), derived from the TAI though synchronized with the *Universal Time*, was developed to maintain the consistency with the *astronomical time* employed for so long (particularly for navigation).

The *gravitational time dilation* -an amazing prediction of General Relativity (Chapter 7)- was corroborated with the different clocks in the satellites comprising the TAI network, which were running at different rates because of their different altitudes (different gravitational fields). Other scales to correct for this phenomenon were conceived with names like *Barycentric Coordinate Time* (TCB), *Geocentric Coordinate Time* (TCG) and *Terrestrial Time* (TT).

As for the hasty increment of accuracy in our measurement of *time*, pendulum clocks -in their time- used to be considered 'accurate' when their accumulated error during a day of operation did not surpass 10 seconds (1 second in 2.4 hours). In February 2008, physicists of JILA, (Joint Institute for Laboratory Astrophysics), developed a new atomic clock based on atoms of Strontium with an error below 1 second in 200 million years! Currently, research is underway to develop technologies which eventually will achieve exactitudes in the order of a ***second*** in several ***thousands of millions of years***, approaching the accepted 'age' of the Universe.

Finally, atomic clocks, artificial satellites, and Relativity Theory all together made the Global Positioning System (GPS) possible, which is already a popular accessory in cars. Besides the already mentioned *gravitational dilation of time* (Chapter 7), there is another prediction referred to as the *cinematic dilation of time* (Chapter 6) and both are essential to track an object on the surface of the Earth with the accuracy needed to allow a driver to efficiently and safely reach his/her destination. Patience! After a few more chapters, we shall be in a position to understand the fundamentals which made such a technological wonder physically possible and, if the reader grasps them without a single mathematical formula, this book main goal will have been accomplished.

3. What we have learned

We started with the most famous equation in history when, actually, it is one of the many remarkable implications of Special Relativity, and any orthodox approach would have presented it only after discussing the theory and its consequences. The rationale behind my didactic heresy was that it allowed me to subliminally introduce modern (relativistic) ideas, maintaining (only temporarily) many classical and intuitive concepts that I assumed we all shared (mainly *space* and *time*). This chapter, instead, was the beginning of a process in which I will gradually try -using reason- to change almost everything that we still erroneously consider as *intuitive* and *natural*. We have given enough time to *time*; next chapter the time will be given to *space*. Let me summarize what I think the reader has learned throughout this chapter:

---

51 The acronym is a compromise between those in English and French. It reminds me the struggle between Catholics and Protestants to accept the Gregorian calendar (which gave Newton two different birthdays!).

• *Relativity* has nothing to do with *subjectivity* and the former can be *objectively* established;

• Human *perception of time* is little more than merely *qualitative*;

• Physics, and the science of physical *measurement* in particular, were developed to avoid the multitude of different *perceptions* experienced by different *individuals*;

• Every *measurement* technique includes *definitions* (conventions) carrying a degree of *arbitrariness* which inevitably will show up in our description of the **real** world;

• Though, of course, *theory* and **reality** are not the same, the former describes the latter *objectively*;

• *Causality* is an observable fact in our macroscopic world which is characterizable **without** reference to the *concept of time*. It is, instead, the *notion of time* that naturally emanates from the *Principle of Causality*.

• *Time*, besides *order*, has *direction* -- but the latter is neither representable by, nor deducible from the Theory of Relativity;

• *Simultaneity* is simply the **absence** of *temporal order* and, ergo, when two *events* are ***distant***, the concept of *simultaneity* is ambiguous unless, besides the definitions of *unit of time* and *uniformity*, we add a third *definition* to resolve the ambiguity.

• Science has learned to ***measure*** *time* with an accuracy which surpasses the unimaginable, but we still do not agree as to whether it is an ***inherent property of the real world*** or a ***mere intellectual creation*** of ours -though extremely effective- to describe the Universe.

In chapters 3, 4, and 5 we shall embark into a similar enterprise in order to ***understand*** and ***demystify*** the concepts of *space* and *motion* so as to, in chapters 6 and 7, present the Special and General Theories of Relativity, and finally comprehend why *space* and *time* are ***relative*** and ***interdependent***.

Additional Recommended Reading

[1] Mccready, Stuart, Editor, *The Discovery of Time*. Illinois: Sourcebooks, Inc., 2001.

[2] Klein, Stefan, *The Secret Pulse of Time*. New York: Marlow & Co., 2007.

[3] Reichenbach, Hans, *The Direction of Time*. New York: Dover Publications, 1984.

[4] Reichenbach, Hans, *The Philosophy of Space and Time*. New York: Dover Publications, 1958.

[5] Nathan, Otto and Norden, Heinz, Einstein on Peace. New York: Simon & Schuster, 1960.

[6] Borges, Jorge L., *Una nueva Refutación del Tiempo*. Revista Sur, No. 115, 1946.

[7] Gimbel Steven and Walz, Anke, *Defending Einstein, Hans Reichenbach's Writings on Space, Time, and Motion*. Cambridge: University Press, 2006.

[8] Hoffman, Banesh, *Relativity and its Roots*. New York: Dover Science Books, 1983.

[9] Shamos, Morris H., Editor, *Great Experiments in Physics*; Firsthand Accounts from Galileo to Einstein. New York: Dover Publications, 1959.

[10] Yourgrau, Palle, *A World without Time - The Forgotten Legacy of Gödel and Einstein*. New York: Basic Books, 2005.

[11] Grümbaum, Adolf, *The Philosophical Problems of Space and Time*. New York: Alfred A. Knopf, Inc, 1963.

[12] Carnap, Rudolf, *An Introduction to the Philosophy of Science*. New York: Edited by Martin Garner, Dover Publications, 1996.

[13] Goldsmith, Donald, *The Ultimate EINSTEIN*. New York: Byron Press Multimedia Books, 1997.

Chapter 3

The Perception of Space...

The sentence 'the earth is a sphere' is an incomplete statement, and resembles the statement 'this room is seven units long'. Hans Reichenbach (The Philosophy of Space & Time).

and its Measurement

In space we know rectilinear triangles the sum of whose angles is equal to two right angles; but equally we know curvilinear triangles the sum of whose angles is less than two right angles. The existence of the one sort is not more doubtful than that of the other. To give the name of straights to the sides of the first is to adopt Euclidean geometry; to give the name of straights to the sides of the latter is to adopt the non-Euclidean geometry. So that to ask what geometry it is proper to adopt is to ask, to what line is it proper to give the name straight! Henri Poincaré (The Foundations of Science).

In Chapter 1, I assumed all of us shared the intuitive and classical notions of *time* and *space*, using the concepts of *instant* (punctual *time*), *duration* (interval of *time*), *simultaneity, position* (punctual *space*), *distance* (interval of *space*), and the derived magnitudes needed to describe *motion*: *velocity* and *acceleration*. In Chapter 2, we maintained the notion of *space* as instinctive and as shared by all of us, while attempted to understand the concept of ***objective*** *time*. After reviewing the classical notions of *space* and *time*, we learned that our ability to ***measure*** *time* starts with our ability to determine when two *intervals of time* are to be considered ***equal***. We further learned that the ***objective*** notion of *simultaneity* between two *events* corresponds to the intuitive idea we all have of it but, to our surprise, only when their locations are ***contiguous*** in *space*.

When two *events* are ***distant***, we concluded (not without dismay) that to say they are *simultaneous* has **no** univocal meaning, unless we adopt a *definition* that, as such, has a certain degree of ***arbitrariness***, but that is suggested and constrained by a notion more fundamental than that of *time*: the *causality* of our macroscopic world. Only through such a convention can we say -without ambiguity- that two *events* separated in *space* have been *simultaneous*. In essence, what this definition does is to offer a unique method of *synchronizing* two ***distant*** *clocks*, so that their respective **local** *times* can be compared **globally**. As I emphatically noted, the need for this *definition* (i.e. the ambiguity without it) exists as the exclusive consequence of the *spatial* separation between the *events*, and has nothing to do with the relative *motion* between different 'observers' (properly expressed: 'frames of reference').

In this chapter, armed with this new knowledge about ***objective** time*, we shall discuss the ontology and epistemology of *space*, contrasting its ***perception*** with its objective ***measurement***. An important conclusion of this chapter will be that, as long as we ***measure*** the *length* of an object from a *frame of reference* in which the object is *at rest*, our intuition and common sense will be reeducated, but not offended. However, we will be again astonished to realize that ***measuring*** that *length* from a *frame of reference* which is in *motion* relative to the object, requires of another *definition* (and ergo, somehow arbitrary) which was ignored by humanity over many millennia. Only after we agree upon this new *definition*, the *length* of an object 'in motion' will have an unequivocal meaning. Moreover, only after we understand **why** this *definition* is necessary, we can rationally digest several assertions within Relativity Theory that, otherwise, the layperson could never comprehend. Get on board to travel throughout *space* and to learn how to ***measure*** it!

1. What is Space? - The Ontology of Space throughout History

The Sumerian, Babylonian, and Assyrian civilizations lived on the plains between the rivers Tigris and Euphrates about 4,000 years before Christ. From their writings on clay tablets, we infer they knew very well how to measure their arable immediate *space*. The concept of *length* or *distance* between two sticks fixed in their land naturally appeared as the *number of times* a rigid object (considered as the yardstick or benchmark) had to be transported in order to cover the *space* from one post to the other, and in such a way as for that *number of times* to be ***minimal***: the concepts of *congruence* between two solid objects, and of *straight line* (with its inseparable intuitive association with ***minimal** distance*) had been born.

They also knew how to quantify their cultivable land by subdividing it in rectangular pieces for which they defined their *area* as the product of the *lengths* of their sides: they soon realized that the number of stomachs they could feed was related to the product of those *lengths* and not to the *lengths per se*. Reckoning that a rectangular piece of terrain could be considered as the juxtaposition of two triangular ones with the same *area* and a common side, they learned to estimate how many souls they could feed plowing pieces of land with triangular and other shapes: the physical science of ***geo*** (land)-***metry*** (measure) had been born. The abstract idea of a *plane* circulated through Babylonian blood, and its extrapolation to the not-yet-explored tridimensional world was inevitable. At the time of harvesting and storing, they extended the concept of *area* to that of *volume*, piling up their produce to save *space*, and soon they discovered that the number of satisfied stomachs was easily estimated if they multiplied the *area* by the *height* of the pile.

The Egyptian civilization (circa 3,000 BC) lived on the plains along the Nile River with agriculture being also decisive for their survival. Repeated inundations forced them to frequently reconfigure their cultivable space. If the plow's groove was the material representation of their idea of a *straight line*, the set of all those grooves, or the abstract idea of shifting one of them uniformly and indefinitely, shaped their abstract conception of a *plane*. The minor undulations left on the terrain by the plow were not significant in the calculations needed for their survival. Their world was viscerally *planar*.

Even though they clearly perceived the third dimension, their possibilities of exploring it were minimal, and when they built the pharaoh's pyramids -as a conspicuous symbol of their power and pioneer spirit- they did it audaciously, displaying their adoration for *plane geometry* as extended to the unknown. If, for them -while they plowed the land- the notions of *straight line* and *plane* could be extended to 'infinity', why could they not extrapolate those notions beyond the pinnacle of the pyramids towards the sky? The mathematical (abstract) science of *plane geometry* extended to our tridimensional *space* had been born.

In the 500s BC, the school of Pythagoras, using the abstract notions of *point*, *straight line*, and *plane* -implicit in the practical geometry of the Babylonians and Egyptians- developed a series of ***abstract*** theorems among which there was the well-known Pythagorean Theorem that establishes the numerical relation between the *areas* of three squares, each one of them with one of its sides being part of a rectangular triangle (Figure 4).

Figure 4. The Pythagorean Theorem

Babylonians and Egyptians knew of that relation from an empirical point of view, even extended to the third dimension. This process of abstraction initiated by the practical necessity of measuring the crop lands continued, and the Pythagoreans discovered something Babylonians and Egyptians could not have empirically ascertained: there existed rectangular triangles whose hypotenuse length could not be expressed as a ***ratio*** between two *integers* (which is what we obtain when measuring a *length* with a *standard* object or *rule*). A ***ratio*** of integers was called a ***rational*** number so this new strange number was naturally given the unfortunate name of 'irrational'.

1.1. The Euclidian Dogma

Two centuries later (323-283 BC), Euclid of Alexandria wrote his celebrated "The Elements", one of the most influential books in history, comprising thirteen volumes, where he masterfully developed in great detail the axiomatic geometry known today as Euclidean geometry. The book clearly abstracted the notions necessary to measure the ***physical*** *space* into axioms and postulates which defined the ethereal concepts of *point*, *line*, *plane*, etc. These definitions were purposely conceived as fully detached from their practical agricultural origin, so as to assign a universal 'eternal' validity to them. In fact, Euclid firmly believed that his axioms reflected evident and indisputable **truths** about the physical universe. With those irreducible premises, he deduced through theorems -proved with an irrefutable logic- the essential properties of *space*, as it had been perceived and conceived in those days. Euclid's work was considered the quintessence of a deductive intellectual construct for over two millennia. The fundamental nature of his five axioms is as follows:

- **Axiom 1**: There is exactly one straight line connecting any two distinct points;
- **Axiom 2**: Every straight line can be continued endlessly;
- **Axiom 3**: Given a segment of straight line and a point, it is possible to draw a circle with the segment as radius and the point as center;
- **Axiom 4**: All right angles are congruent to each other;
- **Axiom 5**: Given a straight line and a point not on it, there is exactly one straight line through the point and parallel to the first line;

It is obvious that the notions of *point*, *straight line*, *circle*, *center*, *radius*, and *plane* are not defined by themselves but by the interrelation established among them by the axioms. It is not so evident that axioms 1, 2, and 5 define the *topology* of the Euclidean plane, while the axioms 3, 4, and 5 define the Babylonian notion of *congruence* of the rigid body and, ergo, provide a *metric* to *space*, i.e. a technique for *measuring distances* with the Pythagorean Theorem. The fact that a circumference is fully characterized by only two numbers (center and radius) implies that the *distance* between any of its *points* and its *center* is unique and, due to the other axioms, has to verify the Pythagorean Theorem. There is another experimental fact hidden in this axiom: the *position* and *orientation* of a rigid body do not affect its *length*, i.e. its *congruence* with the *standard* body (think of a compass while drafting a circumference).

Axiom 3 is valid only on a plane, because in a three-dimensional space, it defines a sphere. Axiom 5 -called the 'axiom of the parallels'- can be expressed in various forms (the one I chose is simpler than the original). Another simpler equivalent form is to assert that the sum of the internal angles of any triangle has to be 180°. This axiom was questioned from the beginning because of the comparative complexity and the limited use Euclid made of it to prove all his theorems. It was thought to be a consequence of the other four axioms, instead of an independent 'truth', and many thinkers dedicated their full lives to conceive a theorem which, using only axioms 1 through 4, could prove the validity of Axiom 5.

According to Proclus (410-485 AD), King Ptolemy I (323-283 BC) of Egypt wrote a book where he proved (erroneously) that the fifth postulate was a consequence of the other four. Proclus tried to prove the same obtaining an equally flawed solution. During the Middle Ages, Arab science flourished giving birth to Algebra. Euclid's book was amply studied in the Arab world and the discussion regarding the fifth postulate continued. By the beginning of the 12th century, an English named Adelhard of Bath, traveled from Asia Minor to Egypt, learned Arab, and crossed the Gibraltar strait towards Moorish Spain (Al-Andalus), disguised as a Muslim student. Once in Cordoba, he obtained an Arab copy of

"The Elements", translated it into Latin, and smuggled it into Christian Europe through the Pyrenees, the book arriving back at its original place. The book in Latin was distributed among the sage people of the time in the West and... they learned the fundamentals of the geometry that the Greeks had developed a millennium before! Once the press was invented, "The Elements" was one of the first published books (Venice, 1482) as a translation into Latin of the Arab version. In 1505, the first translation into Latin (straight from the Greek text edited by Theon of Alexandria in the 4th century) was published.

Babylonians and Egyptians provided the experimental observation and practical application by using the inductive method; the Greeks provided proof of the validity of those empirical procedures by using the deductive method. But the pragmatic origin of the geometric science was gradually forgotten after "The Elements", and the Euclidean geometry became a dogma in which the ***truth*** of its theorems was interpreted as an absolute **truth,** and curiously ***independent*** of the ***truth*** of its axioms. The epistemological problem of how to justify the ***truth*** of those postulates was simply ignored. This fallacious belief that the Universe could be understood by means of an exclusively ***deductive*** system of thought, endured until the 17th century in which the words 'geometry' and 'geometrician' were virtual synonyms with 'mathematics' and 'mathematician' respectively, and the desideratum of every intellectual was to make their work display the solid logic structure and trustworthiness of the 'geometrical method'.

At the zenith of its dogmatic reign, Euclid's geometry was finally employed by Newton to develop his notion of ***absolute*** *space* and, "standing on the shoulders of Galileo and other giants", conceived *time* also as ***absolute*** in order to enunciate his laws of *motion* through *space* (Chapter 4). As a historical curiosity, Newton used -in his calculations of the force of gravity- the French unit of *length* called 'Paris foot' because the French explorers had measured the *size* of the Earth with the best accuracy available at the time[1]. Euclid's hegemony, now strengthened and authenticated by the success of Newton's mechanics, continued and expanded for another two hundred years until the 19th century with the support of Immanuel Kant and his doctrine of the "a priori", which affirms that Euclid's axioms are "synthetic a priori intuitions" -- meaning that they express essential ***truths*** about our Universe, which require neither analysis nor experimental corroboration.

The logical certainty of the geometrical proofs and our evolutionary adaptation to the local properties of the ***provincial*** *space* where we spend our daily lives, made our civilization ignore for two millennia the distinction between ***logic truth*** -that can only assure the veracity of a syllogism- and the **truth** of an assertion by itself in the sense that it affirms something **true** or not with respect to our physical universe. Concisely: a statement obtained through an impeccable logical argument can only be as truthful or as fallacious as the axioms employed to deduce it.

In short, our civilization forgot the agricultural origin of Euclid's geometry, limited to the plains of Mesopotamia, where every property of the **local** *space* was experimentally confirmed by the necessity to survive, and believed that

1 This measured value for Earth size was about 15% smaller than the one currently accepted, and that is why Newton's comparative calculations of objects falling on Earth and the Moon orbit were not consistent.

those concepts, not only constituted absolute *a priori* truths, but that they could be recklessly extrapolated beyond the provincial circumstances of their creation.

1.2. The Modern Ontology and Epistemology of Space

As a product of the marriage between the Euclidean dogma and Newton's mechanics, the classical concept of *space* was solidly established. In Chapter 2, I concluded: "To wrap up the classical notion of *space*, we all subconsciously feel *space* (and its relation with *time*) as a tridimensional scenario -as impalpable as immutable- where *matter* and *radiation* dwell in and events occurred, are occurring, and will occur. We say this conception describes *space* as ***absolute*** and ***independent of time***: *position* is relative to the *spatial frame of reference*, but the *distance* between any two *points* in *space* -at a common instant- is ***absolute***, i.e. the **same** for the whole Universe, independently of the *reference frame*. *Past*, *present*, and *future* are *spatially* ***universal***, i.e. the **same** for us here on Earth, as for any inhabitant of a recondite planet in a remote solar system in a lost galaxy running away from us at thousands of kilometers per second."

The above portrayal is simply Newton's notion of *space* as described in his *Principia*[2]:

> *... Absolute space, in its own nature, without relation to anything external, remains always similar and immovable. Relative space is some movable dimension or measure of the absolute spaces; which our senses determine by its position to bodies; and which is vulgarly taken for immovable space; such is the dimension of a subterraneous, an aerial, or celestial space, determined by its position in respect of the earth. Absolute and relative space, are the same in figure and magnitude; but they do not remain always numerically the same. For if the earth, for instance, moves, a space of our air, which relatively and in respect of the earth remains always the same, will at one time be one part of the absolute space into which the air passes; at another time it will be another part of the same, and so, absolutely understood, it will be perpetually mutable.*[3]

Once again, we see that Newton accepts the relativity of having to select a *frame of reference* in order to ***measure*** *space*, but he thinks this mundane requirement does not affect the ***absolute*** character of *space* in the sense that it is, by its own nature, always "similar and immovable" and, consequently, every *point* in *space* has an ***absolute*** *position* in it, even though, out of convenience, we refer that *position* to a practical *frame of reference*. Newton never accepted that the sheer necessity of choosing a *frame of reference*, besides being convenient, might have been suggestively denying the ***absolute*** character of *space*.

Not surprisingly, Mach -in his *Science of Mechanics*- criticized the ***absolute*** and ***universal*** *space* of Newton with the same ferocity and philosophical clarity he did for the ***absolute*** *time*. Mach said:

> *...Nobody has competence to express judgments about absolute space or*

2 Sir Isaac Newton, *Mathematical Principles of Natural Philosophy*, July 5th, 1687 (known as "The Principia").
3 As cited in *Great Experiments in Physics*; New York: Dover Publications, 1959.

absolute motion, because they are pure things of the mind, pure mental constructions that cannot be produced in experience. All our principles of mechanics are, as we have demonstrated in detail, experimental knowledge referring to relative positions and movements of bodies....[4]

1.2.1. The Fall of the Euclidean Dogma

In the second half of the 18th century, there existed around 30 different proofs of the fifth postulate as a consequence of the other four; all of them incorrect. In 1733, Giovanni Girolamo Saccheri (1667-1733) published a revolutionary book named *Euclides ab ovni naevo vindicatus* (Euclid free of flaws) which remained unread for more than a century to be rediscovered in 1889. In this book, Saccheri attempted to prove the fifth axiom from the other four, using Euclid's preferred deductive method: *Reductio ad absurdum*. He presupposed the fifth axiom was false, hoping to arrive at a contradiction which would reestablish its validity. To his dismay, not only did he not reach any incongruity, but he surmised that the three possibilities the postulate could adopt, namely a) there is only **one** parallel; b) there is **no** parallel; and c) there is an **infinite** number of parallels, could be accepted as a fifth axiom so as to conceive other 'geometries' which -though bizarre- were logically consistent[5] and, ergo legitimate from the mathematical point of view. Notwithstanding, Saccheri rejected many of his own results based on prejudice and aesthetic considerations.

While Saccheri's book was forgotten, three mathematicians, Kart Friedrich Gauss (1777-1855), Johann Bolyai (1802-1860), and Nikolai Ivanovich Lobachevsky (1792-1856) independently discovered the same: allowing the fifth postulate to assume its three possibilities, they could obtain three distinct geometries, all of them perfectly valid from the mathematical point of view, and one of which was the venerated Euclidean geometry -- the only one considered at the moment to be the genuine ***geometry*** of our **real** *space*.

Gauss, according to Guy Waldo Dunnington[6], discovered the non-Euclidean geometries but he never published his work because of the controversy he knew it would unleash. It is believed that Gauss thought of measuring the sum of the angles of a triangle formed by three stars and, in a more modest scale, he used the apexes of three mountains in Germany with the same purpose[7]. Bolyai and Lobachevsky, independently, published the discovery of hyperbolic geometric where the Euclid's fifth postulate is replaced by the one stating that there is more than one parallel or, equivalently, that the sum of the angles of a triangle is smaller than 180° or, equivalently, that the ratio between the length and the diameter of a circumference is greater than the number π (3.1416...). The surface of a saddle is an example of a **bi**-dimensional *space* with ***hyperbolic geometry***. Bolyai concluded his work prophetically affirming that it is impossible to determine, by means of exclusively mathematical arguments, whether our ***physical*** *space* is Euclidean or not -- Physics being the only science with pertinent competence to the effect.

4 Translated into English from *La teoría de la relatividad: sus orígenes e impacto sobre el pensamiento moderno*, Barcelona: Altaya S.A., 1993.
5 An axiomatic system is said to be internally consistent if, given any statement A, it is impossible to conceive two theorems one of which proves that A is true and the other proves that A is false.
6 Carl Frederick Gauss: Titan of Science (ISBN 0-88385-547-X).
7 It is said that Gauss found the sum deviated from 180°, but it was within the experimental error.

The most important contribution of Gauss, from our perspective, was in the field of Differential Geometry with his so-called *Theorema Egregium* (remarkable theorem). With this theorem, Gauss proved that it was possible to define an ***internal** curvature* of a surface by using its ***metric*** intrinsic properties (measuring *distances* and *angles* on the very surface) and, thereby, in a way completely independent of the **tri**-dimensional *space* in which the **bi**-dimensional surface was immersed. *Curvature*, so defined, becomes an invariant or intrinsic property of the **bi**-dimensional *space* the surface is, without reference whatsoever to the **tri**-dimensional *space* that contains it.

In 1853, Gauss asked his pupil Bernhard Riemann to conduct a study on the fundamentals of geometry, a study which turned out to be one of the most important works in the history of geometry. Riemann discovered ***elliptic** geometry* in which Euclid's fifth axiom was replaced by the one saying that there is **no** parallel at all or, equivalently, that the sum of the angles of a triangle is greater than 180° or, equivalently, that the ratio between the length of the circumference and its radius is smaller than π (3.1416...). The sphere is an example of a **bi**-dimensional *space* with ***elliptic** geometry*.

Riemann also extended the *Theorema Egregius* of Gauss, defining what mathematicians call the 'tensor of curvature' for spaces of arbitrary dimension, regardless of their being hyperbolic, elliptic, or Euclidean. And... "what is the use of it all?" my Mother would have asked me in Spanish. Well... it is essential for the General Theory of Relativity (Chapter 7) because, by extending the concept of *curvature* to a **tri**-dimensional *space*, it is concluded that we can talk about the *curvature* of our real **tri**-dimensional *space* independently of the non-existence of a **tetra**-dimensional *space* in which ours could *curve*! Likewise, this new concept of *curvature* allows us to define the *curvature* of *space-time* (**tetra**-dimensional) without the existence of a fifth dimension. Once again: the *curvature of a space*, so defined, is an intrinsic property which does **not** depend upon the existence of a higher-dimension *space*.

It can be proved that the *curvature* of a Euclidean *space* is always ***zero***, regardless of its dimension, which -unfortunately- led some people to apply the adjective 'plane' to a Euclidean *space* of any dimension. In this fashion, ***non-Euclidean** space* is synonymous with ***non-zero curvature** space*. The non-Euclidean geometries (hyperbolic and elliptic), obtained by altering Euclid's original formulation, have *non-zero* but ***constant** curvature* in every point of *space*. The general notion of ***Riemannian space***, instead, includes *spaces* with ***variable** curvature*, i.e. *spaces* containing regions with ***negligible** curvature* (Euclidean for all practical effects), regions with ***hyperbolic*** structure (negative curvature), and others with ***elliptic*** geometry (positive curvature).

Finally, it is remarkable that this generalized notion of *internal curvature* is different from the traditional *curvature* of a curve/surface, which is a measure of their deviation from a *straight line* or *plane* in the ***Euclidean space***: later in this chapter we shall understand that the cities of Necochea (Argentina) and Madrid (Spain) are both built on *surfaces* that, according to our colloquial parlance (subconsciously Euclidean), can be considered as *planar*, i.e. without appreciable deviation from a *plane* and, ergo, with practically **nil *external** curvature* (due to the tremendous size of our planet). Nonetheless, the first city has a **zero** *tensor of curvature* (Euclidean geometry), while the second has a **non-zero** *tensor of curvature* (non-Euclidean geometry). Interesting? Puzzling? At least curious of knowing what all is about? Let us continue demystifying concepts, because this idea of *curvature* and its indiscriminate use and abuse by Popular Science writers has generated a good deal of erroneous folklore. As Rudolf Carnap (1891-1970) said: "All this [confusion] could have been avoided if the term 'curvature' had been avoided" [8].

Hermann von Helmholtz (1821-1894) contributed considerably to the philosophical foundation of the modern notion of ***physical*** *space* and pointed out that Kant's theory of *space*, based on the "synthetic a priori" nature of ***Euclidean geometry*** could not be true to the light of the new ***non-Euclidean geometries*** discovered by Bolyai, Lobachevsky, and Riemann. Helmholtz firmly believed that, had a child experienced since birth rigid bodies behaving differently from what all of us have experienced, the corresponding ***non-Euclidean geometry*** would have become his/her 'synthetic a priori pure intuition' -- to use Kant's phraseology.

1.2.2. Subjective Space (Perception)

"Perception without concepts is blind". I could not agree more with this statement by Kant and, in fact, that was one of the themes behind all my discussions in Chapter 2 on the ***perception*** of *time* and its limitations. However, Kantian philosophy on *space* basically sustains that Euclidean geometry can be visualized naturally and intuitively (without concepts), while the others cannot -- and that is why Euclidean geometry is to be the correct one. Neo-Kantians, motivated by their desire to save Kant's philosophy, assert that our visualization of *space* is purely intuitive and not psychological -- as if intuition were not a manifestation of our psyche.

According to Reichenbach [1], our ability to visualize has two functions: that which produces an image, and the 'normative' one which discovers geometrical relations on the image. The latter is the most important and the cause of philosophical controversy with respect to the epistemological value our power of visualizing has. Kant states that our intuition compels us to believe in, for instance, the truthfulness of Euclid's fifth axiom. Let us imagine: a) a *straight line*; b) a *point* outside of the *straight line*; c) more than one *straight line* passing through the *point* and all of them **parallel** to the first *straight line*. Once we have visualized the first *parallel line*, as soon as we imagine a second *parallel line* passing through the *point*, either we see it identical to the first one or, imagining it slightly different, we will see it as being in a different *plane* or, eventually, as intersecting the *straight line* imagined in a). And... if the reader has -as I do- a poor imagination, it only takes for us to sketch what we just said on paper to conclude the same, with a sheer certainty that reinforces the Kantian point of view.

Besides, it is remarkable that the conclusion we arrive at does not depend on our skill or on how careful we are to draw the *lines* and the *point*: neither the *point* and *straight lines* nor their *parallelism* have to correspond to the abstract idea we have of those concepts. The drawing replaces the function of producing the image 'in the air' and helps the 'normative' function, which is the one reaching the conclusion. We only need to know what we want to draw and not how we draw it; and that is why our conclusion feels so compulsive and inevitable to us, in a 'synthetic a priori' manner -- as if the 'real' ***geometry*** of the *space* where we live in were codified in our DNA.

But... we only need to meditate a little to realize that if the drawing is not clear enough for us to feel the compulsion to conclude, we take the liberty of recurrently 'improving' it until we 'see the truth' of what we want to prove. This trivial observation clearly indicates that we do not deduce the conclusion from the image but that, instead, we modify it until we are convinced it has all the ingredients to help us obtaining the conclusion in an exclusively deductive manner and, ergo, even though we may have been left with the impression that we found a *primal* **truth** about our ***physical*** *space*, all we have done is to achieve a ***logically*** correct conclusion, the ***truth*** of which is to be referred to the ***truth*** of those axioms implicit in our reasoning. ***It is the hidden conditions behind the presentation of a problem, that limit and define the truthfulness of its solution.***

While we apply the normative function, our brain is making sure that we interpret the concepts of 'straight line', 'point', and 'parallelism' à la Euclid and, consequently, we should not be surprised of attaining -as a prize for our intellectual effort- a logical inference in full agreement with Euclidean geometry! As Reichenbach says, the merit of our power of visualization is that it converts the logic compulsion immanent in the Euclidean geometry into a visual compulsion which, I would add, is the result of our subconscious employment of the Euclidean concepts for over two thousand years and of the physical congruence of rigid bodies for over five thousand years.

Now... Stop all this Euclid's bashing for a second! All the above considerations do not imply that the Euclidean geometry does not accurately correspond to the spatial relations between objects we perceive daily in our limited sphere of action. Otherwise, it would not have excelled and survived for millennia! It is our daily experience with the behavior of solid bodies and luminous rays that constitutes the genesis behind our compulsion to imagine and reason in tune with Euclid. Our fallacy resided in having believed that the physical *congruency* we observe with solid objects and light propagation was the only one admissible to devise a way of ***measuring*** *space* and that the ***Euclidean geometry*** based on such ***congruency*** was an inherent property of our ***physical*** *space per se* and, as such, the only 'genuine' and acceptable one for describing the objective **reality**.

As it normally happens, it was the astonishing simplicity, accuracy, and efficacy of the mathematical model we used to describe the physical world that misled us into believing they were the same thing.

1.2.2.1. Necochea's Space and Descartes

As I said, *space* was not the problem. I left home alone heading to school. My theatrical performance was imminent; my Mom would go later. Walking the streets of the little Necochea, even for a six-year-old kid, was almost as easy as it would later be walking from the rim to the center of the stage. René Descartes (1596-1650) would have instantly fallen in love with the divine natural implementation of his immortal *Cartesian Coordinate System*: the 'Quequén Grande' River discharges on the beautiful Atlantic Ocean, the two of them drawing an almost perfect right angle.

I was taught only once how to find my way in Necochea, and it was sufficient to start exploring the city: those streets parallel to the ocean (North-South) are identified with even numbers; those parallel to the river (East-West) are assigned odd numbers; the river is the origin (zero) for the house numbers on even streets; the ocean is the origin (zero) for the house numbers on odd streets; the blocks are square with sides 100 meters long. On every street, in the direction in which house numbers increase (farther away from the sea and the river), the houses on your right have even numbers and those on your left have odd numbers. Finally, the number assigned to a house on a given side of the street is obtained by simply dividing 100 by the number of houses on that side and adding twice the number of blocks from the origin (river or sea).

The street where we lived has the number 54, i.e. it runs parallel to the sea and is at 2.7 km from it ((54/2)x100/1000=2.7); our house had the number 2725=2700+25, so we were 14 blocks from the river (27/2+0.5) and, ergo, 14x100/1000=1.4 km from it. To complete the spatial localization, the house numbers are so carefully assigned that their difference is, with good precision, the distance in meters between the corresponding houses. As a result, the even numbers on one side of the street correspond almost perfectly with the odd numbers on the opposite side: our number was 2725 and our neighbor's across the street was 2726. The *metric* obsession of its designers is conspicuous in several cases on our block: there are houses whose garage and main door have different numbers assigned, with the only purpose of maintaining the *metric* information (distance) characteristic of a *Cartesian Coordinate System*. An address includes everything you need to go there, and more... because given the departure and destination addresses, the *distance* between them and the *time* to cover it can be estimated very easily. The numbers (coordinates), besides indicating the *relative position* (topology) of streets and houses, contain straightforward *distance* information. ***Topology*** and ***metrics*** in the City of Necochea were integrated from scratch during its very foundation in 1881. In mathematical and compact terms: the *coordinates* (house numbers) are the *distances* to the coordinate axes of the Cartesian system (ocean and river). Even though all this sounds extremely complex for a six-year-old kid -and of course, neither was I taught this way nor did I do those calculations- I only needed to venture into a first quick excursion to rapidly grasp how not to get lost. It was joyfully easy for a child to walk around Necochea!

We would attend two hours of class before commencing the celebration meeting. Yesterday, the object of my puerile desire had explained that our planet was round though even in Christopher Columbus's time many people did not believe it! Today, I had to submit as homework a drawing conveying my reaction to such an outlandish idea[8]. No need to mention that whatever Mrs. Albizuri would say had to be true but... even so, it seemed ridiculous to me. Besides, the word 'planet' sounded suspiciously very close to the word 'plane'[9]. I

8 It is believed that Thales of Miletus was the first one to suggest the earth was round (500s BC). Believe it or not, after 2,500 years, the "Flat Earth Society" is still alive and represented in Twitter and Facebook!.

9 My hunch was unfounded: the word 'plane' derives from the Latin 'planum' while the word 'planet' comes from Greek and means wanderer (because the planets moved in inexplicable ways relative to the 'fixed stars').

do not remember what her explanation was of why all of us -my grandparents and uncles in Spain and we in Argentina- believed to be 'standing up'; let alone why those *silly guys in Antarctica* did not fall! -- being the latter the most bewildering idea I wanted to represent on my piece of art. Most surely, I do not remember it because my teacher's 'rational' explanation was naturally incomprehensible to me. Figure 5 depicts the 'PowerPoint' version in 2008, almost identical to my original drawing in 1954.

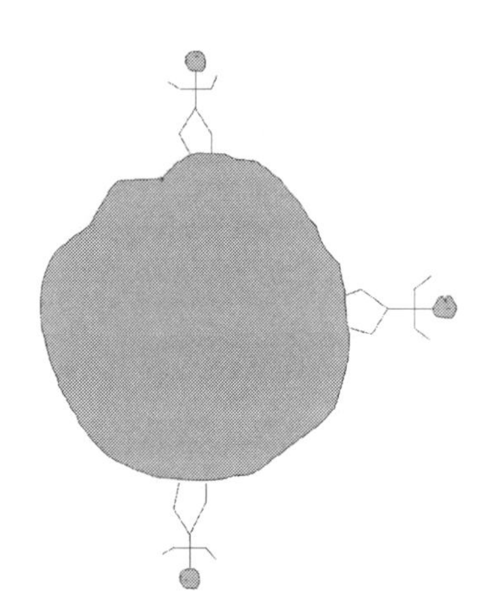

Figure 5. 1954:
"Those silly guys in Antarctica"

After having drawn something so transcendental the night before, and perhaps trying to avoid thinking of the embarrassment I was about to go through in a few hours, today I was in a 'philosophical' mood, and could not stop thinking why the real world of the adults was so different from my most immediate and solid perceptions. The 'General Belgrano' school was on the same street as my house, about 6 blocks of *distance*; I only had to cross a wide and dangerous avenue. All along my walk, the curbs on both sides of the 54th street constituted an almost perfect realization of what adults called two *parallel straight lines*. Looking afar, however, I saw those *lines* converge as if there was no doubt they would eventually meet somewhere not too far away... but my *perception* of the *place* where they would touch seemed to *move* with me. It occurred to me that, perhaps, had the curbs been built *diverging* I would be seeing them as parallel. Another idea: if the Earth was really round as my Angel said, then, no matter how much I struggled not to deviate from my 'straight' path, it would be impossible for me to walk ***straight***. Weird! First, they taught me what a *straight line* is and then... they tell me that it does not exist! Oh! Now I get it: when last week my Dad took me to the local 'aeroclub' and we flew on a tiny airplane, had we wanted, we could have continued flying on a *straight line* forever -- because we were not forced any longer to trail the curvature of our planet... Hmm... I sincerely hope that they do not come now telling me that the 'air' is also round!

I could also see faraway objects (houses, trees, cars, etc.) that seemed *smaller* than those of the same kind but close to me; however, when I got to the place, things had the same *shape* and *size* as they did two blocks behind; and the curbs were still *parallel* in my immediate neighborhood. And these effects had nothing to do with my *motion* -- it was clearly only related to the *spatial* separation between the objects and I. Creepy again! Which were the **real** *size* and *shape* of things, I asked myself; those I perceived when things were ***close*** by me or... those I perceived when we were ***far away*** from each other? I was sensibly sure that, had I used the *meter* tape my Mom employed in preparation to sew the fabric, and measured two alike objects separated by three or four blocks (i.e. had I ***transported*** the tape from one *place* to the other),

I would have concluded that the object had practically the **same** *size*. I never did it, of course, but it was clear to me that my local *perception* so indicated.

But... what kind of physical **reality** was it, if the only way to correctly *perceive* it was by being side by side with it? What was the actual **truth** value, if any, of all those silly statements adults indulged themselves with when talking about other planets, which -according to them- were incredibly farther away from my house than my school was? I could ***transport*** the *meter* tape for a few blocks but, certainly, neither I nor anybody else had ever transported a *meter* tape to Venus. And, why was I so sure, once the car was close to me -without using my Mom's tape- that it was practically the **same** *size* and *shape* as the cars I had seen a few blocks behind (or that it was not)? It was as if I could imagine myself tracing the car contour with the *tape* to correctly obtain its *length* and *shape* without having to carry the tape -together with the 'bolitas[1]'- in my pocket everywhere. This is getting eerie!

I get it; this had to be related to what adults called *perspective* or, simpler, to see the same object from different *points of view*. As per those savvy grown-ups, things had only one ***absolute*** *shape* and ***absolute*** *size*, and all the other different perceptions were simply ***distortions*** due to the *distance* and/or *angle* from which they were observed. This explanation seemed natural and logical to me, though my puerile mind could not stop asking why one of those *perceptions* was bafflingly chosen as ***real***, and the rest as ***distortions***. Would it not be, perhaps, more rational to think that **reality** is the set of all manifestations of an object, instead of just one dubiously selected? But, on the other hand, I had already been told of the so-called optical illusions that certainly could not be considered as objective **reality**. Too complicated for my infantile brain; I better followed the crowd and started worrying about the silly verse I was going to recite, as I was now a few *meters* from the school entrance.

And so I did; I grew up and quickly familiarized myself with that particular interpretation... to the point that my brain stopped asking why ***distant*** things were ***smaller*** or why two ***parallel*** *lines* were not such at a *distance*! Moreover: my brain got accustomed so well to it that stopped seeing ***distant*** objects as 'smaller' or ***distant*** *parallel lines* as 'not parallel'; my intellect subconsciously adjusted to a *space* ***geometry*** that seemed as natural as rational. The definition of ***congruence*** of the rigid body, hidden in the procedure my Mom used to measure, cut, and sew the fabric, had invaded and comfortably installed in my bones forever -- to the extreme of being confused with compulsive ***intuition***. Lucky me! Because now, the *space* that my Mom's *meter* tape would have occupied in my pocket, could be used to carry even more marbles... ***my unconditional indoctrination to Euclidean geometry had been accomplished. Good job, 'adults'!***

But maturation comes always with disappointment; when in the last years of my elementary school I started working as a delivery boy for 'Casa Hidalgo' (a men's clothing store), I had to traverse the city on a bicycle to deliver the merchandise, and pretty soon I learned there existed a few streets which were diagonal to the beautiful *Cartesian Coordinate System* defined by the ocean and the river, that a few blocks were not square, and that their sides were

1 'Bolitas' is Spanish for 'little balls' and is used in Argentina to refer to 'marbles' (as in 'playing marbles').

longer or shorter than the decreed hundred meters. Today, I would say that the city's *topology* was still the same but that its *metrics* was a little different, slightly ***varying*** from *place* to *place*. Things were a little more complicated than expected; accurately estimating the *distance* between two domiciles was not as straightforward... but it was still quite manageable.

By then, I had already been taught about the Pythagorean Theorem asserting that a diagonal street would cover roughly 140 meters -- instead of the 100 meters of a normal street. Already with twelve or thirteen years of age, my piece of art drawn when I was six (Figure 5) was still wandering in my head and I could not help but ask myself: given that I had accepted the Earth was round, would the theorem still be valid for a triangle joining us in Argentina with my grandparents in Spain and with, e.g. the United States? Likewise, would the sum of the three angles of such a huge triangle still be 180° as my teachers assured with messianic wisdom? Luckily, every time these thoughts came to my mind, I had another persistent reflection: that of my pragmatic mother sweetly but obstinately saying to me: "shut up you silly boy and enjoy your adolescence" -- thanks Mom, for helping to maintain my mental sanity and sensibility in this crazy but fascinating world.

1.2.3. Objective Space (Measurement) - Geometry and Field

Physical geometry is the science which studies the *spatial relations* between objects (bodies and radiation) ***independently*** of their composition and/or nature. Even though we are talking about *relations* between objects, we take the linguistic license of abstracting from the matter/radiation needed to establish those relations, and wholly refer to them as the 'geometry of space'. It is clear then that any relation between objects that depends upon their composition or nature, cannot be considered as reflecting a property of the *geometry* of *space*. ***Qualitative*** *spatial* relations define the ***topology*** of *space*; ***quantitative*** *spatial* relations define the ***metrics*** of *space*.

1.2.3.1. Madrid Space and Gauss-Riemann

In December 1982, I returned to my natal country for the first time after my parents had taken me from the City of 'El Cid' (Burgos, Spain), where I was born in 1948, to Necochea in 1949. Twenty days of navigating the Atlantic had converted, after 34 years, into a few hours of flight -- as if the mere passing of *time* had ***dilated*** the present *time* or, equivalently, ***contracted*** *space*.

As in that theatrical morning 28 years ago, I believed again that *space* was not the problem; neither were -for sure- the language, culture, or idiosyncrasy: out of my house I had grown up in Argentina; inside of it, in Spain. I knew both folklores. What a fortune! The problem once again was *time* (or so I believed). As a result of one of my best Quixotic moves (and it would not be the last), I had resigned from my post as Full Professor at the 'Universidad de San Juan' and had to find another job urgently; my daughter Araceli had been born two months ago and... Oh surprise! She apparently did not know the Spaniard proverb which says that 'babies come with a loaf of bread under their arm".[2]

This time, what I wanted for *time* was not to stop (dilate) so I would not have

2 Meaning that parents will somehow manage to feed the new baby no matter what.

to confront the public, but to accelerate (contract) so I could know as soon as possible which would my new *space* be, and what new *times* were awaiting for me and my family. I would spend a week in Madrid seeking for a job, would return to Burgos after 34 years (where I had and still have many relatives), and would continue towards Minneapolis, USA where I had professional relations to continue my search for the job opportunity I desperately needed.

The place in Madrid where I had my first appointment was on the "Paseo de la Castellana" street. I had been explained that a good reference were the 'Ministries of Labor, Social Security, and Public Works' and that, once over there, it would be really easy to find the place just by knowing the number of the portal I was looking for which, if I remember correctly, was '59'. In my natal land for the first time after 34 years and in a totally unknown big city, I expected that reaching the place would take some effort and, most probably, the assistance of some pedestrian. I never was a friend of maps so I did not take the trouble of purchasing one.

Once in the zone of the Ministries, I was on the sidewalk in front of a portal with the even number '104', while the one I was looking for was odd (59). Hoping that the *topological* design of Necochea in which even numbers were used for one side of the street and odd numbers for the other was valid, I crossed the street pleasantly confirming that was the case but... something was wrong: the closest portal to be considered as face to face with the portal '104' across the street had the number '67'; a difference of 37 just by crossing the street! Something was flawed with my Cartesian notion of Madrid's *space*. I became a little suspicious of my ability to reach the *place* on *time*.

Not having a map, I ventured the validity of Necochea's second *topological* rule by which the side of the street with assigned even portal numbers is the one to the right in the direction towards which the numbers increase; in this way, I determined the direction that, most probably (if the rule was valid), I had to walk so as to see the numbers decrease ('67' to '59') and, because these numbers were so close, I expected to get to the place in a few minutes. I walked and walked without seeing any number and I became very doubtful of getting to the *place* on *time*. After more than what in Necochea would have been two blocks, I finally saw the number '63' which confirmed the *topological* rule of my childhood in the New World. But once again I became frustrated by my miserable failure in estimating the considerable *distance* I had had to walk so as to reduce the portal addresses by just three numbers in this strange *coordinate system* of the Spanish capital.

By then, it was clear to me that the portal numbers (coordinates) were nothing but mere indicators of the *topological* properties of Madrid's space, and that they had very little to do, directly at least, with its *metric* properties. Clearly, places ***close*** to one another in *space* had assigned numbers ***close*** to one another; the farther away the portals, the more different the numbers but... to use the portal numbers to estimate their mutual *distance*, required of a formula which, if existed, would not be simple and, even worse, it would have -apparently- to change from *place* to *place*. Naturally, thus, the Pythagorean Theorem was utterly useless to estimate *distances*.

And then I thought: how could this mysterious formula with only ***local*** validity be found? The ideas of Gauss and Riemann came to my mind immediately: we measure with a rule, in the traditional sense of the Babylonians, the *distance* between two ***contiguous*** portals on the 'Paseo de la Castellana' street (e.g. between the portals '63' and '67') and do the same on the 'Raimundo Fernández Villaverde' street (e.g. portals '82' and '85'). In the small (as compared with Madrid) parallelogram defined by the two segments defined by the two pairs of portals in both streets, we draw several segments with different orientations and measure all of their *distances*. Now that we have all the *distances* associated with multiple segments with different orientations, all of them corresponding to a fixed increment of coordinates (67-63=4) on 'Paseo de la Castellana' and to a fixed increment of coordinates (87-81=6) on 'Fernández Villaverde' street, we try to find a mathematical formula which could produce those ***measured*** *distances* out of the two coordinate increments (4 and 6). In my beloved Necochea where Descartes and Pythagoras reign, all those *distances* would have been equal and about 7.21 ($\sqrt{(4^2+6^2)}$); in Riemann's kingdom, all of them may have (and do in Madrid) arbitrary and different values!

Given that Pythagoras' formula is invalid for cases like the one I experienced in Madrid, Riemann generalized it by including not just the squares of the coordinate increments but also their crossed products, with arbitrary coefficients for all squares and cross-products. If we were capable of determining those three coefficients (the one for the square of the increment on the 'Castellana', the one for the square of the increment on 'Fernández Villaverde', and the one for the cross-product between the two increments), we would have given a *metric* to Madrid's *space* in the zone of the Ministries, and the set of those three coefficients in Riemann's generalized formula for the *distance* between two portals is what physicists would call the *metric tensor* in that zone of the city. If we did the same throughout Madrid, we could determine the whole city *metric tensor* which, as we know now, would radically change from one zone to another.

It was about time to stop all this mathematical gibberish and take the bull by its horns: either I continued for who knows how long to reach the darn '59' portal, knowing that I could be wrong, or asked a pedestrian for help. I chose the latter and asked a lady passing by. With my conspicuously Argentinean accent, I explained to her what portal I was looking for and why I thought it had to be very close even though I could not find it. "But... 'majo'[3], it is so obvious you are not from here! That portal is by 'José Abascal' street; keep going this direction two or three more blocks and you will see the house. And so it was: I had to walk more than four Necochea blocks to barely decrease four numbers, from '63' to '59'!

That was my first conscious interaction with a ***non-Euclidean*** *geometry* and, while crossing the Atlantic Ocean towards New York, I could not stop wondering what the intrinsic difference between Necochea and Madrid was. Both cities were built on *planar* and *small* areas as compared to our planet, so the abstract idea of a *plane* had to be valid in both cases. However, Necochea clearly responded to the idea of a ***Euclidean*** **bi**-dimensional *space*, while Madrid was the 'poster child' for a **bi**-dimensional ***Riemannian*** *geometry*! Riemann's *curvature* for Necochea -obtained from its *metric tensor*- was **zero**; Madrid's *metric ten-*

3 'Majo' is a common Spanish adjective for 'nice', 'attractive', 'handsome', etc. (do not take it literally!).

sor gave, instead, a **non-zero** *curvature*! Undoubtedly, the crux of the matter resided in the *metric tensor* which in Necochea was practically the same for the whole city and governed by the Pythagorean Theorem, while in Madrid it varied wildly from one place to the other and responded to a formula much more general than Pythagoras'.

But then... if I was correct in my thinking, the ***geometry*** of a *space* is **not** an absolute inherent property of it, but depends upon how we define the *coordinate system* and its relation with the notion of *distance* between two *points* in it (*metrics*)! And, if this is true... what objective meaning could Einstein's assertion that "space-time is non-Euclidean" have in his General Relativity theory (Chapter 7)? Or, equivalently, what did Einstein mean when he said that "space-time is curved"?

When I felt I was about to lose my mental sanity at 10,000 meters over the Atlantic, once again my Mom's voice 'telepathically' arrived from the Euclidean Necochea with her usual ubiquity and sense of duty: "Shut up you silly boy and figure out how to get a job, so you can feed your wife and daughter" -- thanks mama, for helping me to distinguish the *urgently important* from the ***transcendentally important***.

1.2.4. Topological Properties of Objective Space

Space, compared with *time*, is richer in *topological* properties which are related to our *qualitative* notions of *continuity*, *dimension*, *finiteness*, *infiniteness*, *inclusion*, *connectivity*, *openness*, *closure*, etc. These properties characterize a *space* without a need for a *metric*, i.e. without having defined the concept of *distance* and a technique to ***measure*** it.

As we saw, *continuity* of *time* and *space* is a mathematical hypothesis about the physical world which has proven very fruitful -- even in the microscopic world where the *discrete* nature of matter cannot be ignored. In essence, mathematical *continuity* is an abstraction inspired by our sensorial experiences: we often experience two pairs of *perceptions* (A,B) and (B,C) such that A is ***indistinguishable*** from B (A=B) and B is ***indistinguishable*** from C (B=C); however, our *perception* may allow us to ***distinguish*** A from C (A≠C), imposing a blatant logical incongruence (our sacrosanct **equality** relation would **not** be *transitive*!). This paradox is resolved with the abstract notion of *continuity*, so that magnitudes so small that they cannot be individually *perceived* by our senses or ***measured*** by our instrumentation (infinitesimals) can be accumulated (integrated) to a point that they become *perceivable* or *measurable*[4]. Remember Aquilles and the tortoise?

A *point* is a *space* of ***zero** dimension* without any structure; if we imagine a *point* moving indefinitely and continuously in two opposite directions, we visualize the *straight line* as a *space* of ***one** dimension* with Euclidean *topology*; if we imagine the *straight line* moving indefinitely and continuously in two opposite directions, we envision the *plane* as a *space* of ***two** dimensions* with Euclidean *topology*; finally, if we imagine the *plane* moving indefinitely and continuously in two opposite directions we envisage a *space* of ***three** dimensions* with Euclidean *topology*. If we add a *metric* to these *topological spaces*,

4 Mathematicians call the sum of an infinite number of infinitely small (infinitesimals) numbers 'integral'.

and this *metric* is defined through the Pythagorean Theorem, we obtain ***Euclidean spaces*** (with Euclidean *topology* and *metrics*). The Euclidean *space* of ***three*** *dimensions* is the abstract **model** employed to represent our ***physical*** Universe for millennia.

Besides its *dimension* (three), the Euclidean *space* has other *topological* properties, such as its *infinitude* in the sense that it is logically impossible to fully traverse it (nothing to do with our practical possibilities of doing so), and its *openness* in the sense that traversing it in a given direction will never take us to a place where we have been before. Another *topological* property of the Euclidean *space* is that it is *fully connected* in the sense that given any two points, there is always a *continuous path* between them and, besides, such *space* is also *simply connected* in the sense that it admits for any closed surface (imagine e.g. a sphere) to *continuously* contract to a single *point*. Another property is that of *inclusion*, whereby given three closed surfaces without common points, if the notion of *inclusion* is applicable, then it is possible to say unequivocally that one of the three is between the other two (think, out of simplicity, of three concentric spheres). Equivalently, the surface between both is included in the *exterior* surface and includes the *interior* surface.

In order to understand -without mathematics- the relevance of the *topology* of a *space*, it is useful to envision **bi**-dimensional *spaces* immersed in the Euclidean **tri**-dimensional *space*. The first bombshell we encounter is that the Euclidean **tri**-dimensional *space* contains **bi**-dimensional *spaces* which are not Euclidean! Think of the surface of a ***sphere*** as the Universe for **bi**-dimensional people -let us call them Spheretians- who are condemned to move on (should I say 'in'?) it without any conscience or practical possibility of exploring the ***third*** *dimension* into which their world -we Terrestrians know- is immersed. Clearly, there is a drastic *topological* difference between this imaginary **bi**-dimensional world and the **tri**-dimensional one: it is logically possible to fully traverse it so... it is ***finite*** though ***limitless*** because a traveler will never encounter a sign saying: *Non Plus Ultra* (end of the world). Besides, it is ***closed*** because, had the traveler continued its trip without changing course, s/he would eventually have found her/himself at the departure location. It is, however, ***totally*** and ***simply connected***, as its **tri**-dimensional container is, because a *closed line* on the ***sphere*** can always be continuously reduced to a *point*.

Notwithstanding, there is a subtle difference in the *connectivity* of the Spheretians' universe (clearly non-Euclidean): for any *closed line*, it can be contracted to a point in two ways; for example, the Equator can be contracted towards the North Pole or towards the South Pole. Due to this double possibility, and because *inclusion* is a *topological* property (unrelated to the notion of *distance*), we can legitimately say that either the Equator encloses the Tropic of Cancer because we imagine these two closed lines contracting toward the North Pole, or that the Equator is enclosed by the Tropic of Cancer because we envision the two lines contracting towards the South Pole. Consequently, a ***sphere*** -as a **bi**-dimensional universe- does not admit the concept of *inclusion* in an unambiguous sense or, equivalently, the concept of *inclusion* is ***relative*** due to the *closeness* and *finitude* nature of the Spheretians' world.

Even comparing two **bi**-dimensional *spaces* like a ***sphere*** and a ***plane***, there are dramatic *topological* differences. It is the *topology* of the ***sphere*** that makes it a non-Euclidean *space* trapped in a Euclidean *space* of a higher *dimension* -- while a ***plane*** is certainly Euclidean. As I envisaged that 'philosophical morning' of 1954 when a child, if *terra firma* was really a ***sphere*** as my adored teacher had said, walking 'in a straight line' really meant traversing a *great circle* (like the Equator or a meridian) and, ergo, in a ***closed*** and ***finite*** world, a *straight line* has to be *closed*!

For that reader who viscerally reacts arguing that we cannot call such a line 'straight' because it is obviously *curved* (that's why it is called a 'great circle'!), I cannot but remind him/her that the only reason we have to call it 'curved' -in the colloquial sense of the word- is because we have at our disposal the ***third*** *dimension* to observe the sphere's ***external*** *curvature*[5]. From the point of view of the Spheretian who inhabits such a singular world, his trajectory has been perfectly rectilinear -- as rectilinear as my walk to school (before I learned Earth was round).

This last assertion does not mean that the Spheretians are not capable of realizing that their *space* does not correspond to the Euclidean idea of a *plane*, because -eventually- they will learn that two 'people' walking in opposite directions and keeping the course will eventually meet, and will have to admit (with some reluctance) that what they call 'straight' are, in fact, *closed lines*. In any case, to avoid this kind of confusions, the more general term 'geodesics' was created (which reduces to our mundane *straight line* for Euclidean *spaces*), and then we say that the ***great circles*** of a ***sphere*** are its ***geodesics***. By the way, it is worth noting that the definition of 'straight line' given by Euclid in his 'Elements'[6] does not help to clarify the concept (reinforcing what I just said): "A straight line is one that lies evenly with respect to the points on itself" (?)

Furthermore, if Spheretians' 'straight lines' are our 'great circles', given a 'great circle' and a *point* outside of it, any other 'great circle' passing through the *point* will intersect the first 'great circle' on the poles and, ergo, for the Spheretians there exist **no** lines *parallel* to a given one, i.e. Euclid's fifth axiom is not valid. Equivalently, imagine two *meridians* forming a *non-zero angle* at the North Pole and intersecting the Equator. Because all *meridians* on a ***sphere*** cross the Equator with a 90° angle, the sum of the angles in the triangle formed by those two *meridian* segments and the Equator segment has to be necessarily greater than 180° -- negatively answering another of my childhood questions that morning of 1954. Even Euclid's first postulate is not valid: there are *points* like the North and South poles through which there exist an infinite number of 'straight' *lines* (all meridians are geodesics) -- instead of only one.

Finally, remembering Gauss' *Theorema Egregium*, Spheretians (once they had discovered it) could define and ***measure*** an *internal curvature* of their own space without any reference to a ***third*** *dimension* (of which they are not aware). ***Measuring*** their **bi**-dimensional world on sufficiently extended areas, they would conclude that the *curvature* of their world is *small* but ***non-zero*** and, ergo it is **not** Euclidean. At the same time, they would easily understand why their ancestors believed for millennia in the Euclidean dogma, until they developed the necessary technology to conduct ***measurements*** over big enough

5 External curvature (our colloquial 'curvature') of a surface is a measure of its deviation from a plane.
6 As cited in *The Curious History of Relativity*, Jean Eisenstaedt, 2006.

areas so as to detect the difference.

It is also easy to imagine another **bi**-dimensional 'being' whose *space* is contained in a **tri**-dimensional Euclidean *space* but that is not ***simply connected***, a property that we saw was shared by the *plane* and the *sphere*. Such a world would be the one inhabited by the Toroideans, i.e. the 'guys' living on (in) a *surface* with the *topology* of a doughnut. We say a doughnut has a hole because we, again, see it immersed in our **tri**-dimensional *space*; from the point of view of a Toroidean, there is no such a hole and her observations and trips imply that his/her world is, as for the Spheretians, ***finite*** but ***limitless***, ***closed***, ***connected***, and with ***relative inclusion***. However, a toroid is **not *simply connected*** because there are *closed lines* which **cannot** be *continuously contracted* to a *point*. Picture a *closed line* on the surface of a doughnut which -seen from our perspective- encircles the hole. Imagine it now *continuously contracting* and getting closer to the hole until it reaches a stage at which it stops *contracting* and starts *enlarging* and moving farther away from the hole, returning to the original position. It is obvious, thus, that ***plane***, ***sphere***, and ***toroid*** are *topologically* different.

From the mathematical point of view it is not difficult to extend these **bi**-dimensional *topologies* of ***sphere*** and ***toroid*** to ***three*** or more dimensions, but trying to visualize how a *space* like ours would be like with *spherical* or *toroidal topologies* is not that easy. Einstein, based on his General Relativity theory, believed our Universe had a *spherical topology*. By natural extension of what we discussed in two dimensions, we can infer what properties such a Universe would have. First, as we know from millenary experience, it is Euclidean, but only **locally** which, among other things, means we cannot extend our *straight lines* and *planes* indefinitely. As for the Universe as a whole, if its *topology* were *spherical*, our *space* would be then ***finite***, i.e. with the suitable technology and longevity, there is no place we could not reach, ***limitless*** (the 'Pillars of Hercules' are just mythology), ***closed***, i.e. by traveling without changing course, we would eventually come back to the same place (*geodesics* are *closed*), and finally, it would be ***totally*** and ***simply connected***, i.e. every *closed surface* could be *continuously* contracted to a *point*.

Nonetheless, the state of affairs regarding which *topology* our Universe has is not that simple, because Einstein's Field Equation (Chapter 7), which magically and elegantly condenses all his General Relativity theory, admits multiple solutions with multiple *topological* and *metrical* properties for *space-time* that are as unexpected as fascinating. Einstein, during his visit to CALTEC (California Institute of Technology) in 1930, officially abandoned his belief in a spherical topology for our Universe. Now is the *time* to learn how to ***measure*** *space*.

1.2.5. Metrical Properties of Objective Space - Equality, Unity, Relativity, and Curvature

When we described how Babylonians developed the notion of *length* or *distance* between two posts in their crop fields and the technique to measure them, I said "naturally appeared as the *number of times* a rigid object (considered as the yardstick or benchmark) had to be transported in order to cover the *space* from one post to the other, and in such a way as for that *number of times* to be ***minimal***: the concepts of *congruence* between two solid objects, and of *straight line* (with its inseparable intuitive association with ***minimal*** *distance*) had been born."

Let us converse now about the epistemological problems we encounter when trying to ***measure*** a *distance*. To begin with, let me notice that I abused the language when I said "to cover the *space* from one post to the other" because what Babylonians really did was to compare two bodies or pieces of body: one was the piece of terrain delimited by the two poles and the other the standard body or 'rule'. Had anyone of them been missing, the ***measurement*** would have been impossible (even if we continue thinking that the 'space' between the sticks still exists). In this fashion, we can understand what Henri Poincaré said in his well-known book *Science and Hypothesis* (1902):

> *Experiments only teach us the relations of bodies to one another. They do not and cannot give us the relations of bodies and space, nor the mutual relations of the different parts of space.*

In brief: even though we talk about the 'geometry **of** space' or the 'topological and metrical properties **of** space', or more plainly about the 'available space', we can only discover the *properties of space* through the restrictions it imposes on *matter* and *radiation* (objects) when they interrelate.

We start by defining the concept of ***congruence*** between two *segments* identified by two marks imprinted on two solid bodies: take any bodies A and B; make two marks on A defining a *segment*, and do the same on B; we say that the two *segments* are ***congruent*** when we can orient and position both bodies such that the ***ends*** of both *segments* coincide. Now, ignore B as a body, abstracting the concept of a *segment* materially defined on both bodies, and imagine we use the physical segment marked on A for, abusing the language, defining through ***congruence*** a ***non-physical*** *segment* on *space*. We say that the *extremes* of this imaginary *segment* define two *points* in *space*.

Next step is to define the notion of *length* or *distance* and we start by defining when the *lengths* of two segments are **equal**. If bodies A and B are contiguous in *space* (relative to human dimension), we simply apply the concept of ***congruence*** and decree: two *segments* are **equal** in *length* when they are ***congruent*** (once juxtaposed, their ends coincide). Once again, we ignore body B by loosely talking about the *segment* defined by it, and say that its *length* or, equivalently, the *distance* between its extremes, is **equal** to that of the physical *segment* defined by body A.

It is paramount to understand that with this *definition*, the **equality** between any two *distances* in *space* will be governed by the behavior of the ***congruency*** between solid bodies when they are transported to achieve their *metric* function. It is, then, not difficult to see that *distance* in *space* is the result of a *definition*, convenient and practical but, even so, with a degree of ***arbitrariness***. We say that we have selected a *definition* of ***congruence*** and, in particular, the so-called '***congruence of the rigid body***'.

Now that we know when two *lengths* are **equal** by definition, in order for us to assign a *length* to every object we need to define the ***unit of length*** and we do so by choosing an object or process as ***standard***. Once we have the

standard (A), we repeatedly juxtapose it from one end to the other end of the object being ***measured*** (B), establishing a series of ***partial congruencies***, and annotating how many of them we need to carry out so as to cover B. This number of times which, in general, will not be an integer (it will be rational, i.e. a ratio of integers), is called, *by definition*, the *length* of the body B. It is essential to realize that, even though we certainly need to displace the ***standard body*** throughout the to-be-measured object, in each and every one of the stages where the rule is performing its *metric* function, both bodies (measurer and measured) are in relative ***repose***.

Once more, we abstract the body B and talk about the *length* of the abstract *segment* in *space* or, equivalently, of the *distance* between its *endpoints*. It is evident therefore that the *length* of an object is **not** an intrinsic property of it, but a ***relation*** between it and the ***standard object***. Likewise, the *distance* between two *points* in *space* is not an inherent property of the abstract *segment* defined by both *points*, but a ***relation*** between said *segment* and the ***standard object***.

If objects A and B are ***distant***, then, as when we treated the objective ***measurement*** of *time*, it is necessary to make some epistemological considerations which will be possible for us to understand only after we discuss how to select the ***standard object***.

1.2.5.1. Standard Object/Process - Relativity of Geometry and Universal Fields

Paraphrasing Hans Reichenbach, "It is logically impossible to determine whether the standard meter in Paris is really a meter" [1]. The meter in Paris (Sèvres) is a meter by definition; full stop. The United States had two copies of the ***standard meter*** (21 and 27). Periodically, they were taken to Paris so as to determine if they had changed. These comparisons always included at least three copies so, in case of discrepancies, it would be possible to estimate the probability of these changes occurring in the copies and not in the actual standard [9].

How do we understand all this? When we try to choose a ***standard*** for ***measuring*** *space*, we find ourselves in the same vicious circle we were when trying to select the ***unit of time***. Obviously, if the body is to be used as ***standard of length***, its *length* must not *change*, i.e. the body must be ***rigid***; otherwise measuring the same body will provide different *lengths*, simply because it is the ***standard body*** that ***changed***! Did I say obviously? Not so fast! How do we know the ***standard body*** is ***rigid***? Oh, come on! ***Measure*** it! But... how? if we still do not have the ***standard body*** to ***measure*** it? ***If we measure the standard body with the standard body we will always obtain unity as a result, independently of its variations!***

This vicious circle can be opened when we think in relative terms, instead of assuming that the *length* of a body is an ***absolute*** property. The concept of ***congruence*** allows us to define the concept of *relative* ***rigidity*** between bodies without circularity and without having yet defined a ***standard body***. If we find experimentally that when two *segments* marked on two bodies are ***congruent*** once, they continue being so regardless of how many times we repeat the experiment and, besides, that the same happens for any arbitrary pair of *segments*, we then say that the two bodies are *relatively* ***rigid***. Note again that we have **not *measured*** the *length* of the *segments*, as we have applied the concept of ***congruence*** between them and not between a to-be-measured object and another chosen as ***standard*** (we have not selected the ***standard*** yet). Both bodies found to be ***congruent*** could change their *shape* and *size* dramatically but ***identically***, and they still would continue being *relatively* ***rigid*** as per our definition.

But... is it really possible that two bodies vary their *shape* and *size* while maintaining their mutual ***congruence***? It surely is: if, for example, two bodies are made of the same material (say iron), we only need to heat them up to the same temperature and their ***congruence*** will not be destroyed -- though, if we compare them with neighbor bodies, we will have to conclude that the *relative congruence* (as we defined it) between the two heated bodies and the neighbor ones has changed. We say that both bodies have identically ***dilated*** in a characteristic fashion for iron; but it must be clear that saying that the rest of the Universe has *contracted* while the two bodies remained unchanged is perfectly correct and epistemologically irrefutable, even though it does not fit our intuitive way of seeing things. One thing is unquestionable: a *thermal field* alters the *spatial relation* between bodies and, had we not known that we had applied heat in a uniform way to both bodies (and not to the rest of the Universe), we could not have attributed that change in the *spatial relation* to an *identical* ***dilation*** of the two bodies. Even more perplexing: nothing would have prevented us from interpreting it as a change in the ***geometry*** of *space*!

To discard the last mystifying possibility we ask ourselves: is it possible to know whether a *thermal field* is present in a zone of *space* and, ergo, for us to say it is the *cause* for the observed *spatial relations*? Yes, it is possible because a field of temperatures is a so-called ***differential field***, which means it affects different materials differently. We only need to establish the ***congruence*** between two different materials (e.g. iron and copper), heat them up and observe that the ***congruence*** is destroyed. From this, we deduce that the change in *spatial relation* cannot be explained by the ***geometry*** of the *space* shared by the bodies -- unless we are willing to put up with the complexity of endowing *space* with a *different* ***geometry*** for each *type of material* (in which case ***geometry*** would not be a property of *space* any longer). As we saw in Chapter 1, the thermometer of mercury works thanks to the ***differential*** character of a thermal field, which affects glass and mercury differently so that mercury 'climbs' with our fever.

How do we, thus, choose the ***standard body***? We investigate the great diversity of existent bodies in our accessible Universe and find many which, when compared to one another, are not *relatively* **rigid**, e.g. steel and flexible rubber. But, fortunately, there is a very abundant class of bodies that, even if only approximately, are all *mutually* **rigid**, e.g. all the metals and in general all solid materials; I insist: **rigid** among them, not absolutely **rigid** (a concept not definable). Once the notion of *relative* ***rigidity*** has been elucidated, we take the linguistic license of talking about the 'rigid body' (as we have done in previous sections), as if the adjective had an absolute meaning. As we know by now, it is the tacit assumptions behind our usage of language that blur the concepts it conveys.

Let us deepen our understanding even more: what would happen if we selected a bar of gelatin as ***standard***? Well... epistemologically, it would be perfectly legitimate... but we would find very few bodies *relatively **rigid*** to the ***standard***, and our description of the objective world would be much more complicated because each time we measured the same steel bar, we would get a different length for it. As with the ***measurement*** of *time*, we are epistemologically entitled to choose any body/process we want as standard, but there are choices with maximum practicality and descriptive simplicity (in terms of the laws of Nature we would subsequently discover). For instance, given that the ***unit of length*** intervenes in the *definition* of the concept of *energy*[7], the *Law of Conservation of Energy* would not be valid any longer as such, because the *energy* of a system would vary with the variations of the Universe with respect to the ***gelatin standard*** (which, by definition would always have a length equal to 1).

We, thus, choose a metallic body as ***standard***, say steel, and now we have another problem, previously suggested when talking about ***differential** fields*: if we decree the *length* of this *standard* body simply as the absolute ***unit*** without any considerations as to how it is affected by ***differential** fields* (e.g. thermal fields), we find consequences hard to intuitively accept: if we ***heat*** up the ***standard body***, the rest of the universe ***contracts***! If we ***cool*** it, the rest of the Universe ***expands***! Once again, there would be no logical inconsistency at all, but we would be forced to unnecessarily complicate the laws of Nature (assuming we were smart enough to find regularity in such a chaos).

In order to avoid the above chaos, we complicate a little our *definition* of the ***unit of length*** by taking into account the effect of thermal and other types of ***differential** fields*. We decree that the ***standard body*** will have a *length* equal to 1 only at a **given** *temperature* and -using a mathematical formula derived from the relative *spatial relation* between bodies of the ***standard material*** at different *temperatures*- we specify the *length* of the ***standard body*** at ***every*** *temperature*. We do the same with any other non-negligible perturbation of a ***differential*** character (mechanic, magnetic, etc.). This is why the ***standard meter*** in Paris was made of a platinum-iridium alloy, with an X-shaped cross-section (to minimize deformation), it is maintained at a temperature of 0° Celsius, and is inside an earthquake-proof safe box (even though it is not used any longer as *standard*). With the pertinent semantic and epistemological caveats, we can say that -in this way- we have made the standard body 'more rigid'.

And now the crucial question: are there any ***universal*** (as opposed to ***differential***) *fields*? Where by ***universal*** we mean: a) they affect all objects in exactly the same manner regardless of their composition, mass, etc.; and b) it is impossible to block their effect[8]. Please notice that, by its very definition, a ***universal** field* is very difficult to detect. Because its effect is identical on all objects, and cannot be shielded, the only way to detect a ***universal** field* is by its **variation** in *space* (if it varies). As we saw in Chapter 1, *gravity* is a

7 We defined *energy* as "capacity of doing work" and *work* as the product of *force* with *distance*.

8 Logically, this impossibility stated in b) is inferable from a), but I included it for the sake of pedagogical clarity.

universal** field*[9] because its effect -the *acceleration* imparted to an object- is always the **same** regardless of the object *mass*, *composition*, etc. and, besides, it cannot be blocked as you can shield an electromagnetic field. In anticipation to Chapter 7, *Einstein's Principle of Equivalence* is precisely based on the ***universal nature of *gravity* in sufficiently small zones of space[10] and, besides, it is a direct consequence of the **equality** between *gravitational* and *inertial masses*, a subject we treated with considerable detail in Chapter 1, and that constituted Einstein's predominant muse for the development of his General Relativity.

And... what is the effect of a ***universal** field* on the *spatial relation* between objects (geometry)? If the field affects the *spatial **metrical*** relation at all, by the very definition of ***universal***, it will be **identical** for all objects (the ***standard*** one included) and, ergo, it will be impossible to detect its existence! Think, for instance, that the Almighty has a sense of humor and decides, while we are sleeping, to inflate the Universe in a fixed proportion; will we be aware of his jest after trying all imaginable tests in our repertoire? NO! Because all of us and the entire Universe (including the ***standard*** body) have been enlarged in the same ratio -- leaving all our ***measurements*** **invariant**. Not so evident but equally powerful[11]: if His joke consisted in transmuting the Universe into its specular image, could we know about it? NO! Our 'right' glove (before the hoax) would now neatly fit on our 'left' (after the hoax) hand and... because once 'inside the mirror', we call 'right' to the hand we called 'left' 'out of the mirror', the 'Old Man'[12] would laugh at us for eternity.

Here is the last nail in the ***universal** fields'* coffin: if we cannot prove their existence, and the ***metric** spatial relations* we discovered are **identical** for all objects regardless of their composition or properties, what prevents us from assuming those *fields* do **not** exist, interpreting the existing *spatial relations*, instead, as ***geometrical*** properties of *space*?

The ineludible conclusion is that the term 'geometry of space' is **relative**, even after having chosen a *definition* of ***congruence***, and it only acquires an unequivocal meaning when it is accompanied with the specification of those *fields* affecting the *spatial relations*. But, because the ***universal** fields* are undetectable, the most sensible epistemological decision is to decree them as inexistent! -- attributing those detected *spatial relations* between objects exclusively to the ***geometry*** of *space* as well as to ***differential** fields* (which can be detected). It seems, then, as if the conclusion I arrived at -right before I was 'telepathically' interrupted by my Mother while crossing the Atlantic that Christmas of 1982- was correct -- even though my conclusion was based on completely different considerations (comparing the ***geometries*** of Necochea and Madrid).

To wrap it up, C.D. Broad expressed the gist behind the **relativity** of ***geometry*** very well:

All measurement involves both physical and geometrical assumptions, and

9 *Gravity* is *universal* with respect to the *acceleration* imparted to bodies, but other effects, e.g. the *deformation* it produces on a metal beam, does depend on the beam material and, ergo, it is *differential*.

10 By moving from the Equator to the poles, or leaving the planet surface, gravity changes considerably.

11 The so-called Helmholtz's parable.

12 Einstein used to refer to his God as the "Old Man".

> *the two things, space and matter, are not given separately, but analysed out of a common experience. Subject to the general condition that space is to be changeless and matter to move about in space, we can explain the same observed results in many different ways by making compensatory changes in the qualities that we assign to space and the qualities we assign to matter. Hence it seems theoretically impossible to decide by any experiment what are the qualities of one of them in distinction from the other.*[13]

1.2.5.2. Gravity and Local Curvature of Space - The General Theory of Relativity

These epistemological considerations led Einstein to suspect that *gravity* could be interpreted as a local manifestation of the ***geometry*** of *space* -- instead of the Newtonian interpretation of ***action***, instantaneously and mysteriously, ***at a distance*** (a notion with which Newton never felt comfortable).

As we shall see in Chapter 7, during his herculean task of including the *gravitational field* within the conceptual frame of his Relativity theory, Einstein encountered a serious dilemma. If he insisted on preserving the precepts of Euclidean geometry, he had to change the traditional definition of Babylonian ***congruence*** suggested by the observed *spatial* behavior of the **rigid** body. This was not a small deal as it meant that when the ends of two rules coincided regardless of their *position* and *orientation*, he had to **define** them with **different** *lengths*! Einstein was mindful of the conventional nature of the *definition* of ***congruence*** and, hence, there was no epistemological reason to prevent him from judiciously changing it to suit his needs. But, besides these counterintuitive 'expansions' and 'contractions' of a solid body, he also had to accept that *light* would not travel in a 'straight line' (geodesic trajectory) in a Euclidean *space*.

The other face of the dichotomy he was confronting was to defenestrate the Euclidean dogma but, even though the dogma had been considerably weakened by Gauss, Bolyai, Lobachevsky and Riemann, doing something like that would certainly be considered a scientific heresy. By taking such a bold move, and choosing the ***non-Euclidean geometry*** of Riemann with **variable** *curvature*, he could maintain the millenarian definition of ***congruence of the rigid body***, and *light* would continue traveling in a 'straight line', i.e. on the geodesics of a ***non-Euclidean*** *space*.

No matter which of the two options Einstein chose, the *spatial relations* between objects in the neighborhood of *matter* could be explained as a) there existed a *field* (called *gravitational field*) and a ***geometry*** of *space* so that both jointly explained the objective **reality** or... b) There is **no** *gravitational field* at all, because the *spatial* and *temporal relations* can be explained as the exclusive consequence of the **local *geometry*** of the surrounding *space-time*.

Einstein decided to retain the millenarian simplicity and intuition behind the idea of ***congruence of rigid bodies***, and to include the effects of *gravitation* into the ***geometry*** of *space*. This new geometry would be Riemannian, i.e. with a *variable curvature* from *place* to *place*. Remember that this *curvature* is **not**

13 As cited in A. Grünbaum, *The Philosophical Problems of Space and Time*, 1963.

our colloquial *curvature* (which is Euclidean!) so... please! Do not try to imagine our tri-dimensional *space* curving in our everyday sense! With this approach, an object in **free-fall** does not fall because Earth attracts it instantly and mysteriously at a distance but because, **not** being subjected to any *force* whatsoever, it moves through a *geodesic* ('straight line') of *space-time* -- reminding us of *Galileo's Principle of Inertia* whereby a body free of any influence should move in a *straight line* forever. The classical laws of *motion* and their evolution up to Special and General Relativity theories will be treated in upcoming chapters. Patience, as we are getting there!

For now, let us imagine we consider a reduced zone of *space* close to our planet so that the *acceleration of gravity* is **constant** and, ergo, we are in the presence of a ***universal*** *field*. According to Einstein, there is no *field* (remember the last nail in the coffin for ***universal*** fields?) because the observed *spatial relations* are the exclusive consequence of a different ***geometry***, which is globally ***non-Euclidean*** but that, in sufficiently small regions, it is indistinguishable from the Euclidean one.

That's not true! a perspicacious experimenter would exclaim: I admit that the existence of the *gravitational field* -in a reduced region of *space*- cannot be proved changing the *mass* or *composition* of the body, but you only need to move the object from the Equator to the poles or simply sufficiently elevate it from ground so as to measure a ***different*** *acceleration*; it is the *spatial* ***non-uniformity*** of the *gravitational field* that makes it detectable as an objective **reality** and, therefore, I do not accept your capriciously calling it 'non-existent'.

I agree, Einstein would say, but... what prevents us from interpreting the observed variations in the *spatial relations* between objects when changing places (due to what you call non-uniformity of the gravitational field) as, simply, variations in the ***geometry*** (***metrical*** properties) of *space* while we insist on saying that the *gravitational field* does not exist? Local *gravity* is ***universal*** and, as such, can be ignored; all we have to do is to accept that the ***geometry*** valid in one *place* has not to be strictly valid in another, and to be capable of quantitatively determining how the ***geometry*** changes from *place* to *place*. Does it sound familiar to the variations of Madrid's *metric tensor*? This is the essence of the General Theory of Relativity -Einstein's masterpiece presented in 1916 and the core subject of Chapter 7- so... again, don't despair if this seems incomprehensible as of now. We are just warming up.

And I would add in my imaginary defense of Einstein's position: we are tired of watching on TV the astronauts 'floating' in the International Space Station while at the same time we are 'educated' about the **absence** of *gravity* in the region of *space* where the Station is at only 350 km of altitude. Nonetheless, at that height and from the Newtonian perspective, the Earth *gravitational field* is still very powerful, and precisely because of its strength (according to Newton) the Station continues rotating (and the astronauts float with respect to it). If we accept that the *gravitational field* is an objective **reality** on the surface of the Earth, then we have to accept that it also exists and is very strong where the Station is. Consequently, the folkloric term 'weightless' is incorrect from the perspective of the Earth and helps neither our education nor our understanding of the Universe.

Now, let us change the *frame of reference*: had the astronauts been born in the Station and without any contact with terrestrial science, they would not have had any reason to conceive a *gravitational field* they could not have detected and, hence, the voice 'weightless' would have been neither incorrect nor correct but, simply, non-existent in their lexicon! The *space-time* relation between bodies inside the Station (their natural *frame of reference*) could have been explained exclusively by a ***geometry*** of their **local** *space* and ***differential*** *fields* (which are detectable).

The unavoidable conclusion is that the ***existence*** of some *fields* is **relative** to the *frame of reference* and to the choice of the ***geometry*** of *space*. But... for God's sake! Is there anything at all which has an absolute value independently of our perspective? Of course there is; but we have to be patient and, meanwhile, so as not to anguish, let us meditate upon what Hans Reichenbach wrote in his book *The Philosophy of Space and Time*:

> *The only path to **objective knowledge** leads through conscious awareness of the role that **subjectivity** plays in our methods of research.*

1.2.6. The Length of Distant Bodies

One of the reflections I made during my walk to school that 'philosophical' morning of 1954 was that, despite the distant trees and cars seeming smaller than those close to me, I was convinced that, had I transported my Mom's meter tape with me toward the **distant** *place*, I would have been able to quantitatively confirm that there was no essential difference in their *sizes* and *shapes*. Does this mean that the objects were really of the **same** *size* even though I saw them **smaller** from afar?

Let us analyze the situation in the following equivalent manner. We take two rules with the **same** *length* and transport them to another **distant** *place* by different paths. Once the two rules are together again at the new *place*, we compare them and they still have the **same** *length*. Are we entitled to affirm that the two rules did not change *length* during their respective transportation paths? The answer is NO; given the definition of *length*, the only provable assertion is that both rules have the **same** *length* when they are compared *in situ*. As we know by now, the existence of different ***universal*** fields at each of the ***distant*** *places* would preserve their ***congruence*** in both *places* even though they could have experienced a colossal *expansion* or *contraction* during their respective transportations. The mere *definition* of *length*, through the ***congruence of the rigid body***, does not allow us to compare ***distant*** objects and, due to the supposed ***universality*** of the assumed *fields*, when the bodies get together again, they will continue being ***congruent***.

One more time, the problem is not one of *cognition*, but of *definition*. It is impossible to know whether a rule maintains its *length* when it is transported from one *place* to another and, hence, in order to compare ***distant*** *distances*, we have to introduce another *definition*: given the experimental fact that two bodies *relatively* **rigid** of **equal** *length* are found of **equal** *length* every time they are compared *in situ*, independently of the paths traversed for them to meet, we decree that both bodies also have the **same** *length* even when they are separated.

Remarkable! As we have learned several times now, what we tend to consider as a logical inference bordering on the obvious turns out to be just a sensible *definition* with a degree of ***arbitrariness***. The only restriction this *definition* has to respect is that it should produce the previous accepted result for the case in which the bodies are ***contiguous*** -- which is the only one for which the *definition* of **congruence** can be applied. Clearly, our new *definition* meets this requirement.

From the above arguments and new definition, the answer to my childhood question is: **YES**, the trees and cars seemed ***smaller*** from afar (by ***perception***); **YES**, they had the **same** *size* when I moved from one place to another with my *meter* tape (by ***measurement***); and finally, **YES**, they had the **same** *size* when they were ***distant*** to me, even though I *perceived* them ***smaller*** -- but this latter **equality** of *lengths* was valid by *definition*!

The experimental fact that two bodies which are ***congruent*** in a place keep being ***congruent*** in another place regardless of how we transported them, is very convenient for our ***measurement*** technique: had two bodies been ***congruent*** 'here', but **not** been ***congruent*** 'there', we would have had to define a different ***standard body*** for each and every *place* in *space*, and consider all of them as **equal** in *length* by definition. Only in that way the science of ***measurement*** could have had a global legitimacy; otherwise the *length* of an object would have been a property of it and of its *location*. For the disappointment of the French's pride, there would have been as many 'cities of light' zealously guarding their 'standard meter' as human conglomerations exist on the planet. Luckily, that is not the case, and visiting Paris is a memorable experience. This is the supreme importance and convenience of the definition of ***congruence*** by means of the behavior of the **rigid** body.

These reflections allow us to understand a notion of the highest significance for upcoming chapters: the mere fact of conceiving the idea that we would have had to call two bodies -congruent 'here' but not 'there'- as **equal** in *length*, means that the relation between the concepts of ***congruence*** and *length* is not as straightforward and natural as it seems to be. In the place we call 'here' we would consider that two ***congruent*** bodies have the **same** *length*, and in the place called 'there' we would consider that two ***incongruent*** bodies also have the **same** *length*, i.e. the definition of **equality** would be different for different *places*! Succinctly: the popular assertion "two congruent bodies have the same length", regardless of how natural and logical it seems to be, in reality, it carries (though camouflaged) a very convenient *definition* based on the physical behavior of the **rigid** body -- but this definition is far from being unique!

Once again, we conclude that the different *spatial relation* between bodies in different *locations* can be interpreted by asserting that the ***geometry*** of *space* is a **local** property that can vary from *place* to *place*. Besides, we find out -not without amazement- that the notion of *shape* of an object is also relative to the ***definition of congruence*** chosen to measure it and, consequently, we learn that the statement 'our planet is approximately spherical' (which sounded so ridiculous to me when I was six) had the chosen ***congruence of the***

rigid body hidden in it. With a different ***definition of congruence***, our planet could easily (but weirdly and awkwardly) said to be ***cubic***!

1.2.7. Length of a Body in Motion and its Relation with the Notion of Simultaneity

In Chapter 2 we learned that the concept of *simultaneity* is **relative**. In this chapter, we just learned that both *length* and *geometry* are **relative**. Pithily: both *time* and *space* are ***metrically*** amorphous, i.e. they do not have an intrinsic ***metric***. Undoubtedly, thus, the Newtonian concepts of ***absolute*** *space* and *time* cannot be correct. In this section we shall inquire if *space* and *time* are totally independent à la Newton or, to the contrary, their objective ***measurements*** are so intertwined that, in some cases, the ***measurement*** of *space* depends upon the ***measurement*** of *time* -- in blatant disagreement with our everyday intuition.

Marking a solid body, we defined the notion of a *segment* and, through the concept of ***congruence*** with a *rule*, we defined the *length* of the *segment* or, equivalently, the *distance* between its *endpoints*. Abstracting one of the bodies, we spoke of the *distance* between two *points* in *space*. We also emphasized that the *rule* had to be *in repose* with respect to the *segment* while performing its *metric* function. Let us imagine now two different worlds in relative *motion* (e.g., again, Earth and Mars) and that we want to compare the *lengths* of two *segments* marked on two bodies which are in relative *repose* inside each world.

We already know how to ***measure*** the *length* of our *segment* because it is *at rest* with our *rule*. The same can be said of our Martian neighbors because their *rule* and their *segment* are at relative *rest* as well. We refer to these terrestrial and Martian *lengths* as 'lengths in repose' because the ***definition of congruence*** can only be applied when ***rule*** and ***segment*** are in relative *repose*. Let us make things interesting: taking advantage of the interplanetary cultural exchange program, we make a copy of our *segment* and our *rule* and off we are to Mars. Once physically and socially adapted to the new world, we measure our *segment* with our *rule*, obtaining its 'length in repose' and... what do we get? Well... by now, after our discussion on ***universal*** fields, we can only say what the ***congruence*** ratio between the *rule* and the *segment* is on Mars; any further statement regarding what the *length* of the *segment* is as compared to its *length* on Earth requires serious cogitation to separate ***cognition*** from ***definition***. Now the situation is more complex than my childhood comparison of trees and cars a few blocks away, because those objects were in relative *repose* while Mars and Earth are in very fast relative *motion*. And why does relative *motion* impose a problem? I repeat: because our notion of ***congruence*** is meaningless unless *rule* and to-be-measured *object* are relatively *at rest*.

Once the *rule* is in the world called Mars, it is impossible to compare its *length* with the one it had while on Earth. For the umpteenth time, we learn that what seems to be a sensible and indisputable fact, is simply so because of a convenient definition: we decree that the *length in repose* (better: ***proper*** *length*) of an object measured in *frames of reference* which are in relative *motion*, is -*ceteris paribus*[14]- always the same. Please understand that the to-be-

14 *Ceteris Paribus* is Latin for 'all other things being equal' and commonly used when discussing logical relations.

measured object is *at rest* in the *frame of reference* from which is ***measured***, and that we are defining that the object's ***proper*** *length* will always be the same regardless of the *frame of reference* into which we place it (e.g. Mars or Earth). In brief: the **proper** *length* of an object is independent of the *reference frame* in which it is ***measured***.

It gets even more interesting: we make a copy of our rule and segment and leave them to our Martian host. By definition of 'copy', *rule* and *segment* staying on Mars have the same proper *length* as the originals now back on Earth; and both Martians and Terrestrians will always *measure* (*ceteris paribus*) the same **proper** *length* for the *segment*. If we have the patience of dealing with a 20 minute-delay phone conversation with our Martian friend, we could corroborate our agreement on the ***proper*** *lengths* of original and copied objects, while cheerfully remembering good old times.

But... pretty soon we realize that we would be very limited in our ability to understand the Universe, if we depended upon the existence of intelligent beings in every recondite *place* we are interested in. The only way to progress in our comprehension of the Universe is to be capable of describing the objective **reality** from different *frames of reference* and, hence, we need to devise a technique to ***measure*** the *length* of an object which is in *motion* relative to us.

The crucial question is: what is the *length* of the *segment* we left on the red planet from our perspective here on Earth, i.e. using a *rule* which is *at rest* on Earth (and, ergo in *motion* relative to Mars)? In classical physics this question was never made; it was assumed that the *length* of an object was the **same** regardless of its *motion* relative to the *rule*. This assumption was considered one of those *a priori synthetic truths* which did not admit a scintilla of doubt. Notwithstanding, it should be clear by now that the classical notion of *length* is based on the geometrical notion of ***congruence*** and, if the to-be-measured object is **not** *at rest* with the *rule*, ***congruence*** has no meaning (surprised of how easily can we ignore the obvious for centuries?). We, thus, need to conceive a new *definition* to agree in what we mean by *length* of a *moving* object. As always, this new *definition* will have some ***arbitrariness***, but it has to be a consistent extension of our classical *definition* based on the ***congruence of the rigid body***. In short: however we define the *length* of a *moving* object, when it is *at rest* with us, the new *definition* has to become the classical one.

Here is a natural and consistent extension: we will call ***moving*** *length* of the Martian *segment* ***measured*** from Earth, the classical *length* of a new *static segment* defined on Earth through a ***snapshot*** -based on our **local** *time*- of the *moving* Martian *segment*. In other words: we call *length* of an object in *motion* with respect to us, the *distance* we ***measure*** between two *simultaneous positions* -in our *reference frame*- of its *endpoints*. We simply converted a *moving segment* into a *static* one so we can apply the ancient notion of ***congruence***. Obviously, in the special case in which the relative *speed* between the *segment* and our *rule* is ***zero***, this new *definition* becomes the classical one, giving what we already know as the *length in repose* or, better, the ***proper*** *length*.

Even though this definition seems quite natural and simple, the attentive reader who remembers the relativity of *simultaneity* we discovered in Chapter 2, must be anticipating the arrival of a huge headache (more like a migraine!):

because the *simultaneity* between ***distant*** *events* is definable with a degree of ***arbitrariness***, how will our new *definition* of *length* affect the *length* of the Martian *segment* when we choose our *definition* of *simultaneity* here on Earth at leisure?

Let us define two *events* A and B; A is defined by the coincidence of our observation of one of the *endpoints* of the *moving segment* with a fixed *position* here on Earth. This *event* is defined by a *position* and a **local** *time*. B is likewise defined by the coincidence of our observation of the other *endpoint* with another *position* here on Earth, in general *distant* from that of A. B is defined by another *position* and the **same** *time* as for A, i.e. A and B are *simultaneous* according to our clocks.

The *segment* defined by the *positions* of *events* A and B is *at rest* with respect to us so we can ***measure*** its *length* and the result will be, by definition, the ***moving*** *length* of the Martian *segment* on Earth. Now, due to the leeway we have in defining when two *distant events* in relative *repose* are *simultaneous* (the greater the latitude the larger the *distance* between A and B), had we changed our *definition* of *simultaneity* for another permissible one, the same pair of observations (*events* A and B) will have the **same** *spatial* coordinates but **different** *metrical times* (though, of course, they are still *topologically simultaneous*) and, therefore, the **same** *distance* between A and B is not any longer the ***moving*** *length* of the Martian *segment*, because A and B are not *metrically simultaneous* for us anymore. This is telling us that the *distance* from A to B will have to *change* so as to make their *times metrically* ***identical*** and, ergo, the *length* of the *segment* on Mars, as ***measured*** here on Earth, depends upon our ***measurement*** of *time* on Earth. It is also not difficult to conclude that, had Earth and Mars been *at rest* with respect to each other, the *length* would have simply been the object's ***proper*** *length*, and independent of our *definition* of *simultaneity*. This is why in classical physics, *time* and *space* have always been independent.

In summary, the particular *definition* of ***metrical*** *simultaneity* we adopt affects the *length* for an object in *motion* and, ergo, the ***measurement*** of *space* depends, in general, upon the ***measurement*** of *time*! Once we adopt a definition of *simultaneity*, e.g. Einstein's (previous chapter), the ***moving*** *length* will be different from the ***proper*** *length* and will depend upon the *speed* with which the body is *moving* with respect to the *reference frame* -- in the same manner we saw that the *energy* of an object depends on its *speed* relative to the *frame of reference*.

At this point, I bet the reader must be wondering if all this philosophical mumbo jumbo is worth her/his time and effort to follow me. YES! I assure you it is worth your time. Why? Because another folkloric assertion of the great majority of Popular Science books on Relativity theory says that "a body in motion is shorter the faster it moves, with its length reducing to zero as its speed approaches that of light in vacuum". So expressed, the statement sounds as fantastic as unintelligible, because nobody is prepared to believe that a block of steel will ***shorten*** until disappearing by the mere fact of ***moving*** faster and faster. Let alone when, on top of it, these authors tell us that the steel *mass* becomes ***infinite*** right when its *length* becomes ***zero***!

Remember that, in Chapter 1, once we got rid of the semantic confusion associated with the word 'mass', and understood it means 'inertia' (**not** quantity of *matter*), its increase with *speed* relative to the *reference frame* stopped being unintelligible and counterintuitive. The way to maintain our mental sanity with respect to the decrease of an object *length* with *motion* is similar: no matter how fast two objects are *moving* with respect to each other, their ***proper*** *lengths* are always the same, so that our common sense and intuition are not offended. Remember that ***proper*** *length* means the *length* when the object is *at rest* in the *reference frame*. It is the ***moving*** *length*, i.e. the *length* as ***measured*** from a *reference frame* in which the object is *moving*, that changes with their relative *speed*; and the reason is found in the very *definition* of ***moving*** *length* which involves the ***measurement*** of time. Besides, the effect is fully ***reciprocal***: the *segment* we left on Mars, when measured from Earth, is ***contracted*** (compared with the one we brought with us), and when our Martian host ***measures*** our *segment*, s/he finds it to be ***contracted*** (**not** expanded!) with respect to his copy.

Asking 'which is the real length' is pointless. All of them are different manifestations of the same objective reality, and the effect is due to the relative motion between both frames of reference.

1.3. Historic Evolution of the Measurement of Length up to the current Unit

As Herbert A. Klein says in his historical treaty on the science of physical measurement [9]: "The metamorphoses of measurements form a special chapter in the annals of illogical human behavior". For millennia, humanity sought ways of defining the ***unit of length*** in a reproducible manner, using parts of our anatomy (inch, foot, finger, nail, digit, etc.), botanical units (barley seeds), our planet dimensions, processes like the pendulum, a boat on water (marine knot), and others much more arbitrary, giving birth to an extremely erratic, anarchic evolutionary process, plagued with inconsistencies which, fortunately, ended up in the metric system and its extension, the current International System of Measures (SI) [9].

In the 18th century there were two candidates to become the ***standard body/process*** for the ***unit of length***: the *length* of a pendulum with a half-period of a second, and the ten millionth of the *distance* of the meridian from the North Pole to the Equator passing through Paris. In 1791 the French Academy of Sciences chose the meridian definition discarding the pendulum (because gravity variations on the planet affect the relation between its period and length). An expedition lasting from 1792 till 1799 measured the distance between Dunkirk and Barcelona so as to establish the ***standard body***. In the interim, in 1795, France built the first ***standard bar*** made of brass as the official **unit** which was called 'meter', based on the preliminary measurements of the meridian (afterwards found to have an error of 0.2 mm). Even so, the original brass slab was adopted as the **unit** and, as a result, the circumference of our planet on that meridian is a little more than 40,000,000 meters. In 1799 the ***standard bar*** was made out of platinum.

In 1875 the International Office of Weights and Measures (BIPM) was founded in Sèvres (outskirts of Paris), defining the international ***meter*** as the *distance* between two marks made on a bar built in 1889 with an alloy of 90% platinum and 10% iridium at 0° Celsius. A sign of great progress was that the United States and England defined their 'yards' in terms of the ***meter***, instead of constructing their own bars of platinum-iridium.

In 1893, the ***standard meter*** was measured with an optical interferometer by Albert A. Michelson who proposed the idea of defining the ***unit of length*** through the *wavelength*[15] *of light*, because it offered an extremely precise method to measure lengths through the interference[16] of two light rays. In 1960, the proposal was officially accepted and the International System of Units (SI) defined the meter as "equal to 1,650,763.73 wavelengths of the 'orange-red' light emitted by the atom of Krypton-86 in vacuum". The original bar is still preserved for historical reasons in Sèvres under the conditions strictly established in 1889.

With respect to the precision and reproducibility of the ***standard meter***, in 1889 it was possible to realize comparisons of *length* with a precision of one part in 1 million (0.0001%). In 1960, using Krypton-86 light, precision improved 100 times (0.000001%). The lack of coherence[17] in the light emitted by the Krypton-86 lamp only allowed to accurately measure *lengths* between 0.1mm and 0.4mm. The invention of the laser would solve these problems. In the 1970-1980 decade, the National Bureau of Standards (USA, today NIST) utilized a laser beam emitted by gaseous Helium-Neon allowing to accurately measure several tens of meters without coherence problems, though with a reproducibility lower than that of the Krypton lamp.

Finally, because of the dramatic improvements in laser technology, and with the objective of maximizing the reproducibility of the ***standard meter***, and the precision of *length* measurements, the Resolution 1 of the 17th General Conference on Weights and Measures (CGPM, 1983) defined the standard meter as follows:

The meter is the length of the path travelled by light in vacuum during a time interval of 1/299,792,458 of a second. The old definition is abrogated.

It is worth noting that, by defining the ***meter*** in this way, the *speed of light in vacuum* is fixed without uncertainty at the value 299,792,458 meters per second. The foundation behind this decision will be clear in Chapter 6 (Special Relativity). To improve the reproducibility of the ***standard meter*** among different laboratories in the world, the recommended radiation was the laser of Helium-Neon whose wavelength in vacuum is 632.99139822 nanometers with an uncertainty of 2.5 thousand millionth per cent while, as we saw in the previous chapter, the uncertainty in the reproduction of the second as a unit of time is about 100,000 times better.

The definition for the ***standard meter*** was improved in 2002 recommending it be limited to "wavelengths sufficiently short to minimize the effects predicted by General Relativity with respect to the uncertainty in its realization".

15 Let us remember the *wavelength* is the distance that light traverses during a cycle of the electromagnetic field. For visible light, this distance is between 380 nm (violet) and 750 nm (red).

16 The phenomenon of *interference* between waves shall be described in Chapter 5.

17 *Coherence* is a property of a luminous ray related to the precision with which *interference* can be measured.

This recommendation will be understood in Chapter 7.

2. What we have learned

Let us sum up what I think the reader has learned in this chapter:

• Physical ***geometry*** is the science which studies the *spatial relations* between bodies and radiation (objects) independently of their composition or nature;

• The term 'rigid body' is the result of a linguistic license, because ***rigidity*** is a concept ***relative*** to the ***standard body/process***;

• Mutual ***congruence*** of ***rigid*** bodies and their behavior when *transported* were historically employed to *define* when two *segments* in *space* have the same *length* or, equivalently, when two *distances* between two *points* are ***equal***. Nonetheless, this *definition* of equality between *distances* is not unique;

• It is impossible to know whether a rule maintains its *length* when *transported* to another *place* and, ergo, two rigid bodies with the **same** *length*, when together, are assigned the **same** *length* when afar by *definition*;

• The ***geometry*** of *space* is a ***relative*** term which only acquires unequivocal meaning after a) assuming there are **no *universal*** *fields*; b) specifying the definition of ***congruence*** to ***measure*** *distances* or, equivalently, specifying the *metric tensor*;

• The notion of *curvature* of a Riemannian *space* is different from the intuitive idea (subconsciously Euclidean) of *curvature* as deviation from a straight line or plane. Both cities, Necochea and Madrid, according to our colloquial language are built on a *plane*, but their Riemann *curvatures* are different with the former's being **zero** (Euclidean space) and the latter's being **non-zero** (non-Euclidean or ***curved*** *space*);

• Classical physics assumed that the *length* of a *segment* was an ***invariant***, i.e. independent of the *reference frame*, even when the latter was in relative *motion* with the former. Science ignored for millennia that the very definition of *length* implied that object and rule had to be in relative *repose*;

• The extension of the notion of ***proper*** *length* to the notion of ***moving*** *length* shows that, when there is relative *motion*, the ***measurement*** of *space* depends upon the ***measurement*** of *time* -- demonstrating that the ideas of ***absolute*** *space* and *time*, one independent of the other à la Newton, could not be correct.

• From using our anatomy, our planet dimensions, and through the fabrication of a **rigid *standard body***, humanity ended up choosing the process of *light propagation in vacuum* to define the ***unit of length*** named 'meter' as the international ***standard***. By means of this *definition*, the value for the *speed of light* is not obtained by measurement but adopted so as to ***define*** the ***meter***. The standard prototype of platinum-iridium was abrogated and the ***meter*** was defined as the ***distance*** traveled by light in a ***time interval*** equal to 1/299,792,458 seconds. What we understand by a 'second' was defined in Chapter 2.

In the next two chapters (4 and 5), we shall embark on a similar endeavor of understanding and demystifying the concept of ***relative motion***, so that in

Chapter 6 we can tackle the Special Theory of Relativity, and comprehend in depth why *space* and *time* are ***relative*** and ***interdependent***.

Additional Recommended Reading

[1] Reichenbach, Hans, *The Philosophy of Space and Time*. New York: Dover Publications, 1958.

[2] Gimbel, Steven and Walz, Anke, *Defending Einstein - Hans Reichenbach's Writings on Space, Time, and Motion*. Cambridge: University Press, 2006.

[3] Hoffman, Banesh, *Relativity and its Roots*. New York: Dover Science Books, 1983.

[4] Grümbaum, Adolf, *The Philosophical Problems of Space and Time*. New York: Alfred A. Knopf, Inc., 1963.

[5] Poincaré, Henri, *Science and Hypothesis*. New York: Dover Publications, 1952.

[6] Rucker, Rudolf v. B., *Geometry, Relativity and the Fourth Dimension*. New York: Dover Publications, 1977.

[7] Eisenstaedt, Jean, *The Curious History of Relativity*. Princeton: University Press, 2006.

[8] Carnap, Rudolf, *An Introduction to the Philosophy of Science*. Edited by Martin Garner, New York: Dover Publications, 1996.

[9] Klein, Herbert Arthur, *The Science of Measurement - A Historical Survey*. New York: Dover Publications, 1974.

[10] Goldsmith, Donald. *The Ultimate EINSTEIN*. New York: Byron Press Multimedia Books, 1997.

Chapter 4

Who was Right: Ptolemy or Copernicus?

Had I been present at the Creation, I would have given some useful hints for the better ordering of the Universe. Alfonso X of Castile 'The Wise' (1221-1284), referring to the Ptolemaic View of the Universe used in the Alfonsine Tables.[1]

None... and... Both

...there is no difference between the two statements: 'the earth rotates once a day' and 'the heavens revolve about the earth once a day'. The two mean exactly the same thing, just as it means the same thing if a say that a certain length is six feet or two yards. Bertrand Russell, ABC of Relativity, 1925.

In Chapter 1, I supposed all of us shared the intuitive and classical notions of *time* and *space*, employing the concepts of *instant*, *duration*, *simultaneity*, *position*, *distance*, and their derived magnitudes to describe *motion*: *velocity* and *acceleration*. In essence, thus, *motion* appears to us as ***change*** of *position* in *time*. But we also learned that, when there is *acceleration*, *motion* is accompanied with the existence of *forces*. Galileo and Newton's laws of *motion* were explained in classical intuitive terms, using the semantics mostly provided by the Merriam-Webster dictionary.

In Chapter 2 we kept the notion of *space* as intuitive, while we attempted to understand the modern concept of ***objective*** *time*. We learned that the concept of *simultaneity* corresponds to the instinctive idea we all have of it, but only when the *events* are ***contiguous*** in *space*. When *events* are ***distant***, we concluded that saying they are *simultaneous* has not an unequivocal meaning, unless we agree in adopting a *definition* which is suggested and limited by a concept even more fundamental than *time*: the *Principle of Causality*. This recognition of the conventionality of *simultaneity* will allow us -in Chapter 6- to understand the ***relativity*** of *time* and its role in the conception of *space-time*.

In Chapter 3 we discussed the ontology and epistemology of *space*, contrasting its *perception* with its *objective* ***measurement***. We learned that the so-called ***congruency of the rigid body*** is not the only possible *definition* to ***measure*** *space*; that the ***geometry*** of *space* is a ***relative*** notion; that the concept of ***curvature*** of *space* is different from our intuitive (subconsciously

1 The Alfonsine Tables contained data to compute the position of the Sun, Moon, and planets relative to the fixed stars. They were based on the observations of Ptolemy and others, and prepared in Toledo (Spain) in 1252-1270.

Euclidean) idea of it as deviation from a *plane*; and that to ***measure*** the *length* of an object, using a *reference frame* in *motion* relative with the object, requires of another somehow arbitrary *definition* which makes it dependent upon the ***measurement*** of *time*. This conventionality of *length*, and its dependence upon *time*, will allow us -in Chapter 6- to understand the ***relativity*** of *space* and *time*, as well as whether there is anything in the mysterious *space-time* that is ***absolute***.

Even employing the intuitive and classical notions of *time* and *space*, it was clear in Chapter 1 that both traversed *distance* and *velocity* are concepts ***relative*** to the *reference frame* and, caricaturing this solid **truth**, I traveled from the Mormon Temple in Salt Lake City to the Caesar's Palace in Las Vegas maintaining a constant **nil** *velocity* and having traversed a **zilch** *distance*! What? How? Well... I decided to go rogue and chose my own automobile as the *frame of reference*. Does it mean I saved my gasoline money? Of course not because, with respect to my capricious *reference frame*, our planet (and the rest of the Universe) was *moving* in the opposite direction during my whole trip! Otherwise, how could I have been across from the Mormon Temple at the beginning and in the heart of the 'Sin City' at the end?

But... did the planet ***really*** move? Or... did my car ***really*** move? Why did I add "and the rest of the Universe" between parentheses? Would it not be sufficient with just our planet *moving* in the opposite direction? If *motion* is simply ***change*** of *distance* in *time*, then all *motion* is evidently ***relative***. But... is it only that? Why, almost a hundred years after the publication of the General Theory of Relativity in 1916, our folkloric 'educated' view is Copernican, i.e. that the Earth is the one that *moves* and not Ptolemaic, i.e. just the reverse, as people believed in ancient times?

Who was right? Copernicus or Ptolemy? Maybe none? Maybe both? If it is so easy as I described it during my trip to Las Vegas, why is it so difficult for us to think in relative terms? Well, for the good reason that human nature loves absoluteness, and erroneously considers it as a state of higher knowledge[2]. And this difficulty reaches its apex when *motion* involves ***rotation***: seven years after his *annus mirabilis*, Einstein was still struggling with ***rotation*** and, in 1912, wrote to his friend Michele Besso: "... As you can see, I am still very far from conceiving rotation as relative...". He took another four years to complete his masterpiece.

The classical concepts of *inertia* and *inertial mass* were also intuitively treated in Chapter 1 as a mysterious humanoid resistance of *matter* to ***change*** *speed* -- and this ***change*** of *speed* is called *acceleration*. I said: "... if the body is at rest, it resists starting to move and, if it is moving, it resists being stopped. Because of that reluctance, we hold tight while standing on a bus since upon its sudden stop we would be violently and mysteriously pushed in the direction the bus had before the unexpected brake. Contrariwise, while the vehicle is moving at a constant velocity on a road without bumps, we can freely move without danger of falling and even hold a glass of water without spilling it." I am sure that there is not a single reader who is not identified with this traveling experience and who does not think of my assertion as perfectly natural and

2 In fact, the first acceptation in the dictionary for 'absoluteness' is "free from imperfection"!

logical. Notwithstanding, in this factual statement we find the *Gordian Knot* for the long-lasting philosophical clash between the champions of ***absolute*** *space* and *motion* and the defenders of their ***relative*** nature.

This and the next chapters are dedicated to review that philosophical struggle between intellectual giants and to try understanding the idea of *motion* and its ***relative*** character. If you survived the last three chapters, it is probable that -with everything we have learned from them- you will find this chapter -despite its epistemological depth- more pleasant and easier to digest so as to, at the end of the next chapter, be enabled to comprehend the Special Theory of Relativity (Chapter 6) and later the General Theory of Relativity (Chapter 7). Let's enjoy the adventure of *motion* together!

1. What is Motion? -The Ontology and Epistemology of Motion throughout History

I shall divide this brief historical review until the arrival of Einstein in two eras characterized by their conception of *motion*: the ***kinematics*** era and the ***dynamics*** era. ***Kinematics*** is the discipline which studies the ***motion*** of objects as a simple ***change*** of *position*, without any reference to the notion of *force*. **Dynamics**, in turn, studies *motion* in relation with the *forces* which modify it. Both disciplines constitute the science of ***Mechanics***.

From our first months after birth, it is obvious to us that one way of making an object *at rest* to *move* is, besides letting it fall, ***pushing*** or ***pulling*** it, and that those two actions are always associated with a special muscular sensation in our body. Soon we learn to refer to that sensation as 'applying force'. In this way, the notions of *force* and *motion* are instinctively associated in our brain, reinforcing another basic intuitive rendering of our surrounding world: the notion of ***causality***. We rationalize that *force* produces *motion* because we think (erroneously) that *repose* is something totally different from *motion* and that the object -before being subject to our *force*- was entirely free of other *forces*.

From these primitive experiences -essential for our development- it is only natural that we hastily and subconsciously conclude that every *motion* implies the existence of a *force*. Even when we see an object coming toward us in the air without any apparent *force* acting on it, we reason that some *time* before, it was subjected to a *force* which started its *motion*, without reflecting on why the body continued its *motion* given that the *force* disappeared as soon as the object started its *trajectory*. With respect to our experience with *falling* bodies, given that we obviously do not need to ***push*** them to start *falling*, we early in life intellectually (though subconsciously) assimilate Newton's legacy, accepting the existence of a mysterious, invisible, and omnipresent *force* which grown-ups call 'gravity'.

If the *moving* bodies are electrically charged, then the discipline which studies them is named ***Electrodynamics***. As we saw in Chapter 1, the increment of the electron *mass* with its *velocity* is a phenomenon which had been observed, without a coherent explanation, before the most famous equation in history predicted and explained it. If the *mass* of the electron *changed* with its *speed*, the particle known as electron obviously did not behave according to Newton's precepts regarding the intrinsic constancy of an object *mass*.

We also learnt in Chapter 1 that the phenomenon of *radiation*, which transmits *energy* from place to place without transporting *matter*, shows the characteristics of *motion* as well, because it propagates with a certain *speed*. Furthermore, we proved that -according to the most celebrated equation- *radiation* has *mass* (*inertia*). *Electromagnetic waves* (including *light*) have, thus, all the characteristics of *motion*. *Motion* is, manifestly, more fundamental than *matter* because it can ***transmit*** *energy* and *mass* without ***transporting*** *matter*. However, we should not forget that the only way of generating or detecting *radiation* is through its interaction with *matter* - because generation is achieved with a ***transmitter*** and detection with a ***receiver*** (these devices being natural or man-made).

It is a curious fact of history that ***Electrodynamics*** was the discipline which forced the science of ***Mechanics*** to restructure so as to adjust to the *Principle of Relativity* inherently respected by the former -- even though the notion of ***relativity*** of *motion* appeared first during the study of the latter. In fact, what *a posteriori* was coined as the 'Special Theory of Relativity', had been presented by Einstein in his famous publication entitled "On the Electrodynamics of Moving Bodies". The *motion* of electrically charged bodies and of *radiation* will be discussed in detail in Chapter 5.

1.1. The Kinematics Era - Ptolemy, Copernicus, Brahe, and Kepler

Aristarchus of Samos (310-230 BC), according to Plutarch and Archimedes, was the first to propose the heliocentric view of the Universe against the Aristotle's geocentric paradigm of the time. His writings were lost during one of the fires in the Library of Alexandria. The Greek-Egyptian astronomer, chemist, geographer, and mathematician Claudius Ptolemy (circa 140 AD) empirically systematized -in his astronomical treatise named *Almagest*- the vision of the Cosmos shared by Plato and Aristotle. In contrast with these two great philosophers, Ptolemy was a great empiricist and studied the mass of experimental data that existed on the *motion* of planets and stars, with the objective of conceiving a *kinematic* model which could order their *positions* in the past, so as to allow him to predict their *positions* in the future. The Ptolemaic model can be synthesized by saying that our Earth is the center of the Universe and that the Sun, the Moon, and the planets revolve around it. The 'fixed stars' were named that way because they did not show any *motion* beyond their diurnal rotation[3] and, ergo, they were considered as ***fixed*** on a sphere which rotated as a whole around the Earth. What is more evident than that when their *positions* are observed from our planet?

Ptolemy described the *motion* of celestial bodies by means of the epicycle technique of Apollonius (circa 262-190 BC), which represented their *trajectories* as circles whose centers traversed other circles. It is evident that Ptolemy knew the Earth was round but he thought that it was immobile arguing that, if it rotated, it would leave the birds behind and, if it displaced, it would get away from the 'celestial sphere' and we would notice it by comparing what we saw during the day and during the night[4]. He also explained that if the

3 The apparent lack of motion is due to the tremendous distances between them and us.

4 The interstellar distances are so immense that Earth displacement is comparatively negligible.

Earth rotated, its surface would develop speeds in the order of 2,000 km/h, creating violent hurricanes and dust storms which would destroy everything. His astronomical theories influenced the scientific world until the 16th century when Copernicus (1473-1543), one of the Renaissance exponents, looked at the planetary trajectories imagining he was on the Sun instead of on the Earth, i.e. he changed the *reference frame*.

At the age of 33, Copernicus had already conceived his revolutionary heliocentric theory, but very little of his doctrine was published while alive. His masterpiece, entitled "On the Rotation of Celestial Bodies", was published posthumously in 1546. Revising his final manuscript on his death bed, he did not realize that his friend Andreas Osiander had inserted a cautious preface trying to minimize the differences between the heliocentric theory and the official position of the Church.

In comparison to the multiplicity of Ptolemy's epicycles, Copernicus's description was much simpler and exact. The so-called *Ephemeris Tables* were superior to the *Alfonsine Tables*. The complexity of his homonymous tables motivated Alfonso X 'The Wise' to pronounce his famous offering of advice to the Deity at the moment of Creation so as to simplify the *motion* of the skies. Was ***descriptive complexity*** implying lack of **reality**? Copernicus had said: "It is more *convenient* to suppose the Earth rotates because, then, the astronomical laws are expressed in a simpler language". Was ***descriptive simplicity*** an indicator of **reality**?

Copernicus calculated the radiuses of the planetary orbits with an accuracy better than 1%[5] and knew that the Sun had to be a little shifted from the center of the solar system, to avoid discrepancies in the experimental data. The definitive triumph of the Copernican vision was going to require better instrumentation for astronomical measurement. Experimental data employed by Copernicus had an angular accuracy of 10' (minutes of arc) -- which is approximately the *angle* defined by a 2.5 cm (about an inch) diameter coin, when we look at it from a *distance* of about 6 m (about 20 feet). This means that two objects 6 m away from the observer were not distinguished unless they were separated by more than 2.5 cm. Such a poor resolution relative to current telescopes[6] can only magnify the intellectual stature of Copernicus.

Tycho Brahe (1546-1601), financially supported by the King Frederick II of Denmark and Norway, increased the exactitude of his sextant to better than half a minute, i.e. 20 times better than what Copernicus had had available. An angle of 0.5' corresponds to that covered by the same coin now seen from a distance of 120 m. When Tycho Brahe died, his experimental data went to his assistant Johannes Kepler (1571-1630). Brahe's data for Mars and its retrograde[7] motion together with Kepler's own observations, were essential for him to formulate his famous three *Laws of Planetary Motion*, leaving out no doubt about the elliptic nature of planetary motion.

Both Copernicus and Kepler thought that the solar system was the totality of the Universe, and Kepler vehemently opposed to Giordano Bruno (1548-1600)

5 1% is the uncertainty in the measurement so the greater the accuracy, the lower this percentage.

6 The Hubble Space Telescope, in orbit at about 590 km since 1990, has an optical resolution of 0.1", i.e. 6,000 times greater than Copernicus' sextant.

7 Planets displace along the Zodiac from West to East, with brief periods of retrograde motion from East to West.

who thought that the stars were independent solar systems. Bruno believed in the uniformity of the Universe in contrast to Aristotle's doctrine of the two worlds: the *sublunary* (imperfect and variant) and the *superlunary* (perfect and with immutable circular *motions*).

1.2. The Dynamics Era - da Vinci, Galileo, Newton, and Mach

Astronomical accuracy had to be again substantially improved, and another giant of the Renaissance applied his genius to go farther from Kepler accomplishments: the telescope was invented circa 1590 and Galileo -our hero in Chapters 1 and 2- built one in 1609. Galileo and Francis Bacon (1561-1626) are considered the founders of the scientific method, in which systematic experimentation is an essential component in the quest for understanding Nature. Even so, a century before, Leonardo da Vinci (1452-1519) had accurately identified both deductive and inductive mental processes embedded in the scientific method, when he said and acted in accordance with his teachings:

> *When tackling a scientific problem, I first plan experiments, because I pretend to determine the problem in accord with experience, showing then why objects are forced to behave that way. That is the method we have to follow when investigating natural phenomena.*

Defying the Authority of the time, Galileo said: "I do not feel obliged to believe that the same God who gave us the senses, reason, and intellect, expects us not to use them". He also said: "That person who tries to solve a scientific problem without the assistance of Mathematics, attempts the impossible. We should measure all that is measurable and convert into measurable what is not". He wrote to Kepler: "You would laugh after hearing some of our most respected philosophers at the University denying the existence of the new planets through purely logical arguments, as if the planets were magical apparitions... another scientist did not want to look with my telescope because that 'would only confuse him'...". We already know the great adversity Galileo suffered because of his intellectual bravery.

But, as we showed in Chapters 1 and 2, it is the investigations and revolutionary conclusions of Galileo on the *motion* of bodies on Earth, and not his multiple other astronomical achievements, that really interests us and now we resume, so as to start exploring the epistemology of *motion*.

1.2.1. The Giants of the Road

Traversing by bus the 360 km of humid pampas between Necochea and Bahía Blanca was a good part of my routine between 1966 and 1974. The 'Universidad Nacional del Sur' is the most important university in the Argentinean south. During the long four-hour trip, my mind tried hard to assimilate the great amount of theoretical and practical information necessary to formally move forward as a student of Electrical Engineering though, more than sporadically, I could not help unleashing my penchant for philosophy. Having understood the concept of convergence of a sum of an infinite number of summands, and thus being able to thrust aside all the pseudoparadoxes of Zeno, made me feel very proud. Consistently with the University's motto -*ardua veritatem* (the truth is attained through effort)- after a great deal of intellectual hard work, I had gradually become versed in the subtleties and power of *differential and integral calculi*, and little doubts remained in my mind regarding the **reality** of *motion*... or... so I believed.

Due to my compulsive punctuality, it was not uncommon that, once seated, ten or fifteen minutes could have elapsed before the bus departed heading to Bahía Blanca. Not less common was that, in the interim, we were trapped between other *Giants of the Road* intermittently arriving and departing. While my bus waited at the Terminal ready to begin moving back, I was absorbed in my innumerable epistemological inquiries, and the ghost of the ***relativity of motion*** fearlessly appeared before me, once and again, defying my capacity of comprehension and my ability to avoid being deceived.

Not being able to see the solid and huge *frame of reference* constituted by the bus Terminal (or any other reference rigidly attached to *terra firma*), together with the smoothness with which those giants initiated their *motion*, and my constant rumination over a variety of subjects, more than once for a brief moment while looking through the window, made me feel as if... either we were backing off when, 'in reality', other bus was arriving (and we were static on the platform)... or we were *moving* forward when, 'in reality', another bus was departing in the only way possible, i.e. *moving* backwards. *Mutatis mutandis*, sometimes it seemed as if a giant bus by my side was arriving when, 'really', mine had smoothly started to *move* backwards leaving the platform.

Was I being deceived? Was my transient *perception* an illusion with the *motion* opposite to my *perception* the only 'real' (or vice versa)? What was the meaning of 'terra firma'? Is there something 'real' behind our *perception* of *motion*? Why did I feel awkward each time I used the word 'real'? Is there something else other than the mere *temporal* ***change*** of the *distance* between two objects? i.e. does absolute *motion* 'exist'? Why did I feel discomfited each time I used the word 'exist'? Or, paraphrasing Mach: is ***absolute*** *motion* nothing more than a useless mental construct?

Please notice that, besides my not being able to use *terra firma* as reference, I mentioned three other conditions potentially causing my delusion: a) the ***smoothness*** with which *motion* was initiated; b) my frequent state of ***cogitation***; and c) my ***interaction*** with the ***external*** world by "looking through the window". Regarding *terra firma*, from our very birth, we have learned to use it as our subconscious default *frame of reference*; our infant physical integrity depended upon that default. Why? Well... much before we intellectually accept that our Earth 'moves', every sensible kid in his cradle concludes during the first month of life that ***ground*** and ***walls*** are 'stagnant', while people and objects 'move' so that within a year he wants to emulate them, learn how to walk, and confirms with physical pain the 'solidity' and 'immobility' of ***floor*** and ***walls***.

But, what was going on when, eighteen years after my crawling days, surrounded by huge buses, that 'solid' and 'immobile' *terra firma* disappeared from my ***perception*** field? If ***absolute*** *motion* existed as a **real** phenomenon, I should be capable of correctly discerning the 'real' from the 'illusion' even without *terra firma*. Perhaps my brain -not having its usual reference- was using a ***wrong*** *frame of reference*? Are there ***wrong*** *reference frames*? Isn't the ***objective reality***, by its very conception, independent of the *frame of reference*?

With respect to "looking through the window", I felt sometimes I was 'moving' when I was still 'in repose', and other times I thought I was 'stopped' when I was 'moving'. With the windows open, my subconscious *reference frame* was alternating from my own bus to the other *giant of the road*. But, then, if I eliminated all ***interaction*** with the ***exterior*** (closing the windows), the only available *reference frame* would be my own bus and, ergo, it would be impossible for me to detect any *motion*! Not so fast; there were still two more conditionals: the "smoothness of startup" and my "state of meditation" which -together with the natural noisy environment of a bus terminal- conspired against my *perceiving* any noise/vibration of the bus engine or any other subtle effect indicative of our 'real' motion.

The summer of 1967 had ended and I was returning to Bahía Blanca. During those vacations in Necochea, precisely because of my tendency to be distracted, more than once I had been terrified for a few seconds when, driving my girlfriend's 'fitito'[8] around the seaport area and being stopped at a traffic light on an uphill street, the motion of a giant cereal transporter passing by my side had made me feel that my car was the one *moving* downhill, braking to the max without apparent success, because my mistaken sensation of *movement* continued for barely another creepy second.

As a result of those scary experiences, this time I had decided to be fully conscious of the ***external*** world during my stay on the Terminal platform, so as to clearly identify those moments at which *terra firma* disappeared from my visual field, and -when that happened- to detect any effect that could objectively indicate whether I was *moving* or not. My conclusion invariably was that, in the great majority of cases, even when what I saw through the window indicated the contrary, when I sensed vibration/noise from the vehicle engine, or a slight *spontaneous* decrease of the ***contact pressure*** between my back and my seat, my bus was the one 'really' moving backwards and, vice versa, when those two effects did not occur, I could infer with great certainty that my vehicle was *at rest* on the platform regardless of what I saw through the window. Everything seemed to indicate that visual data from the ***exterior*** only could inform me about ***relative*** *motion* but, if I ignored it completely, I was able to detect the 'real motion' or, as Newton called it, "absolute", by what he meant *relative* to ***absolute*** *space*. From then on, every time I wanted to investigate the nature of *motion*, I closed my window and fully ignored those which were open.

But I soon realized that the engine vibration and noise were effects not associated with *motion per se* but with the internal combustion technology, and that I could easily imagine an engine with a level of vibration and noise below human sensorial threshold (or any detection technology). The ontology and epistemology of *motion* certainly could not depend upon the current state-of-the-art in automotive technology. There only remained one effect apparently associated with our *motion*: that slight *decrease* of ***contact pressure*** on my back when moving *backwards*, which I had already discovered that became an

8 In Argentina, in those days, the Fiat 600 -because of its being so small- was colloquially called 'fitito'.

increase of ***pressure*** when the bus started moving *forward*. This effect was doubtlessly related to the vehicle startup, and was more intense the less dexterous the driver was or, equivalently, the more jerky the startup.

Once on the road at *cruise speed* (about 90 km/h), the ***pressure*** on my back was exactly as when on the platform before departing; the orange juice was perfectly leveled and stable in its glass while we played 'truco'[9]; and people moved through the aisle as if they were on *terra firma*. If the 'truco' cards were in *repose* on the tray, they continued at *rest* unless we exerted a *force* on them. If I vertically launched a little object so as to reach as high as possible without touching the ceiling, the object went up and down on a perfectly ***vertical** trajectory*, exactly as if it was on solid ground. If I threw it horizontally, it underwent the typical ***parabolic** trajectory* a body would have had outside the bus. The value of the vehicle's cruise *speed* did not seem to be relevant and, unless I looked through the window and saw the immense wheat crops, cattle, or a sporadic house or tree, there was **no** conceivable *mechanical* experiment which I could carry out inside the bus, so as to discern that we were cruising at 90 km/h -- or at any other **constant** *velocity*.

Summarizing: it was impossible to determine, without ***external** reference*[10], the speed at which we were traveling. Assuming that a) we were *moving* and *terra firma* was *static*, b) we were *at rest* and *terra firma* was *moving*, or c) both *terra firma* and the *bus* were *moving* with an ***arbitrary*** but ***constant*** relative *speed* were all completely equivalent situations and, ergo, indiscernible. It is remarkable to understand that in order to prove b) and c) it was not necessary to conduct an experiment in which 'terra firma' was 'really moving' corroborating everything was identical; not even a) could be done so as to be certain which the one 'moving' was. Situations a), b), and c) were all proved as equivalent at once, by proving there was **no** *mechanical* experiment we could do on solid ground that would give different results when aboard the vehicle while at ***constant** speed*. Now, I finally could understand why the first little book about Relativity I had read at the age of fourteen, confused the notion of *frame of reference* with the notion of *repose*, tacitly assuming that choosing a *frame of reference* was equivalent to assume it was 'in repose' -- as if being 'in repose' could be meaningful without the adoption of another *frame of reference*! It was simply a linguistic license, totally innocuous (as long as we were conscious of its conventionality).

It is also paramount to understand that this equivalence between both *reference frames* (*terra firma* and *bus*) when their ***relative** speed* was **constant**, meant that any one of them was valid for observing *mechanical* phenomena, i.e. the *laws of motion* were **identical** in both *frames*. The object -going up and down inside the bus- was going up and down when launched on solid ground under the same conditions. However, if the object going up and down inside the bus were observed from *terra firma*, its *trajectory* would not be ***vertical*** but ***parabolic***, because in the latter *frame of reference* the object would have started its *motion* with an initial ***horizontal** speed* equal to the bus *speed* with respect to ground (while when using the bus as *reference*, the object initial ***horizontal** speed* was **zero**). In both *frames of reference*, the object is ***verti-***

9 'Truco' is a game of cards, very popular in Argentina, which is played with Spanish cards.
10 The tachometer on the vehicle dashboard requires the physical contact between the wheels and the road.

cally subjected to *gravity* and thrown ***vertically*** but depending upon which one is used, there exists or not a ***horizontal initial speed***. Once again, my virgin mind confirmed that the concept of *spatial trajectory* of a projectile was **relative** to the *frame of reference*. The laws of Nature were the same for both *frames of reference*, but some observations, far from being ***absolute***, were simply different ***perspectives*** of the **same** physical **reality**.

Notwithstanding, all the above considerations were valid under quite precarious circumstances: the mere presence of a bump on the road, or -more dramatically- when the driver was forced to suddenly brake to avoid disaster, my back violently separated from the seat, and my orange juice went forward like a bullet. These were obviously the same symptoms I had identified during bus departure though now drastically amplified. And, what was the difference between a backward startup by a seasoned driver and an unexpected brake to avoid a collision? The obvious difference was that the ***change*** of *velocity* that the bus experienced in a given *period of time* (its *acceleration*) was much larger than during the startup, and that was why the effects were augmented.

Using *terra firma* as the vantage point, while braking, the bus had evolved from ***uniform*** *motion* (**constant** *speed*=**null** *acceleration*) to highly ***decelerated*** *motion* in a very brief *interval of time*, and this ***accelerated*** *motion* had prolonged until we reached full stop (another form of ***uniform*** *motion*). Changing references, during that quick transition from one form of ***uniform*** *motion* to another (as seen from *terra firma*), inside the vehicle and with all windows closed, objects that had been in *repose* flew without a *force* to explain their ***change*** of *motion*, the orange juice went on its way to paint the seat in front of me without an apparent *force* to *cause* it, the 'truco' cards got scattered all over decorating seats and floor for no apparent mechanical *cause*, and an object ***vertically*** thrown would not have traversed an up and down ***vertical*** path any longer.

Galileo and Newton's laws of *motion* (Chapter 1), taught to me in my first Physics course in college, did not seem to be valid inside the bus during that brief chaotic experience, not even knowing the driver had pressed a pedal named 'brake'. If, instead, the events inside the bus were observed from *terra firma*, the explanation I gave in Chapter 1 would satisfy mostly everyone: the objects flew forward simply due to their *inertia*, because they were, before braking, moving at 90 km/h and, ergo, what seemed chaotic inside the vehicle (with the windows closed), from outside (with the windows open) was interpreted as the natural reluctance (*inertia*) of those objects to change their *motion*, with the consequent tendency to keep traveling at 90 km/h as long as they did not find other objects or effects reducing their *speeds*. In the absence of those interfering effects, the flying objects would draw the typical parabolic *trajectory* due to *gravity* until reaching the floor. Furthermore, from *terra firma*, we do not even need the notion of *inertia* to qualitatively understand what was going on because, being the bus and objects not rigidly attached, no wonder that when the bus braked and the objects did not, their *spatial relation* changed.

But, inside the vehicle, with all its windows closed, the objects were all *in repose* right before the driver pushed the brakes and, hence, their *inertia* could not explain why they started flying like a bullet given that without a *force* -according to Newton- they should have remained *in repose*. In this *reference frame* (temporarily *decelerated* with respect to ground), the only manner I could explain the apparent and momentary 'chaos' was through the spontaneous emergence of a *field of forces* that could change the state of *repose* in which bodies were before the bus braking. This field had to be ***variable*** in *time* so as to disappear once the bus fully stopped or re-attained ***uniform*** *motion*. What could be explained ***without*** the existence of a *force* from ***outside*** (although we needed the peculiar concept of humanoid reluctance (*inertia*) to *motion*), to explain it from ***inside***, we needed the existence of such a *force*! ***Inside*** the bus -while braking- in order to make sense of any phenomena, it was necessary to contemplate such a mysterious *force*. But... was this *force* **real** or fictitious? And, if it was **real**, who or, better, what exerted it? And...that outlandish humanoid reluctance... What in the world was that? Where did it come from? Why could it not be considered as a *force* as well? Or... maybe... couldn't both concepts be as artificial as unnecessary to explain all these mechanical phenomena?

The object flying forward, observed from *terra firma*, had a ***horizontal*** component of **constant** *velocity* equal to the one the bus had before braking, simply because *Galileo's Principle of Inertia* (first Newton's Law) teaches us that the object 'natural behavior' (free of any external influence) would be simply to continue with the bus original speed 'forever'. Due to *gravity*, the same object had a ***vertical*** component of ***variable*** *velocity* (with the *acceleration* of *gravity*). Observed instead from inside the bus, the object had started in *repose* to *move* with a ***horizontal*** *acceleration* equal to the *deceleration* (brake) of the vehicle with respect to ground. This is so because the flying object *moved* relative to ground at **constant** *speed* and hence (think about it a little), the *acceleration* of any third body (the bus) had to be the same with respect to any one of them.

But... then, if the *acceleration* of all flying objects (from inside the bus) was determined by the *deceleration* of the bus relative to ground, all flying objects had the same acceleration regardless of their *masses* or chemical *composition*! Hmm... is this not what we called a ***universal*** *field*? And... is this not what characterizes a *gravitational field*? I had to conclude that, if this mysterious ***horizontal*** *field* 'really' existed, it behaved as the ***vertical*** *gravitational field*!!! By the same token, I concluded that observing the bus braking from *terra firma* was not equivalent to observing the events from ***inside*** the bus: without ***external*** *reference*, before braking our *motion* was undetectable; while braking, the 'chaos' could be interpreted as the hallmark of ***accelerated*** *motion*. Did this mean that *speed* was **relative** but *acceleration* was ***absolute***? Are they not complementary attributes of *motion*? Why this duality?

Assuming Newton was right and ***absolute*** *space* exists, and that *terra firma* was *in repose* while the bus was *moving*, started to brake, and the strange *field of forces* appeared, would this *field of forces* have been present also, had the bus been the one *in repose* and the Earth the one that was *moving* in the opposite direction at 90 km/h and, when braking, started to *decelerate* until full stop? If in this opposite case, in which it is Earth that is *moving* and *decelerating*, the bodies inside the bus would not be shot as bullets when braking, evidently we would have to conclude that the 'chaos' we experience when braking cannot be due to the ***relative*** *motion* between bus and Earth, i.e. that ***accelerated*** *motion* is **not** ***relative***.

But... is it enough to imagine the Earth ***moving*** and the *bus* ***in repose*** to prove/refute the ***relativity of*** *motion*, or do we have to **move** the *rest of the universe* together with our planet while we maintain the bus ***at rest***? From the point of view of the mere *distance* between Earth and bus, it is obvious that the observed *distance* is totally equivalent no matter which (Earth or bus) is the one that *moves* irrespective of the *rest of the universe*. However, in order to achieve a perfect equivalence so that we can proclaim the relativity of *motion*, we cannot just imagine that the Earth ***moves*** while the bus and the *rest of the universe* stay ***in repose*** because, in that case, the *distance* between the bus and the *rest of the universe* would **not change**, while when, during our everyday life, we imagine the only one that *moves* is the autobus, that *distance* will **change**. The absence of 'chaos', or the existence of a different 'chaos' inside the omnibus, could be explained with the lack of symmetry.

The full symmetry necessary to accept the ***relativity of motion*** requires that we compare the *bus* ***moving*** while the *rest of the universe* is ***in repose***, with the bus ***in repose*** while the *rest of the universe* is ***moving*** in the opposite direction. Furthermore, the mere *distance*, i.e. the ***kinematic*** criterion, cannot be the only one to check: all observable phenomena (mechanical, electromagnetic, optic, etc.) have to be the **same** in both perspectives; otherwise, that one which is not the same becomes the proof that ***absolute*** *motion* is detectable because we observe something different. That is why it is critical that the 'chaos' within the bus while braking exists regardless of whether it is the *bus* or the *rest of the universe* that brakes. But... what the heck am I saying? There is no way of experimentally proving that the 'chaos' appears in both cases, because... how can we make the *rest of the universe* ***move*** while we keep the bus ***in repose***? And, by the same token, can we say that it is the *bus* that ***moves*** while the *rest of the universe* is ***at rest***? Of course not! "Cosa 'e loco" (madman's business), the Argentinean comic narrator Luis Landriscina (1935-) would say.

And my brain kept boiling: I could not but wonder whether there was a relation between that mysterious *horizontal force* necessary to explain the havoc within the bus and the well-known *centrifugal force* of ***rotational*** *motion*. Our professor of Physics had taught us that when we made a little rock (attached to a string) ***rotate***, "the centripetal force was real and kept the rock rotating, while the centrifugal force was fictitious and did not exist". How in the world could something fictitious make so much damage when releasing the string or when the omnibus abruptly braked? Was the professor perhaps tacitly using a *frame of reference* of which he did not make mention, i.e. using one of the many possible languages to describe the same physical **reality**? Was there any relation between this and my childhood experiences with the *Carrousel by the Church*? My brain was about to explode. Next time I went to Necochea, I would go to the playground across the church where I had spent so many joyful hours when a little kid.

After braking, once in repose, everything went back to the 'normality' we had experienced before the incident -- when we were suavely cruising at 90 km/h. Based on our everyday experience, during the abrupt braking, we did not need to look through the window to reckon we were *decelerating*, which compelled me to accept that there had to be an intrinsic difference between *uniform motion* (undetectable unless we looked through the window and, even so, without knowing who was moving), and ***accelerated*** *motion* (easily identifiable without looking out to the exterior). Was this what Newton meant when he talked about ***absolute*** *motion*? How could ***uniform*** *motion* be **relative** while ***accelerated*** *motion* be **absolute**? In such a case, I imagined Alfonso X 'The Wise' offering -one more time- advice to the Deity at Creation.

Even though -during the exposition of my experiences- I was referring to the 'real' motion without indicating my *frame of reference*, it was apparent that I subconsciously was talking about my *motion* with respect to *terra firma*. So much for the concept of 'reality'! Because... what is it, then, that 'uniform' and 'accelerated' mean? I only needed to imagine a *frame of reference moving* in perfect unison with our braking autobus to convert what we considered the epitome of ***accelerated*** *motion* into the 'poster child' for ***uniform*** *motion*! Or, vice versa, we only had to imagine a *frame of reference moving* with **non-zero** *acceleration* relative to our bus while traveling at a ***constant*** *speed* of 90 km/h (according to the tachometer!), to convert our ***uniform*** cruising into a ***non-uniform*** (*accelerated*) one! Nonetheless, in spite of our now ***accelerated*** *motion*, we were still pleasantly and safely playing 'truco'! In fact, when I said "inside the bus with all windows closed" I was tacitly using the bus as *frame of reference* and, hence, I had been as 'at rest' when on the platform waiting to depart, as when crossing the pampas at 90 km/h, and even while braking -- had I had an ideally perfect safety belt. That is how, as I already described, I would travel to Vegas from Salt Lake City **without** *moving* many years later. No wonder that Newton desperately needed to believe in ***absolute*** *space*; otherwise, he would have been convinced that nothing makes sense in this crazy world.

From all those disquisitions, it seemed ridiculous to me that we used our planet as the *frame of reference* in order to detect a *motion* that could only be considered ***absolute*** when its *acceleration* was **not** zero -- while our very planet, in turn, was rotating and displacing with respect to the Sun at about 30 km/s, as the solar system was traveling within the Milky Way, and the latter was moving with respect to another galaxy!

Back to reality; in a few days I would have a test on Calculus. I would certainly return to Necochea in a few months and, then, after making sure nobody looked at me, I would turn around on the carrousel over and over, in my last attempt to understand Newton. My beloved *Carrousel by the Church* was the closest thing to the celebrated water bucket that Newton had sagaciously employed to prove the existence of ***absolute*** *space* and, besides, it would allow me to experience 'in the flesh' the -as famous as supposedly fictitious- *centrifugal force*.

1.2.2. The Principle of Inertia and Galileo's Principle of Relativity

Galileo was the first who experimentally studied the *motion* of bodies so as to determine the precise relation between *force* and *motion*, demonstrating the so-called *Principle of Inertia* and, as we saw in Chapters 1 and 2, formulating the law of a **common** *acceleration* for all bodies in our *gravitational field* -- regardless of their *mass* and chemical *composition*. Archimedes (287-212 AC) writings -translated into Latin by the mathematician Tartaglia (1500-1557)- as well as those of Leonardo da Vinci, had a great influence on Galileo. In fact, da Vinci had surmised the *Principle of Inertia* before Galileo demonstrated it by way of experiment.

As extensively discussed in Chapter 2, Galileo measured -with his water clock- the *time interval* that polished little balls took in rolling down a polished inclined plane. His instinct compelled him to suspect that the temporal variation of speed (*acceleration*) was **constant** and challenged himself to experimentally confirm it. First, he mathematically proved that, if the *acceleration* was **constant**, then the traveled *distance* had to be proportional to the ***square*** of the *time interval*. For example: if after **one** *second* the little ball had traveled **one** *centimeter*, then after **two** *seconds* the ball would travel **four** *centimeters*, after **three** *seconds*, **nine** *centimeters*, and so forth. Mathematicians say that "when distance is proportional to time squared, its second derivative[11] is constant".

As we know, Galileo ***measured*** *time* by means of ***weighing*** the water collected during each experiment. Based on the assumed proportionality between elapsed *times* and water *weights*, he experimentally found that doubling the *times*, *distances* were quadrupled, tripling the *times*, *distances* were nine times greater, etc. and, from that experimental finding, using a sheer mathematical relation, Galileo concluded that -as he had intuited it- the ball *acceleration* was **constant** throughout the fall. Subsequently, he conducted similar experiments but gradually ***increasing*** the *slope* of the plane as close as possible to ***verticality***, measuring the different *accelerations* for different *inclinations* and, in this fashion, despite the ball would -of course- stop rolling once falling vertically, Galileo correctly deduced the ***law of free-fall*** on our planet.

Unleashing his experimental spirit, Galileo smoothly connected two planes so that the ball could go down and then up finding that, ***neglecting*** *friction* effects, the ball always climbed the second plane (before starting going down) up to exactly the **same** *height* it had started on the first plane (Figure 6). He then made the second plane ***longer*** and ***longer*** while gradually ***reducing*** its *slope* towards ***horizontality***. And... what did he find? That the ball traversed ***longer*** and ***longer*** *distances* even though it had started from the **same** *height* on the first plane! At the same time, the ***constant*** (during each experiment) *acceleration* was being reduced down to nothing!

As depicted in Figure 6, in an instant of profound inductive inspiration, Galileo extrapolated his measurements and concluded that, had the second plane been ***horizontal*** without friction or any other foreign influence, given that the ball once on the second plane would have never reached the ***initial height*** (on the first plane), it would have continued rolling 'forever' with **constant** *speed* (**constant *zero*** *acceleration*) equal to the *speed* it had at the instant it entered the ***horizontal*** plane.

11 The 'second derivative' is the rate of change of the rate of change of distance with time, i.e. the *acceleration*.

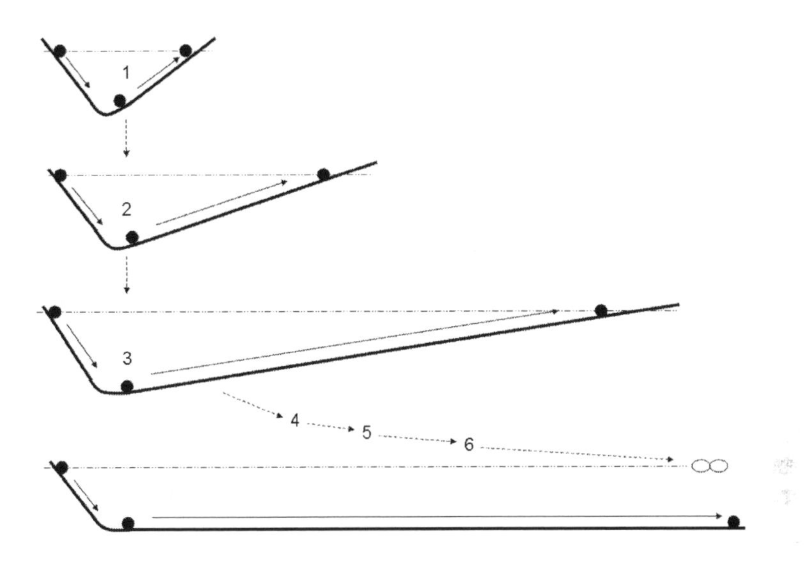

Figure 6. Galileo's Principle of Inertia

In this way, Galileo concluded that the 'natural' state of a body (i.e. when it is free from any *forces*) was not **repose** (as Aristotle taught) but the so-called ***uniform** motion*, which is defined as being **constant** in both *magnitude* and *direction*. Of course, **repose** is a special case of ***uniform** motion* for which the magnitude is ***zero*** and, ergo, it is also a 'natural' state. The genius of Galileo replaced the ancient question of 'why do bodies move?' with the question 'why do they stop?' revolutionizing the science of Mechanics.[12]

The *Principle of Inertia,* thus, can be expressed as: 'When a body is free from all external influence, it will persist in its state of repose or uniform motion'. But there was an irony not unperceived by Galileo: for those advocating the dichotomy between the *sublunary* and *superlunary* worlds, ***repose*** was the natural state in the former, and ***circular** motion* was the natural state in the latter; Galileo now was replacing ***uniform** motion* (i.e. rectilinear with constant speed) for *repose*, but he knew that by extending and reducing the slope of the second plane, the little ball was following the shape of our planet and, consequently, would eventually trace a ***circular** trajectory* instead of a *straight line*!

Evidently, at least for bodies moving on the surface of the Earth, the *Principle of Inertia* had only a **local** validity. What would happen if the body could move in the tri-dimensional *space* while being completely isolated from external interactions? Would it move eternally in a *straight line* at **constant** *speed*? Or... would my childhood fear of being told the "air was also round" become an eccentric reality? Remembering our discussions in Chapter 3, we know that our intuitive notion of 'straight line' is implicitly Euclidean. In a Riemannian space,

12 Plutarch surmised the Principle of Inertia circa 100 AD, when he said: "Everything moves with its natural motion, when it is not affected by something". Nicholas of Cusa (1401-1464) believed the Earth moved continuously in space without us being conscious of it. Descartes unequivocally conjectured the Principle by saying: "A body, when in repose, has the power of staying in repose resisting every change. Equally, when it is in motion, has the power to continue in motion with the same velocity and in the same direction".

instead, the 'straight lines' (geodesics) can be closed. Had Galileo known of non-Euclidean spaces, most probably he would have expressed his principle as: 'When a body is free of all external influence, it will persist in its state of repose or motion along the geodesics of space'. There were still two and a half more centuries to go before Einstein would conceive the principle in those terms.

Galileo knew that his *Principle of Inertia* had been obtained by induction as a generalization of limited terrestrial experiments and, hence, its validity could depend on the *state of motion* of his *frame of reference*; hence, the question of whether his Principle was valid or not in a given *reference frame* could be determined only by *ad hoc* experimentation. A *reference frame* within which the *Principle of Inertia* is valid is called an *Inertial* or *Galilean System of Reference* or, simply, an *Inertial Frame*. With the experimental accuracy available in Galileo's time, our planet certainly qualified as an *Inertial Frame* -- at least **locally**.

It is worth noting that a wrong folkloric definition of an *Inertial Frame* in the Popular Science literature (even in textbooks) reads that "it is a frame in uniform motion". We know very well by now that the idea of *motion* requires a *frame of reference*, so that such a definition of an *Inertial Frame* has no meaning whatsoever, confusing the reader because it tacitly reaffirms the idea of ***absolute*** *motion* -- when the goal of every didactic exposition of Relativity Theory should be precisely the opposite. Besides, as my experience with the *Giants of the Road* clearly showed, ***uniform*** *motion* in a *reference frame* does not have to be ***uniform*** in another *reference frame* (both *frames* only need to be in relative **accelerated** *motion*). This is the proper definition: we call *Inertial or Galilean Frame* any *reference frame* in which the *Principle of Inertia* is valid within an accuracy that suits our purposes; full stop.

Considering different *reference frames* (*terra firma* and a ship in calm seas), Galileo realized he could enunciate a more general principle: the *Principle of Galilean Relativity* which reflects no more and no less than my experience with the *Giants of the Road* crossing the humid pampas of Buenos Aires. Remember that, with the windows closed, there was **no** *mechanical* experiment I could execute to distinguish between any two different states of ***uniform*** *motion* (all referred to *terra firma*), regardless of what those different cruise *speeds* could be. As soon as the abrupt braking ended and we were *in repose*, things were back to a 'normality' as normal as when we were cruising at 90 km/h. Any other *speed*, as long as it was ***constant*** relative to ground, would have been indistinguishable for us -- unless we looked through the window. Galileo expressed it much better in his masterful *Dialogue concerning the two chief Systems of the World, the Ptolemaic and the Copernican* (1630):

> *Shut yourself up with some friend in the main cabin below decks on some large ship, and have with you some flies, butterflies, and other small flying animals...hang up a bottle that empties drop by drop into a wide vessel beneath it... have the ship proceed with any speed you like, so long as the motion is uniform and not fluctuating this way and that... the droplets will fall... into the vessel beneath without dropping toward the stern, although while the drops are in the air the ship runs many spans... the butterflies*

and flies will continue their flights indifferently towards every side, nor will it ever happen that they are concentrated toward the stern, as if tired out from keeping up with the course of the ship. Therefore, somebody conducting experiments below decks could not know whether the ship is moving or in repose.[13]

But... what about Galileo's linguistic and epistemological precision? What did Galileo mean when he said "so long as the motion is uniform"? Obviously he was using *terra firma* as his tacit *frame of reference* and what his maritime experiment proved was that two systems in **relative** ***uniform*** *motion* (ground and ship) are indistinguishable from the mechanical point of view. Notice please, though, that what Galileo did **not** prove was that either the Earth or the boat were *moving* ***uniformly*** -- both assertions without meaning whatsoever. In summary, I would articulate the *Principle of Galilean Relativity* in this way:

If we find in Nature an Inertial Frame of Reference, then we automatically have an infinite number of them: all those frames which move ***uniformly*** *with respect to the former, independently of their actual speeds.*

With a minimum of mathematics, this Principle can be deduced once we assume the existence of an *Inertial Frame* that we will call the *System of Coordinates* A (SC-A). Remember that a *System of Coordinates* is a mathematical representation for a *Frame of Reference*[14], i.e. the abstraction of a physical system formed by three mutually perpendicular rigid bars defining three perpendicular planes, thus allowing us to determine the *position* of a *point* in *space* with the three *distances* to those planes. These *distances* are the coordinates (X,Y,Z) for the *point*. Remember as well that in order to define an *event* we need to know the *time*. In brief, an *event* is univocally determined with a quartet comprising three *spatial* and one *temporal* coordinates (X,Y,Z,T). As usual, if what follows is excessively mathematical for the reader, it can be circumvented without missing the core of our message.

If SC-A is *inertial*, any object free of all external action, if *at rest* will continue *at rest*, and if in ***uniform*** *motion* will continue doing so. Pithily: the *velocity* of an **isolated** object does **not** change. Now imagine another system SC-B in ***uniform*** *motion* with respect to SC-A. We want to prove that SC-B is also an *Inertial Frame*, i.e. that the *velocity* of an **isolated** object does **not** change in it either. For the sake of simplicity we assume that *space* has only two dimensions instead of three, and that the ***uniform*** *motion* of SC-B with respect to SC-A, as well as that of any object *motion*, occur along the X coordinate (Figure 7).

As we know, in order to determine the *velocity* of an object in SC-A we need to know how its spatial coordinates change with *time*, and the same to determine the *velocity* of the same object in SC-B. But because SC-A and SC-B are in relative ***uniform*** *motion*, i.e. they are separating/nearing with a **constant** *velocity*, the object coordinates (position) in SC-A cannot be the same as its coordinates in SC-B. We already know that *position* is relative to the *reference* frame. In our case, only the X-coordinate will change. We call $V_{A/B}$ the velocity of SC-B measured in SC-A and $V_{B/A}$ the velocity of SC-A measured in SC-B. It is clear that $V_{A/B} = -V_{B/A}$. On the other hand, for Galileo, the time (T) and the

13 As cited in Roger Penrose's The Road to Reality, New York: Alfred A. Knopf, 2004.

14 The same *Frame of Reference* can be mathematically represented with different Coordinate Systems.

distance between two points (at the instant T) were absolute, i.e. the same in both *frames of reference* so that, if SC-A and SC-B coincide in their origins when we start measuring time ($T_A = T_B = T = 0$), their *distance* will increase in time being the product of the speed with which they separate or approach and the elapsed *time* ($V_{A/B}.T = -V_{B/A}.T$). Likewise, the object X-coordinate (X_B) in SC- B, according to Galileo, must be equal to the *distance* between the object and the Y-axis in SC-B measured from SC-A ($X_{A/B}$).

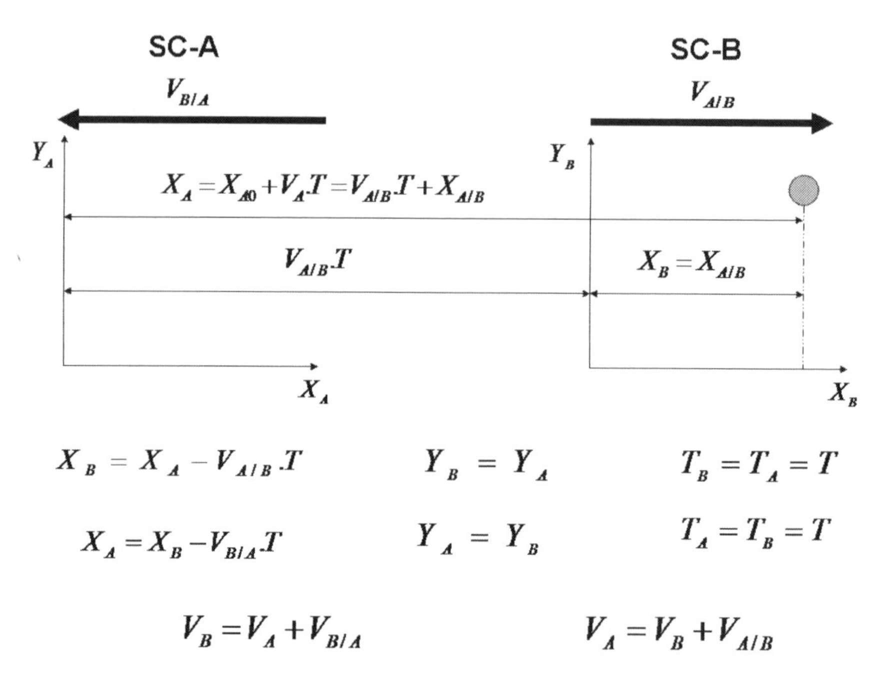

Figure 7. The Principle of Inertia is Covariant under Galileo's Transformation

Let's assume the object is free of any external influence and in ***uniform** motion* in SC-B with velocity V_B; will it continue in ***uniform** motion* with the same *velocity*? Because SC-A is an *Inertial Frame*, if the object is isolated, it is either *in repose* or moving with ***constant** speed* (V_A) and, ergo, its coordinate (X_A) shall be the sum of its initial position X_{A0} plus the product of its *velocity* with the elapsed *time* ($X_A = X_{A0} + V_A.T$). Besides, it is easy to see (Figure 7) that X_B can be calculated as its X-coordinate in SC-A ($X_{A0} + V_A.T$) minus the *distance* between SC-A and SC-B: $X_B = X_{A/B} = X_{A0} + V_A.T - V_{A/B}.T = X_{B0} + (V_A - V_{A/B})T$. Finally, from the equality of the first and last terms, we see that the object velocity in SC-B is $V_B = V_A - V_{A/B}$ and, given that V_A and $V_{A/B}$ are constant by hypothesis, we conclude that the object velocity in SC-B is also constant, i.e. its motion in SC-B is uniform -- as we wanted to prove.

This transformation between the *coordinates* of different *reference frames* in relative ***uniform** motion* is called *Galileo's Transformation*, and what we just proved, i.e. the *Principle of Galilean Relativity*, can be succinctly expressed as follows:

> *The Principle of Inertia is* ***invariant*** *under Galileo's Transformation because, given any two reference frames linked by the transformation, if one of them is* ***inertial****, the other is also* ***inertial****. The* ***positions*** *and* ***velocities*** *of an object are* ***different*** *in each frame, but the Principle of Inertia is valid in both. More specifically, we say that the Principle of Inertia is* ***covariant*** *with respect to Galileo's Transformation.*

Before continuing, I have the hunch that the alert reader must be wondering: if we could logically deduce it, why is it called a 'principle', i.e. why is it considered an ***axiom*** (with an inductive component, e.g. Galileo's experiments), instead of a ***theorem*** (purely deductive)? The answer to this question is that, in order to logically prove the Principle, I said "...for Galileo, the time (T) and the distance between two points (at the instant T) were absolute, i.e. the same in both *frames of reference...*". It was indisputable for Galileo that both *time* and *distance* between two *points* at a **common** *instant* would be the same regardless of their being ***measured*** on *terra firma* or inside the ship. His experimental observation of what was going on below decks, identical to what he was used to observing on solid ground, allowed him to enunciate his *Principle of Relativity*, and the sheer fact that his *Transformation* was consistent with his *Principle*, did in turn reaffirm the validity of his *Transformation* and the hypothesis behind it. Nonetheless, in the following chapter we shall see that this *transformation* is not the only one under which the *Principle of Relativity* remains ***covariant***!

It is crucial to realize that the equality $X_B = X_{A/B}$, i.e. the hypothesis that the *distance* between two *points* in *space* -at a given instant- is the same measured either from SC-A or SC-B, was something of an "a priori synthetic" truth for Galileo (and Newton). Notwithstanding, in the light of Chapters 2 and 3, after understanding the relative character of *simultaneity* and that the *length* of a *segment* -***measured*** from a *reference frame* in relative *motion* with it- depends upon the ***measurement*** of *time*, it should be clear to us that the assumption that *time* and *distance* do **not** ***change*** with the *frame of reference* could be mistaken.

Given those possibly unfounded hypotheses, the *Principle of Galilean Relativity* continues being a ***principle*** and not a ***theorem***. My article of faith, camouflaged under my malicious expression "Besides, it is easy to see (Figure 7)..." consisted in assuming that the *length* of a *segment* was the **same** regardless of its being ***measured*** from SC-A or from SC-B. In summary, given that *Galileo's Transformation* implicitly contains the notions of ***absolute*** *space* and *time*, we have to admit that this transformation could be wrong in its philosophical core even though, of course, we know it works with good exactitude for the great majority of our everyday experiences.

Concisely: what I proved was the consistency of *Galileo's Transformation* with the *Principle of Galilean Relativity*, and not the *Principle per se*, because for a mathematician (and with the benefit of hindsight), it is easy to see that there are other transformations consistent with the *Principle*.

Crossing the Argentinean pampas, according to Galileo and our everyday experience, if I walked toward the bus rear at 1.5 km/h, I was moving at 90-1.5=88.5 km/h with respect to ground. If I walked toward the front, my speed relative to ground was 91.5 km/h. If we call SC-A *terra firma* and SC-B the autobus, then, in the former case V_B = -1.5 km/h and V_A = 88.5 km/h, while in the latter V_B = 1.5 km/h, and V_A = 91.5 km/h. Of course, $V_{A/B}$ = 90 km/h. This arithmetic sum is what we call the *Galilean Composition of Velocities*. It is worth noting that, vice versa, had my walking speed been measured from inside the bus and from *terra firma* (i.e. had we measured V_A and V_B), I could have calculated the bus velocity with respect to ground ($V_{A/B} = V_A - V_B$).

The first line of formulae in Figure 7 shows how to calculate the object *coordinates* in SC-B, knowing its *coordinates* in SC-A; the second line teaches us how to calculate the object *coordinates* in SC-A, knowing its *coordinates* in SC-B; the third line depicts how to use the *Galilean Composition* to calculate the object *speed* in SC-B or in SC-A. It is critical for our understanding to realize that the *mathematical structure* of formulae in the first and second lines is identical. Likewise, first and second equations in the third line are *structurally identical*. All these equations are equivalent and one is obtained from the other through algebraic manipulation and using the equality $V_{A/B} = -V_{B/A}$.

It is this identical structure or reciprocity between SC-A and SC-B that assures that an observer situated in one reference frame or the other is impotent as to determine whether s/he is moving or not.

Now, what happened while the bus was braking with all our windows closed? Given that our omnibus was in *accelerated* motion with respect to *terra firma*, and assuming the latter is an *Inertial Frame*, the *Principle of Galilean Relativity* did not entitle us to conclude the bus was -while braking- an *Inertial Frame* but... was it or was it not? Given that the objects that had been *in repose* became bullets without an apparent *force* which could explain the situation, obviously the *Principle of Inertia* was not valid (the objects should have remained *in repose*) and, hence, the bus did not constitute an *Inertial Frame*. However, had we had the audacity of assuming that, while braking, a *field of forces* had appeared explaining the behavior of the flying objects, the *Principle of Inertia* would have remained intact. Without such a courageous move, I can enunciate -temporarily and in accordance with my experience crossing the pampas- the following statement:

***Uniform** motion (repose included) is relative from the mechanical point of view, i.e. undetectable -without external reference- by mechanical experiments. **Accelerated** motion, instead, presents -prima facie- peculiarities which make it mechanically detectable without external reference and, ergo, absolute.*

Please note the qualifier 'mechanical' I included in my statement, because all my experience so far had been mechanical and therefore it was conceivable that there could exist other phenomena (e.g. of an *optical* nature) allowing us to discriminate between two Galilean (Inertial) *frames of reference*. In fact, that possibility was the holy hope that advocates of ***absolute*** *space* and *motion* had.

1.2.3. Newton's Laws of Motion and Principle of Relativity

Galileo, using his experimental and theoretical knowledge acquired with inclined planes, oriented his efforts to describing the *motion* of a cannonball horizontally fired: he imagined that the horizontal plane which supported the little ball in his previous experiments disappeared and, consequently, the Earth *gravitational field* was now playing a predominant role. He conceived the ball *motion* as comprising two independent *motions*: a) a ***horizontal*** *motion* purely *inertial* (*gravity* acted ***vertically***), i.e. the ball would -in the absence of any horizontal force- move with a **constant** *speed* equal to its **initial** *speed*); and b) a ***vertical*** *motion* -caused by the *gravitational field*- whose *acceleration* (inferred by him by gradually increasing the slope of his inclined planes) had to be **constant** (the ***vertical*** *distance* would vary with the **square** of *time*). Combining the ***horizontal*** *trajectory* with the ***vertical*** one in a *Cartesian System of Coordinates*, Galileo theoretically demonstrated that the *spatial trajectory* of the cannonball (neglecting, as always, air friction effects), had to be what mathematicians called a parabola.

In the same year that Galileo (who wrote in Italian) died, Newton (who would write in Latin) was born. He would be the genius behind the unification of the ideas of Copernicus and Kepler about the *superlunary* world with those of Galileo about the *sublunary* world, in a single theoretical framework capable of describing and explaining the whole Universe with only one set of laws -- proving the reality of the uniformity and unity of the Universe that Giordano Bruno had imagined.

Newton imagined that Galileo's cannon: a) was operated from a place at a high enough altitude so that air friction would be negligible; b) had more and more power (the bullet with more and more ***initial*** *speed*); and c) was taken away once it was fired. Under these assumptions, he concluded that since the ***horizontal*** *speed* would be **constant** (no friction) and the 'going down' (vertical) *motion* could eventually -for a high enough **initial** ***speed***- be compensated by the sphericity of the planet, the bullet would reach the departure place with exactly the same original speed and, ergo, would become an 'eternal' artificial satellite -- exactly like our natural Moon.

Combining the *superlunary* laws of Kepler with the *sublunary* law of **constant** *acceleration* of Galileo, Newton also deduced that the ***magnitude*** of the *gravitational force* had to decrease with the **square** of the *distance* and that its ***direction*** had to be defined by the *straight line* joining both bodies, thereby arriving at the *Law of Universal Gravitation* we saw in Chapter 1. In this fashion, the union of both sidereal and terrestrial worlds had been conceptually and quantitatively achieved. In my opinion, this course of events is much more probable than the legend about the apple falling on Newton's head. Newton presented his masterpiece named *Mathematical Principles of Natural Philosophy* to the Royal Society in three parts between 1686 and 1687. His three Laws of Motion (*Axiomata Sive Leges Motus*) were[15] :

First Law (the law of inertia)

Every body perseveres in its state of rest, or of uniform motion in a right line unless it is compelled to change that state by forces impressed there-

15 As first translated from Latin into English by Andrew Motte in 1729.

on.

Second Law (the law of acceleration)

The alteration of motion is ever proportional to the motive force impressed; and is made in the direction of the right line in which that force is impressed.

Third Law (the law of action and reaction)

For every action, there is always opposed an equal reaction: or the mutual actions of two bodies upon each other are always equal, and directed to contrary parts.

The First Law is essentially the *Galilean Principle of Inertia*, but the great difference is that Galileo enunciated it as **locally** valid on our planet and employing **local** *reference frames* which were in ***uniform*** *motion* with respect to *terra firma*. Newton, instead, had the vision and nerve to postulate their cosmic validity. But *terra firma*, according to Copernicus, was rotating like a spinning top while quickly navigating around the Sun. What sense would it make, thus, to say that the *motion* of an object was a *straight line*, if we did not specify a *frame of reference*?

Newton knew, of course, of *Galileo's Principle of Relativity*, and the potential that ***non-uniform*** *motion* had to prove the existence of ***absolute*** *space*. Precisely because of this potential, he knew that the *Principle of Relativity* was not valid for all possible *frames of reference* (e.g. my *Giant of the Road* while braking). How could he, then, give his *Laws of Motion* a cosmic validity? The answer resided in the ideas of ***absolute*** and ***Euclidean*** *space* and of ***absolute*** *time*; only then the notions of *repose*, *uniform motion*, *accelerated motion*, and *straight line* would also become ***absolute*** (all referred to ***absolute*** and ***immobile*** *space*) and, ergo, would have univocal meaning.

The Second Law (which includes the First), is the well-known "Force is equal to mass times acceleration" we discussed in Chapter 1. Once assumed the existence of ***absolute*** *space*, *time*, and *geometry*, *acceleration* was also ***absolute***, and Newton could introduce the notion of ***absolute*** *force*. This law, in conjunction with the third one, allowed him to define *force* beyond our anthropomorphic 'pull' and 'push' and, most importantly, to ***measure*** it in terms of *inertial mass* and *acceleration*. Assuming that the *inertial mass* employed by this law was equal to the *gravitational mass* he included in his *Law of Universal Gravitation*, Newton proved that two bodies -sublunary or superlunary- with **different** *masses* had to 'fall' with exactly the **same** *acceleration* -- as Galileo had demonstrated on Earth. At last, humanity was in a position to predict not only the *trajectory* of a bullet on our planet but on any other, as well as to predict the existence of unknown planets, the return of a known comet, etc. The Universe was finally uniform; the Deity had not needed Alfonso The Wise's advice.

But Newton could not argue against *Galileo's Principle of Relativity* and had to acknowledge that ***uniform*** *motion* was irrefutably **relative** -- at least from the *mechanical* point of view. Like it or not, he had to accept what we now call *Newton's Principle of Relativity* which is the generalization of *Galileo's Principle* to all three *laws of motion* and to the entire Universe. Extending the definition of *Inertial Frame of Reference* to all frames where the three laws of motion are valid, *Newton's Principle of Relativity* can be enunciated as follows:

> *If we find in Nature an Inertial Frame of Reference, i.e. one in which the three Laws of Newton are valid, then we automatically have an infinite number of them: all those frames which move* ***uniformly*** *with respect to the former, independently of their actual speeds.*

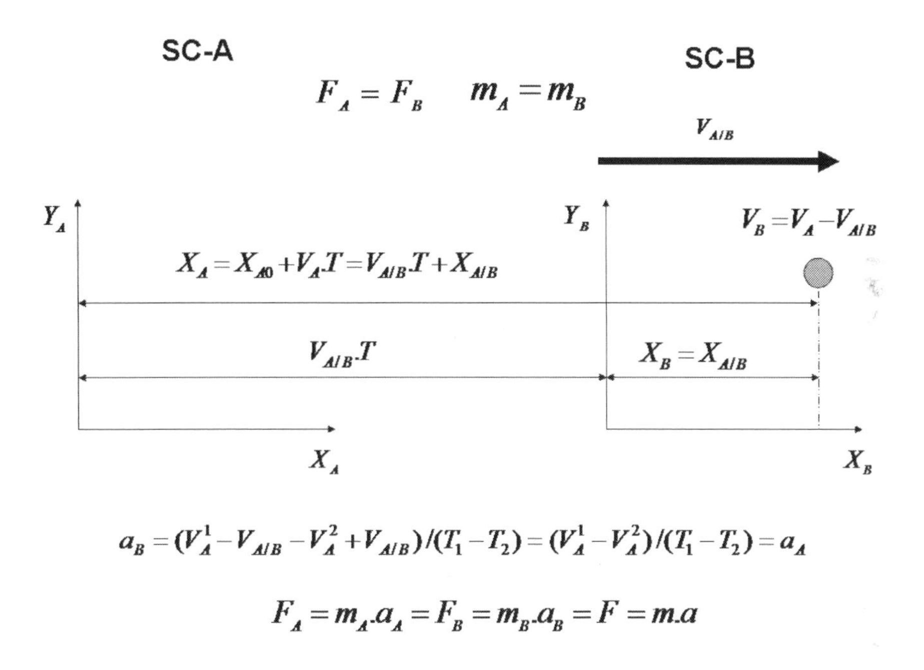

Figure 8. Newton's Laws are Covariant under Galileo's Transformation

We already saw that the *Principle of Inertia* (First Law) is valid in any two *frames of reference* with relative ***uniform*** *motion* (assuming it is valid in one of them). Proving that the second and the third laws are also valid (Figure 8) is again easy if we assumed -as before- that *time*, *distance*, and now *mass* are ***absolute***. Besides, if *acceleration* and *mass* are ***absolute***, then the *force* at a given instant is also absolute. Let us then assume that a force F_A acts on an object with mass m_A in an Inertial Frame SC-A and therefore $F_A = m_A.a_A$. As we know, the relation between the velocities of the object in both coordinate systems is $V_B = V_A - V_{A/B}$. Besides, the acceleration is the change of speed per unit time, i.e. it is obtained (in SC-B) by subtracting two velocities V_B^1, V_B^2 at two times and dividing it by the change in time. Finally then:

$$a_B = (V_A^1 - V_{A/B} - V_A^2 + V_{A/B})(T_1 - T_2) = (V_A^1 - V_A^2)(T_1 - T_2)$$

But the last expression is simply the *acceleration* of the object in SC-A, i.e. a_A and hence the *acceleration* of an object in two *reference frames* with relative ***uniform*** *motion* is the same and, given that for Newton the *mass* and *force* were absolute (independent of the reference frame), we conclude that the

Second Law is also valid in SC-B, i.e. $F_B = m_B.a_B$.

In the same way we said that the *Principle of Inertia* was ***covariant*** under *Galileo's Transformation*, we can now say that the *acceleration* is also ***invariant*** under the same *transformation* and, thus, *Newton's Laws of Motion* are ***covariant*** under ***Galileo's Transformation***. This explains very clearly why, while my bus crossed the pampas with **constant** *velocity* relative to *terra firma* (where *Newton's Laws* were valid), our activities and our ability to predict the *motion* of the orange juice and the deck of 'truco' cards, were identical to those when we were at home. As we saw, this was not true while the omnibus was braking.

The same epistemological elucidation regarding the character of ***principle*** or ***theorem*** I made for the *Principle of Galilean Relativity* -easy to do with the benefit of a five century hindsight- can be made for its generalization, i.e. for the *Principle of Newtonian Relativity*. Reiterating, *Galileo's Transformation* assumes that *time* and *distances* are ***absolute***, i.e. the same in all *Inertial Frames* and, hence, to the light of Chapters 2 and 3, we have to concede that it might be wrong.

Newton dedicated all his energy to defend the ***absolute*** nature of *space* and *time* using *accelerated motion* as its embodiment, and restrained himself to say that, even though ***uniform*** *motion* was clearly **relative** in experience, it was ***absolute*** in its essence -- as he had demonstrated ***accelerated*** *motion* was, in both essence and experience. This attitude was intensely criticized by Leibniz. We saw in Chapters 2 and 3 the philosophical rivalry of both giants through the famous correspondence between Samuel Clarke (on behalf of Newton) and Leibniz, whose central argument was the *Principle of the Identity of Indiscernibles*: what is ***indiscernible*** is ***not different*** and, hence, it makes no sense to speak of ***absolute*** *motion*.

Another weak point in Newton's theory was the notion of a mysterious *force* he called *gravity* acting *instantly* and at a *distance* even in vacuum, without any mechanism to account for it. If the Sun suddenly disappeared, our planet would instantly go astray into the deep sidereal space. Nobody was more aware of the Achilles' heel of Newton's theory than Newton himself. In fact, he wrote:

> *That gravity should be innate, inherent and essential to matter, so that one body may act upon another at a distance through a vacuum, without the mediation of any thing else, by and through which their action and force may be conveyed from one to another, is to me so great an absurdity, that I believe no man who has in philosophical matters a competent faculty of thinking, can ever fall into it.*[16]

1.2.4. The Carrousel by the Church

How could something fictitious have made so much damage had the string holding the rock got broken or the *Giant of the Road* abruptly braked? The Christmas of 1968 was approaching; it had been 10 years since the last time I had played on that peculiar carrousel in the park across the church in downtown Necochea. It was a miniature merry-go-round with four little triangular compartments -one for each kid- with such a negligible mechanical friction

16 As cited in Banesh Hoffmann's *Relativity and its Roots*, New York: Dover Publications, Inc., 1983.

that a tiny impulse given before jumping inside would propel it with very high angular velocities for a few minutes.

My memory of the happiest moments of my childhood remained vivid: once rotating, my visual field would include the Church 'del Carmen', the 'Bank of the Argentinean Nation', the 'School of Nuns', the 59th Avenue, the 'José Manuel Estrada' High School (in the background), the statue of the 'General Mariano Necochea', the City Hall, the 'El Faro' bakery, the 58th Avenue, the Museum of Nutrition, and back to the 'del Carmen' Church.

A guy with stubble would not mix very well with the kids, let alone be understood in his riding the merry-go-round so as to resolve his philosophical queries; I had to wait for them to leave. Six years had passed since I had read that little book on Relativity Theory and had not understood a thing. Even though, for now, I was planning on committing a municipal infraction so as to understand Newton, I knew that most probably sooner or later I would conclude this intellectual giant was wrong, and I would end up thinking like Mach and Einstein. A few months before, I had been greatly impressed by the nobleness and respect for Newton's intellect that Einstein had shown in his *Autobiographical Notes*:

> *Enough of this. Newton, forgive me; you found just about the only way possible in your age for a man of highest reasoning and creative power. The concepts that you created are even today still guiding our thinking in physics, although we now know that they will have to be replaced by others farther removed from the sphere of immediate experience, if we aim at a profounder understanding of relationships.*

But, at the same time, Einstein's conclusions seemed to me as preposterous now -being twenty years old- as the sphericity of the Earth when I was six. The famous 'twin paradox', with which that incomprehensible little book ended, appeared utterly ridiculous and impossible of having any connection with **reality** to me. Nevertheless, my instinct plus the idealism characteristic of youth were telling me that a person who had asked for Newton's forgiveness while demolishing his philosophical edifice, deserved utter respect from me, as well as the necessary effort to understand him -- even if it would take my lifetime.

While -in 1968- I waited for kids from three to ten years of age to finish their joy and leave the playground, I imagined that, if the 'twin paradox' was not a paradox, i.e. if it described **reality** truthfully, then had I in 1954 -instead of flying a little airplane with Dad at 200 km/h- departed on a spaceship at speeds relative to our planet nearing the *speed of light in vacuum* (~300,000 km/s), I could have been here now, in 1968, ***looking at*** exactly the same kids but ***looking like*** them: in short pants and with hair only on my head! Is this not what people call traveling to the future? Even today at sixty-two, I get goose bumps just by thinking about it. Curious to know how is that possible? Of course! Could the reader imagine that its explanation has to do with the concepts of *inertia* and *gravitation*? I couldn't... but so it does -- as we shall see in Chapter 7.

And I kept killing time: would humanity ever develop the needed technology to travel in *space* at those high *speeds* so as to confirm such an incredible phenomenon? The Soviet Union had recently (September 1968) sent, in its *Zond 5*, turtles and other little animals orbiting the Moon and returning safe and sound; President John F. Kennedy had promised in 1961 that the United States would put a man on the Moon before the decade ended and, at this very moment, Frank Borman, James Lovell, and William Anders (Apollo 8) were revolving around our natural satellite -- though without plans for landing. Radar and laser technologies had made possible sending signals to other planets and back so as to experimentally corroborate many other extraordinary predictions of Relativity Theory. But, even so, all these marvelous technological achievements were insignificant with respect to what was needed in order to travel to the future. Would there possibly be other ways of proving such a far-fetched prediction -the so-called *time dilation*- without having to wait for the necessary *space* traveling technology?[17]

Dusk would become dark in a couple of hours; the last kid had left, and the place was ready to be turned into my lab. To start with, I had to clear out my ideas: what was the difference between my experiments on a *Giant of the Road* and those I could perform here on a mini-carrousel? On the former, choosing *terra firma* as reference, the bus could be in an ***infinite*** number of states of ***uniform*** *motion*, i.e. it could be *in repose* at the Terminal or cruising the pampas at any **constant** *speed*. All those *states of motion*, according to the *Principle of Relativity*, were indistinguishable from inside the vehicle with the windows closed and, because *Newton's Laws of Motion* were valid in *terra firma*, they had to be also valid on the bus while cruising at an ***arbitrary*** but **constant** *speed*. That is why we had so much fun playing 'truco' as if at home. All those *Reference Frames* constituted *Inertial Frames*.

Would there exist an analogous set of indistinguishable *states of motion* for a merry-go-round? I gave the carrousel my best impulse without riding on it, so as to see it gyrating from solid ground. Watching attentively any piece of the carrousel except one on its *rotation axis*, say on its periphery, it was obvious that the *velocity direction* (tangent to the carrousel circumference) of such a material piece -even if its *velocity magnitude* was constant (covering equal distances in equal times)- was continuously ***changing***. Consequently, the *velocity* of any piece of carrousel -except for the *center*- could **not** be constant because at least its *direction* was changing all the time and, if it was not constant, then its *motion* with respect to *terra firma* was ***accelerated*** (non-uniform). All rotation was then by its very essence ***non-uniform*** when observed from a *reference frame in repose* with its *rotation axis*. According to Newton, every piece of matter kept gyrating because of the *centripetal force* exerting its action towards the center of rotation, so that the piece could continuously change its *direction of motion*. Now at least I had surmised why Newton had focused on ***rotation*** as the key argument to prove the existence of ***absolute*** *motion*.

My first conclusion, thus, was that the *Principle of Relativity* did not entitle me to presume the carrousel was an *Inertial Frame* like *terra firma* was but... was it or was it not? I gave it my strongest impulse and quickly jumped on it. Given that I was about to abruptly change from ***repose*** to ***accelerated*** *motion*, I expected a sensation similar to the *pressure* ***change*** between my back and the seat in the bus, so I held tight from the metal beams that defined my

17 I did not know by then that a few years before (in 1963) the famous 'Experiment of Mount Washington' (Chapter 6) had experimentally confirmed -with cosmic radiation- that *time dilation* was an irrefutable fact.

compartment, in an attempt to reproduce the function of a seat belt. This time the feeling was that 'somebody' or 'something' was persistently ***pushing*** me towards the *periphery* and out of the carrousel. Of course that I knew all this from the hours and hours that I had spent playing when a child and, even then I had already being told about the *centrifugal force*[18]. But the difference between this experience and that with the autobus was that now there was **no** *cruise speed* at which I could stop holding tight, relax, and play 'truco'. Any insinuation of doing so would have been a recipe for disaster.

Had I released my hands from the metal bars, I would have been shot like a bullet, unhappily ending on the floor, and that would be because in the carrousel, being in ***non-uniform*** motion, the situation was *permanently* **equivalent** to the one inside the omnibus while **braking**. Seen from *terra firma*, my tangential launch could be easily explained again with the concept of *inertia*: I had a *velocity* relative to ground equal to that of the carousel's; letting my hands loose, carrousel and I would become independent objects and I would start a *trajectory* with an **initial** *velocity* equal to the carousel *velocity* and only governed by *gravity* (and a minimal air friction) until reaching the floor.

However, seen from the carousel, and closing my eyes in an attempt to ignore that I knew I was rotating with respect to ground, I was *in repose* in the merry-go-round and, upon releasing my hands, I would be launched without an apparent *force* causing that motion. But, as when the bus started *braking*, according to the *Principle of Inertia*, if there was **no** *force*, I should have stayed *in repose*! In close parallel with my experience crossing the pampas, I hastily concluded that the carrousel could **not** be an *Inertial Frame* ever (the bus could, but only while at **constant** *speed*!). And which was the real difference between the two experiences? In the case of the omnibus, and seeing everything from inside, in order to explain the chaos we had to accept the action of a temporary *field* of gravitational-like *forces* that seemed fictitious to me; in the case of the carrousel, and seeing everything from inside, the analogous explaining *force* already had a name: *centrifugal force*, and I could not sensibly consider it fictitious given that I had to permanently counteract it to avoid killing myself. It was the necessity of having to counteract the *centrifugal force* what gave it the status of 'real', while it was harder to do the same with the *force* in the bus because it only appeared during the transitions between *inertial* states (before and after acceleration). But... during the braking, I also had to hold on tight so as to counteract that force and not hurt myself, so at least temporarily, it was as 'real' as the one I felt constantly in the carrousel.

In summary, if I chose the carousel as the *reference frame*, there existed a ***radial*** *field* of *centrifugal forces* whose intensity was **zero** at the *center*, and the greater the farther away from it. The farther away I was from the rotation axis, the tighter I had to hold on so I could keep rotating without falling. If I chose *terra firma* as the *reference frame*, when I released my hands, my body did what it had to do because of my *inertia*: fly with a *trajectory* determined by my **initial** *speed* (the carrousel *speed* before getting loose), by *gravity*, and by air friction. They were just two different languages to describe the same physical **reality**. When my professor of Physics had said that the *centripetal force* that held the rock rotating was 'real' while the *centrifugal force* was

18 Besides the centrifugal force, there exists the so-called Coriolis force

'fictitious', she had been using the language associated with using *terra firma* as the natural *frame of reference*; had she chosen the gyrating object as reference, then the *centrifugal force* would have been 'real' and the centripetal 'fictitious'. Once again, as in Chapter 3, we learn that the existence of *forces* is **relative** to the *frame of reference*. These *forces* which would explain the chaos seen from inside are called (because the notion of *inertia* explains it from outside) inertial forces, and their associated *accelerations* are called *inertial accelerations*. The remarkable fact with these *inertial accelerations* was that they were independent of the *mass* and chemical *composition* of the objects, i.e. they were as ***universal*** as a *gravitational field*!

But... if a force exists or does not exist depending upon the *reference frame*, what kind of objective **reality** does it have? Given that we have been able to maintain a perfect parallelism between my experiences with the *Giants of the Road* and the *Carrousel by the Church*, I am compelled to conclude that the *inertial forces* that explained the chaos in the bus during its braking are as ***real*** or as ***fictitious*** as the *centrifugal forces* that explained my need to hold on tight on to the merry-go-round. Apparently, the abrupt ***braking*** of the bus allowed me, without external reference, to infer that we were in ***accelerated*** *motion* due to the reigning havoc in comparison with the normal behavior we experienced either at home or while in the bus before braking. As for ***rotation***, by its very nature, it made a *centrifugal force appear*, which allowed us to detect the ***accelerated*** *motion* without external reference. Once again, without further cogitation, everything seemed to indicate that ***accelerated*** *motion* was **absolute** while ***uniform*** *motion* was **relative**.

I could not give up. As with the *Giants of the Road*, if we supposed that Newton was right and ***absolute*** *space* existed, the final proof of the relative nature of *rotation*, depended upon our ability to prove that the *centrifugal force* would also appear in the case in which the carrousel was *in repose*, being now the church, the bank, the school of nuns, the high school, the statue, the City Hall, the bakery, and the museum all together, what was rotating in the opposite direction. Doubtlessly, from the ***kinematic*** perspective, the sequence of images in my visual field would be exactly the same. However... would I need to firmly hold on to the carrousel body, or could I pleasantly be playing 'truco' while enjoying my beloved view? i.e. would the *centrifugal force* be present? The spontaneous answer to this question -due to our everyday experience- is that if the carousel does not gyrate, there is **no** *centrifugal force...* though we do not imagine the *rest of the universe* turning around the merry-go-round. But... what in the world am I saying? There is no way to experimentally prove that the *centrifugal field* appears in both cases because, how can we make the *rest of the universe* ***rotate*** while keeping the carrousel ***in repose***? And, by the same token... can we say that the merry-go-round ***turns*** while the *rest of the universe* stays ***in repose***? Of course not! "Cosa 'e loco!" (madman's business!).

If Einstein had taken ten years to conceive ***rotation*** as **relative**, I was not going to understand it so easily. I had to reread and deeply meditate about Newton's famous experiment with the water bucket, which he had used to believe in the existence of ***absolute*** *space* and *motion*.

1.2.5. Newton's Pail with Water

Only the genius of Newton could have concluded so much out of so little. He took a pail and hung it from the ceiling with a rope attached to its handle; he twisted the rope as much as possible to attain a winding effect; held the bucket, filled it with water, confirmed the water surface was ***flat***, and released the pail so it would start revolving. Initially, the water remained ***static*** relative to *terra firma* and its surface continued ***flat***; the pail *angular velocity* relative to the water was ***maximal*** and equal to its *angular velocity* relative to ground. The water surface continued ***flat***.

The water -being in contact with the pail wall- gradually started to ***rotate*** with its *velocity* relative to the pail decreasing while its surface turning ***concave*** (water closer to the walls higher than that closer to the center). With high enough initial *speed*, eventually water and pail were gyrating at the **same** *speed* relative to ground, i.e. their mutual relative *speed* was ***zero***; at that moment, water concavity was ***maximal***. Newton then grabbed the bucket with his hands to abruptly stop it with respect to ground; the water continued ***rotating*** for a while. Immediately after, the relative *velocity* between water and pail was again ***maximal*** like at the beginning, but it was the water that was ***turning*** relative to ground. However, while at the beginning the water surface was ***flat***, now the ***concavity*** was as intense as when water and pail were *turning in unison*.

What did Newton conclude? That the relative *speed* between water and pail could not explain the ***concavity*** of the surface, because the latter did not exist when the former was ***maximal*** (beginning) but the ***concavity*** was, instead, ***maximal*** when both the ***relative*** *speed* was ***zero*** (right before Newton stopping the pail) and when the ***relative*** speed was ***maximal*** (right after stopped the bucket). And, what was the cause of the ***concavity***? The cause was the *centrifugal force* on the water that, due to the presence of the walls could not move radially, climbing them instead. In a few words, the ***concavity*** was a manifestation of the existence of a *centrifugal field*, and its action was as conspicuous when the ***relative*** *speed* between water and pail was ***maximal*** as when it was ***zero***. Clearly, the latter could not be the *cause* of the ***concavity***. On the other hand, the *centrifugal force* went hand in hand with the *angular velocity* of water relative to ground gradually ***increasing*** until right before the pail was stopped and, when that happened, the ***concavity*** showed complete indifference to the event though gradually ***decreased*** as the water *angular speed* relative to ground was ***decreasing*** due to its friction with the stopped walls. Evidently, it was the ***water rotation*** with respect to ***ground***, and not with respect to the ***pail***, that was related to the water ***concavity***.

So far Newton's reasoning was impeccable. What was his not-so-impeccable next epistemological step? Given that *terra firma* was everything but 'firma' and that the ***relative*** *motion* between water and pail was irrelevant, the *centrifugal force* had to be the consequence of the water *motion* with respect to ***absolute*** *space*, i.e. of ***absolute*** *motion*. His ***absolute*** *space* was immutable, affected by nothing: the archetype of an arena containing all *matter* in the Universe; but, on the other hand, his ***absolute*** *space* did affect *matter*. How? Giving *inertia* to it, because Newton believed that a body *inertia* was a measure of its resistance to be *accelerated* with respect to ***absolute*** *space* and, ergo, it emanated from an action of ***absolute*** *space* upon all material objects.

It is the last conclusion about ***absolute*** *space* and *motion* that contemporaries of Newton like Berkeley, Huygens, and Leibniz fervidly criticized, but it was Mach in 1882 who focused his critique with the specificity and epistemological lucidity necessary to call Einstein's attention. Mach first opposition was that the notion of ***absolute*** *space* was incongruent with *Newton's Principle of Action and Reaction* (Third Law) because ***absolute*** *space* acted on *matter* but *matter* could not act on *space*. But Mach's second argument was the strongest and explains all my questions during my experiments with the *Giants of the Road* and the *Carrousel by the Church*.

According to Mach, the water ***concavity*** did not occur because of its ***absolute*** *motion*, but because of its *motion* ***relative*** to the *rest of the universe* (to which he used to refer as the 'fixed stars'), and Newton's mistake had resided in having considered only the relative *motion* between *water* and *pail* when, for the situation to be entirely symmetrical and equivalent, he should have considered the relative *motion* between *water* and the *rest of the universe*. It was the *mass* of the rest of the universe while ***rotating*** with respect to the *mass* of *water*, or vice versa, that produced the *centrifugal field* which, in turn, produced the ***concavity***. In the same way an electrical charge in *motion* produces a *field* which is different from the one produced when the charge is *in repose*, Mach argued that a body in ***accelerated*** *motion* produced a time-varying *gravitational field* different from the one produced by a static object. Remember I had discovered that (crossing the pampas, and on my mini-carrousel) the mysterious *inertial force* behaved like a *gravitational field*.

The *mass* of the *pail*, in relation with that of the *water*, was insignificant to produce the ***concavity***. And Mach continued: if the ***pail*** could be built with such a thickness that its *mass* were sufficiently large, the *water* surface would insinuate a ***concavity*** at the beginning, when the *water* was *in repose* relative to ground, and the *pail* gyrated at ***maximum*** *angular speed*. When *pail* and *water* were turning at the ***same*** *speed*, i.e. when its relative *motion* did not exist, there would still be a ***concavity*** due to the *rest of the universe*, while the ***concavity*** due to the *pail* mass would be *nil*. When the *pail* was abruptly stopped and the *water* continued ***rotating*** for a while, the opposite to the initial situation (with the same relative speed between pail and water) was in place, and the slight ***concavity*** due to the *pail* would reappear together with the effect of the *rest of the universe*.

Stretching our imagination and letting the bucket grow up until becoming the 'rest of the universe', our mental experiment would start with the *rest of the universe* **rotating** with the *water* ***in repose***; it would go through *water* and *rest of the universe* being in *relative* **repose**, and would end with *water* gyrating and the *rest of the universe* ***in repose***. But... now the state of affairs would be entirely different from the one described by Newton: the ***concavity*** would be ***maximal*** at the beginning, ***zero*** when in the intermediate state (*water* and *rest of the universe* in *relative* **repose**), and ***maximal*** again at the end, i.e. now the *centrifugal field* would go hand in hand with the ***relative speed*** between the *water* and the *rest of the universe* and, hence, ***rotation***

-*accelerated* by its very nature- would be also **relative.** If -assuming the existence of **absolute** *space* so that we can suppose either *water* or the *rest of the universe* as *in repose* (as we did at the start of this paragraph)- we conclude that in both assumed scenarios the *centrifugal field* **exists**, so that the latter cannot be indicative of *motion* with respect to ***absolute*** *space* of either the *water* or the *rest of the universe* and, ergo, the hypothesis of an ***absolute*** *space* is unnecessary!

The advocates of **absolute** *rotation* adduced that all the above was a sheer useless mental experiment ("madman's business"), because it was impossible to make the *rest of the universe* ***rotate*** so as to prove the *centrifugal force* also existed, as when the *water* ***gyrated***. But... to sustain that in our real experiment (the only one possible) it is the *water* that ***rotates*** is simply a linguistic license because the objective **reality** is that what we observe is the **relative** *motion* and a *centrifugal force.* If we assume it is the *water* that ***rotates***, we say that the *centrifugal field* is produced by its ***rotation***; if we assume the *water* is *in repose* while the *rest of the universe* ***gyrates***, we say that the *centrifugal field* is produced by the *rest of the universe* ***rotation***. The objective **reality** is that the *centrifugal field* appears when there is **relative** *rotation*; full stop. Besides, remember that the notion of *centrifugal force* is necessary only when we describe **reality** from a *reference frame in repose* with the 'rotating' object (water in pail, carrousel). If we choose *terra firma* (more properly 'the rest of the universe'), the notion of *water* ***inertia*** explains the observed **reality**: relativizing ***non-uniform*** *motion* requires relativizing the concept of *force.*

Poincaré agreed with Mach: he said -in his book Science and Hypothesis (1902)- regarding the Copernican system: "... these two propositions, 'the earth turns round' and, 'it is more convenient to suppose that the earth turns round' have one and the same meaning". But neither Mach nor Poincaré could offer a consistent theory treating *rotation* as **relative.** Little more than a decade was needed for Einstein to present his masterpiece, in which the *Principle of Relativity* was extended to ***non-uniform*** *motion.* Mach's philosophical position was the seed for Einstein's General Theory of Relativity (Chapter 7), with which he expected to abolish **absolute** *space* forever. Notwithstanding, even today, the ghost of **absolute** *space* seems to sporadically emerge in unexpected apparel.

In accord with Einstein, even from the mechanical point of view (i.e. kinematic and dynamic), every object can be considered *in repose* by convention and, ergo, Ptolemy's and Copernicus' points of view are equally acceptable. The former, describing reality using Earth as the *reference frame*, is extremely complex, and *motion* of celestial bodies requires the interaction of *forces* and *masses* which do not obey Newton's Laws; the latter, describing the same **reality** with the Sun as the *reference frame*, is immensely simpler, and celestial *motion* can be explained in terms of *forces* and *masses* obeying Newton's Laws (as a very good approximation). As Reichenbach [1] said: "The theory of relativity does not say that the conception of Ptolemy is correct; rather it contests the absolute significance of either theory."

Let me finish this section reflecting on the following Arthur S. Eddington excerpt [8]:

Ptolemy on the earth and Copernicus on the sun are both contemplating the same external universe. But their experiences are different, and it is in the process of experiencing events that they become fitted into the frame of space and time - the frame being different according to the local circumstances of the observer who is experiencing them.

1.2.6. Is our Planet an Inertial Frame of Reference?

To wrap up, I need to emphasize how subtle the concept of *Inertial Frame of Reference* is. Once I clarified the folkloric error of assigning a ***uniform*** state of *motion* to an *Inertial Frame* without any *reference* (something impossible), I defined it first as a frame in which the *Principle of Inertia* is valid. Finally, when treating Newton's Laws, I generalized the definition of *Inertial Frame* to one in which *Newton's Laws of Motion* are valid. Equivalently, after introducing the notion of *inertial forces*, we can say that an *Inertial Frame* is one in which there are **no** *inertial forces* (I know: coherent semantics is not scientists' forte).

Besides, the reader probably noticed that, in contrast to other expositions of the *Newtonian Principle of Relativity*, I enunciated it starting with "If we find in Nature an *Inertial Frame*..." In short, both Newtonian Mechanics and Special Relativity (Chapter 6) describe Nature as seen from *Inertial Frames* and, hence, their validity depends upon the existence of such *frames of reference* in the Universe. However, the only way to find such a *frame* is to prove that *Newton's Laws of Motion* are valid in it with the exactitude we need or, equivalently, that there are no *inertial forces* in it. Once we find such a *frame*, then we have an infinitude of them at our disposal.

On the other hand, during my experiences with the *Giants of the Road* and the *Carrousel by the Church* I assumed that *terra firma* was an *Inertial Frame*. I said: "With the experimental accuracy available in Galileo's time, our planet certainly qualified as an *Inertial Frame* -- at least locally". Based on that premise, I concluded -experimenting with the carousel- that *rotation* was a type of *accelerated motion* with respect to *terra firma*. But, then, I cannot but imagine you telling me: "Gotcha! Because our precious planet is nothing but a giant merry-go-round, as clearly indicated by our bulged Equator (due to the *centrifugal force*), the *Coriolis force* that controls the winds, and the varying oscillation plane of Foucalt's Pendulum so... if the planet *rotates*, it **cannot** constitute an *Inertial Frame*'.

You are right: independently of whether the Earth or the *rest of the universe* is the one that *rotates*, if we use our planet as the *reference frame*, the *centrifugal* and *Coriolis fields* do exist and, hence, Earth is not a perfect *Inertial Frame*; in fact, we already know that we cannot use it to describe celestial *motion*, if we want Newton's Laws to be valid. But there are many phenomena for which the effects of Earth *rotation* and *translation* around the Sun are insignificant and, for the desired accuracy, *terra firma* behaves like an *Inertial Frame*[19]. As we saw, Galileo discovered the *Principle of Inertia* and the *law of constant gravity acceleration* employing *terra firma* as reference. Nonetheless, the reproduction of Galileo's experiments with current accuracy would

19 In particular, the effect of daily rotation can be fully eliminated by conducting our experiments on one of the poles.

show deviations with respect to Newton's Laws, attributed to the *non-inertial* nature of our planet as a *reference frame*. Sometimes, the lack of exactitude in our observations may be the motor behind our progress in gradually understanding Nature.

2. What we have learned

As usual, I would like to summarize what I think the reader has learned along this chapter:

• *Motion* manifests to us ***kinematically*** (*temporal* change of *position*) and ***dynamically*** (existence of *forces*);

• Ptolemy, Copernicus, Brahe, and Kepler were the exponents of the *kinematics* era; da Vinci, Galileo, Newton, and Mach combined the *kinematic* and *dynamic* aspects of *motion* giving birth to the science of ***Mechanics***;

• If we find a *reference frame* where *Newton's Laws* are valid, we automatically have an infinite number of them. All move at **constant** *velocity* with respect to one another, and we call them *Inertial Frames of Reference*;

• It is impossible, without *external reference* and with purely *mechanical* experiments, to know the *velocity* with which an *Inertial Frame* moves with respect to another. To assume that it is *in repose* is as valid as to assume it is *moving* at any other ***constant*** *velocity*;

• *Galileo's Transformation* relates *space* and *time coordinates* between two *Inertial Frames*. *Time* and *distance* are assumed **absolute**, i.e. the same in both *frames*, and the *spatial coordinates* easily obtained because *distance* between both *frames* changes proportionally with *time* (mutual ***uniform*** *motion*). The relativity of *simultaneity* and of *distance* we discovered in Chapters 2 and 3, suggest that *Galileo's Transformation* may not be correct;

• If the relative *motion* between two *reference frames* is ***accelerated*** and *Newton's Laws* are valid in one of them, they are **not** valid in the other, i.e. if one is *inertial* the other is **not**;

• In a *non-Inertial Frame*, there appear the so-called *inertial forces* and *inertial accelerations*. The *centrifugal force* is the epitome of an *inertial force*. This *inertial field* is as ***universal*** as a *gravitational field*;

• Inclusion of these *inertial forces* in the *dynamic* description of *motion* inside a *non-Inertial Frame* could make *Newton's Laws of Motion* valid again;

• Our planet (*terra firma*) can be used as an *Inertial Frame of Reference* to describe many local phenomena, but there are other phenomena for which the *inertial forces* due to Earth *rotational motion* cannot be ignored;

• The relativity of ***uniform*** *motion* has been irrefutable for millennia, while ***accelerated*** *motion* showed apparent signs of identification without ***external*** *reference*. Mach was the first one to present a logical/epistemological solid argument to consider ***non-uniform*** *motion* as relative as the ***uniform*** type. Einstein would put Mach's argument into mathematical terms and would make it part of his General Theory of Relativity in 1916. Unfortunately, the ghost of Newton's ***absolute*** *space* was not completely eliminated.

In the next chapter we shall analyze the concept of *motion* when bodies are electrically charged (Electrodynamics), as well as when there is no transport of *matter*, i.e. during propagation of *undulatory radiation*. In order to be able to gallop with ***light***, we should first gallop with ***sound***. By the end of Chapter 5, it is my hope that the reader will have assimilated the philosophical fundamentals and historical background that gave birth to Relativity Theory. With that background, s/he should be prepared to understand Special Relativity in Chapter 6, and General Relativity in Chapter 7.

Additional Recommended Reading

[1] Reichenbach, Hans, *From Copernicus to Einstein*. New York: Dover Publications, 1970.

[2] Einstein, Albert and Infeld, Leopold, *The Evolution of Physics*. New York: Simon & Schuster, 1966.

[3] Russell, Bertrand, *ABC of Relativity*. London: Routledge, Taylor and Francis Group, 1997.

[4] Gimbel, Steven and Walz, Anke, *Defending Einstein, Hans Reichenbach's Writings on Space, Time, and Motion*. Cambridge: University Press, 2006.

[5] Hoffman, Banesh, *Relativity and its Roots*. New York: Dover Science Books, 1983.

[6] Poincaré, Henri, *Science and Hypothesis*. New York: Dover Publications, 1952.

[7] Carnap, Rudolf, *An Introduction to the Philosophy of Science*. Edited by Martin Garner, New York: Dover Publications, 1996.

[8] Eddington, Arthur S., "*The Theory of Relativity and Its Influence on Scientific Thought*" (The Romanes Lecture, 1922). Oxford: The Clarendon Press, 1922, pp. 3-6, 11-12 y 31-32.

[9] Goldsmith, Donald, *The Ultimate EINSTEIN*. New York: Byron Press Multimedia Books, 1997.

Chapter 5

Galloping with Sound

Grammar and ordinary language are bad guides to metaphysics. A great book might be written showing the influence of syntax on philosophy. Bertrand Russell

The Grand Cosmic Conspiracy

When it had become clear that light was of an undulatory nature, physics argued that if there were undulations, there must be something to undulate - one cannot have a verb without a noun. And so the luminiferous ether became established in scientific thought as 'the nominative of the verb undulate', and misled physics for over a century. Sir James Jeans, *Physics and Philosophy*, 1943.

In Chapter 4, I presented the ontology and epistemology behind *motion* of *matter* with its ***kinematic*** and ***dynamic*** aspects. I defined the concept of *Inertial Frame of Reference* as a frame where *Newton's laws of motion* are valid. The philosophical premise behind these laws was the existence of ***absolute*** *space* and *time* and hence of ***absolute*** *motion*. Even though none of my discussions about *Inertial Frames* included the notion of ***absolute*** *space* (precisely to show it was not necessary), the latter was the *Inertial Frame par excellence* for Newton. Ironically, Newton had to accept that there existed a class of *motion* -***uniform*** with respect to ***absolute*** *space* and *time*- that had all the characteristics of being **relative** and none of being **absolute**. He had, thus, to extend the *Galilean Principle of Relativity* to his three *laws of motion*, and we refer to it as the *Principle of Newtonian Relativity*. In sum, if we find an *Inertial Frame*, we have -*ipso facto*- an infinite number of them: all those whose relative *speed* to the former (and between one to another) is ***constant***.

Given any two *Inertial Frames*, I presented *Galileo's Transformation* that relates the *space* and *time* coordinates between both *frames*. As we know, Galileo and Newton assumed that *time interval* and *distance* between two *points* at a given *instant* were **absolute**, i.e. the **same** for all *Inertial Frames*. Likewise, they postulated that the *mass* of a body and the *force* actuating on it were **absolute**, i.e. the **same** for all *Inertial Frames*. We also know -from Chapters 1, 2, and 3- that these hypotheses are dubious and shall be declared erroneous in Chapters 6 and 7. For now, we shall continue describing the course of history, while at the same time trying to understand why such a drastic philosophical change took place at the beginning of the 20th century.

The unavoidable question was: would there exist other types of experiments (non-mechanic) that would allow us to prove that ***uniform** motion* was as 'absolute' as the ***accelerated*** one? Rephrasing: does a non-mechanical phenomenon of *motion* exist for which we know, or believe to know, or for which we could measure its ***absolute** velocity*? The study of electricity and magnetism phenomena had made great advances during the 19th century and seemed to offer the possibility of saving the Newtonian concept of ***absolute** space*. The advocates of ***absolute** space* (there still exist [9,10]) were willing to sacrifice classical mechanics, if necessary to save ***absolute** space*. Why sacrifice? Because Nature had already shown that mechanical phenomena displayed the duality of *velocity* being **relative** while *acceleration* being **absolute**, and if it could be shown that it was possible -through electromagnetic experiments- to determine both ***absolute** velocity* and *acceleration* of an object, then the Universe would be revealing a new level of schizophrenia: Mechanics was relative and Electromagnetism was absolute. Alfonso 'The Wise' would jump out of his tomb once again to offer tardy advice to the Almighty. The scientific mind would not have rested until discovering the beauty of the supposed unity and internal consistency of the Universe, and classical mechanics would have been eventually modified/perfected (something that did happen anyway).

On the other hand, if there was no such non-mechanical phenomenon, the *Principle of Newtonian Relativity* could be extended to all Physics, though -of course- only for *Inertial Frames of Reference*, and the unity of Nature would be partially recovered. This chapter historically reviews the numerous failed theoretical and experimental attempts during the 19th century to determine the *motion* of our planet in ***absolute** space* (disguised first as 'luminiferous ether' and then simply as 'ether'), efforts that put the scientific community in a huge logical/epistemological crossroads, and prompted Einstein to enunciate in 1905 his *Principle of Special Relativity* as well as his *Principle of the Constant Velocity of Light in Vacuum*.

When electricity and magnetism phenomena were seriously investigated, given the success of Newtonian Mechanics, attempts were made to explain them through mechanical processes. As a results of the calamitous failure of those attempts, *classical mechanics* had to surrender to *electromagnetism*, and to be considered as a good approximation to physical **reality**, but only in those cases where the relative *velocities* of the bodies/radiation being studied were much lower than the *speed of light in vacuum*, and in which the intensity of *gravitational fields* was negligible. Given that those two conditions are precisely the circumstances defining our provincialism with respect to the microcosm and macrocosm between which our habitat is entrapped, it is not difficult to understand the superlative past and present success of Newtonian Mechanics.

1. From the Atlantic Ocean to the Andes... and then to the Wasatch Front

In August of 1974, the marvelous time as a student had ended, and I moved from Bahía Blanca to San Juan, very close to the majestic Andes mountain range. Now, I had to concentrate on a very specialized technical field and make all the mistakes possible, so as to acquire a diploma much harder to get: that of 'expert'. Because I was a Professor/Researcher at an academic institution for automatic control in the 'Universidad Nacional de San Juan', I began looking for 'something that needed to be controlled' and, after a few failed attempts, in 1978 I started working on optimal control of mineral grinding systems.

Around 1980, I finally found what I was looking for: there was no instrumentation available for measurement and control of particle size in suspensions/ emulsions at industrial concentrations. Controlling the ore particle size during grinding is crucial to maximize liberation of the precious metal from waste -in preparation to concentration processes like flotation- while minimizing energy consumption. Two physical principles seemed to have a good chance of success: the interaction of the suspension with ultrasonic (*mechanical*) waves, and the interaction with *electromagnetic* waves. From my perspective, I wanted the potential inventions to meet three basic criteria: a) to be intellectually challenging for me in the short, medium, and long terms; b) to have commercial value in the medium and long terms; and c) to have an acceptable chance of obtaining the needed funds to develop those new technologies.

Criterion 'a' was met: the depth of theoretical physics, mathematics, and engineering necessary to develop those technologies was precisely the kind of intellectual challenge I was looking for. Besides, in the long term, an entirely virgin field for basic research -clearly open to me- was that of developing a new fundamental theory to accurately predict multiple-scattering phenomena (reflection, refraction, diffraction and absorption) occurring at the typically high industrial concentrations -- either with optical or with ultrasonic radiation. The sheer possibility of working theoretically and experimentally with *mechanical* and *electromagnetic* waves was highly stimulating for me, as it would give me the opportunity of fulfilling my teenager promise of understanding the Theory of Relativity "even if it took my lifetime".

With respect to condition 'b', besides the need I had already identified in the mining industry, a great majority of the industrial products, either they are produced as suspensions or emulsions, or they go through that state in some stage of their production and, in addition, particle size control is critical not only for the final product properties, but also to minimize production costs. The potential invention had high commercial value in numerous industries, e.g. chemical, mining, cosmetics, food, pharmaceutical, semiconductors, etc.

As for the third criterion, i.e. how to get the monies involved in a research and development project of such a magnitude, was not as apparent as I would have liked it to be. Between 1980 and 1983, I made multiple efforts of getting Argentinean companies interested, though without any success. At the same time, during that period, after almost a decade of academic activity, I had been little by little developing the internally conflictive feeling that I was teaching how to do things that, in reality, I had never done! And, consequently, it was about time for me to spend some time out of Academia entering industry and commerce where the probability of getting those funds needed to concrete my ideas would be, undoubtedly, much higher. The Quixotic decision of resigning from the top academic position I had reached from so much hard work, was silently taking shape inside me. I would end up in another majestic place: the

Wasatch range of Utah, USA, from where I am writing this book more than a quarter of a century later.

2. The Epistemological Crossroads for Physics at the end of the 19th Century

In Chapter 4, during my experiences with the *Giants of the Road*, we learned that *terra firma* (i.e. our planet) can be considered as an *Inertial Frame* which moved ***uniformly*** with respect to the bus -- when the latter was crossing the pampas at ***constant*** *speed*. Let us remember that our planet *rotation* is relatively very slow with respect to its translational *speed* and the *length* of its solar orbit, so that the ***inertial*** *forces/accelerations* affecting a ***small*** zone on our planet can be ignored; and that is why -in the experiments I described in previous chapters and shall describe in this one- Earth can be considered as an *Inertial Frame of Reference*.

My discussion in Chapter 4 only considered three real protagonists: autobus, *terra firma*, and myself. The first two were alternatively used as *frames of reference* to describe my observations. In this chapter, so we can understand my exposition from a historical perspective, I have to include a new hypothetical protagonist: Newton's ***absolute*** *space*. Reiterating, ***absolute*** *space* is, by definition, in ***absolute*** *repose* and constitutes the *Inertial Frame of Reference par excellence*.

So as to not get trapped in a semantic tangle, let us repeat that we call ***absolute*** *velocity* of an object (body or radiation) the *velocity* it has with respect to the ***absolute*** *space* of Newton. Tersely: the **absolute** is, by definition, **relative** to a universal standard called ***absolute*** *space*. With this clarification, if ***absolute*** *space* and *terra firma* are *Inertial Frames*, applying the *Principle of Newtonian Relativity*, the relative *velocity* between both has to be **constant**. But then, given that I experimentally proved that when the bus was cruising at a ***constant*** *speed* with respect to *terra firma*, it constituted an *Inertial Frame* (Newton's laws were valid), so it is *terra firma* which, thus, has to move at an unknown but ***constant*** *speed* with respect to ***absolute*** *space*.

On the other hand, once again, Earth -besides its daily rotation- revolves around the Sun and, if we accept it does so in ***absolute*** *space*, its *velocity* **cannot** be constant as it continuously changes its direction during its daily and annual cycles. Even though these inconsistency could very well be indicative of the epistemological difficulties of the notion of ***absolute*** *space*, in this case, Newtonian mechanics is out of trouble by simply noticing that during the **short** *time* period of, and the **small** *distances* traversed during, my experiments with the bus, our planet *velocity* can be considered as practically **constant** in magnitude and direction. In this fashion, all the terrestrial experiments to be described in this chapter can be considered as in different *Inertial Frames*, one for each *position* of our planet on its daily and annual trajectories.

2.1. The Problems of the Galilean Composition of Velocities

In classical mechanics, as we saw, the *velocity* of an object is **relative** to the *reference frame*, while its *acceleration* is **absolute**, i.e. the **same** for all *Inertial Frames*. While crossing the pampas, according to Galileo and our everyday experience, when the tachometer read 90 km/h, if I walked toward the bus rear at 1.5 km/h, I was moving at 88.5 km/h (90-1.5) with respect to *terra firma*. If I walked towards the front, my *speed* relative to solid ground was 91.5 km/h (90+1.5). Using the symbolism of Chapter 4 (Figure 6), and calling SC-A *terra firma* and SC-B the bus, then in the first case V_B =-1.5 km/h and V_A = 88.5 km/h, while in the second case V_B = 1.5 km/h and V_A = 91.5 km/h. Of course $V_{A/B}$ = 90 km/h. This arithmetic summation of velocities is the *Galilean Composition of Velocities*, which we know works wonders for all situations in our daily lives.

It is paramount to understand that this arithmetic sum involves three *velocities* which are measured in two different *reference systems*: V_A which is the object velocity as measured in SC-A, $V_{A/B}$ which is the velocity of SC-B as measured in SC-A, and V_B which is the object velocity as measured in SC-B. Remember that the *Principle of Newtonian Relativity* implies that $V_{A/B} = -V_{B/A}$, i.e. the bus *velocity* relative to *terra firma* is equal in magnitude but opposite in direction to the *velocity* of *terra firma* relative to the bus, and its absolute value is a characteristic constant for any two *Inertial Frames*. With this caveat, the *Galilean composition* is, then, a relation between two *velocities* measured in two different *Inertial Frames of Reference* in relative ***uniform** motion*.

Up to the 19th century, the experimental legitimacy of this *composition of velocities* reaffirmed the validity of *Galileo's Transformation* relating the *space* and *time* coordinates in both frames and, ergo, the soundness of the hypotheses about ***absolute** time* and *distance* -- as these were the basic pillars of the *transformation*. As a reference, remember that *Galileo's Transformation*, for the simple case of *motion* along the X-Coordinate, reads:

$$X_B = X_A - V_{A/B}.T \; ; \qquad Y_B = Y_A ; \qquad Z_B = Z_A ; \qquad T_B = T_A = T ;$$

Let us now imagine another frame SC-C solidly attached to my body, so we can describe all the experiments from my private perspective. Imagine as well that, as I walk towards the bus front with speed $V_{B/C}$=1,5 km/h, I fire a gun. The initial velocity of the bullet relative to the gun (i.e. SC-C) is defined by gun technology, and it will be the same regardless of my state of motion. Let us assume it is V_C = 1,100 km/h. Which will the bullet velocity V_A relative to *terra firma* be? It is not difficult to see that the answer is obtainable by applying the Galilean Composition twice:

1. Bullet relative to myself (SC-C): $V_C = 1{,}100$

2. Bullet relative to autobus (SC-B): $V_B = V_C + V_{B/C}$ = 1,100 + 1.5 = 1,101.5

3. Bullet to *terra firma* (SC-A): $V_A = V_B + V_{A/B} = V_C + V_{B/C} + V_{A/B}$ =1,101.5 + 90= 1,191.5

We see that, given that bullet, bus, and I are all *moving* in the same direction, using the *Galilean composition*, the bullet *speed* with respect to solid ground is the sum of the *speed* relative to the gun (SC-C) plus the gun *speed* relative to the bus (SC-B), plus the bus *speed* relative to *terra firma* (SC-A). Now we imagine the bus is *moving* forward inside a Jumbo 747 whose *speed* relative to ground is about 1,000 km/h. The reader most probably already figured out that all s/he has to do, so as to find the new bullet speed, is summing up the Jumbo speed to the previous value, and obtaining 2,191.5 km/h for the bullet *speed* relative to *terra firma.*

Finally, suppose we discover that Copernicus was literally right in the sense that the Sun is in ***absolute** repose*; in this case, our planet would have an ***absolute** translational speed* of about 30 km/s or 108,000 km/h and, then, the bullet ***absolute** speed* would be either 105,808.5 km/h or 110,195.5 km/h depending upon the Earth ***absolute** direction* with respect to the ***absolute** common direction* we assumed for all bodies under consideration. In words: one of the two velocities would occur in a given position of the planet while the other -assuming the ***absolute** direction* of the objects has not changed- would occur 12 hours later.[1]

It is evident then that, if the *Galilean Composition of Velocities* is correct, the bullet *speed* with respect to any chosen *Inertial Frame* may, in principle, be as large as we please because, for the purpose, we only need to pile up a large enough number of mobile 'platforms' moving in the **same** *direction* while firing the gun from the top platform. Thus, according to Galileo and Newton, there is **no** limit of *speed* in the Universe, regardless of the *reference frame* we use. As an interesting observation, it is also easy to see that, in classical mechanics, only an ***infinitely** large* velocity would be **absolute**, i.e. common for all *Inertial Frames* because, then, the relative *speed* between the two *Inertial Frames* ($V_{A/B}$) would be negligible, with V_A practically equal to V_B.

An infinite *velocity* is equivalent to Newton's instant *action at a distance* and, until 1675 when Ole Roemer studied the eclipses of Jupiter moon 'Io', people still believed that *light* had all the characteristics of an *instant action at a distance*. Also notice that the possibility of a signal with ***infinite** speed* is consistent with Newton's ***absolute** time*. However, Newton was one of the few who immediately believed in Roemer's conclusion regarding the ***finite** speed* of *light*.

Another way of seeing that there is **no** *speed limit* in classical mechanics is realizing that if we applied a ***constant** force* to an object, because Newton assumes that its *mass* is a ***constant***, its *acceleration* (ratio of *force* over *mass*) also has to be a ***constant*** and, ergo, its *speed* would increase beyond any limit. But, as I already pointed out, experiments conducted between 1901 and 1905 by Walter Kaufmann (1871-1947) showed that the electron *mass* did **not** stay *constant* but increased with its *velocity*, giving one more reason to suspect the soundness of *Newton's Second Law*. And, if the *mass* (inertia) increased without limit with *speed*, the *acceleration* (*force* over *mass*) would gradually reduce to ***zero*** and, hence, there would be a ***maximum** speed*! ... and the *Galilean Composition* would **not** be valid. So much for piling up *moving* platforms.

Talking about piling up *moving* platforms, the famous experiment of Hippolyte Fizeau (1819-1896) in 1851 had proved that a *light beam* traveling in and with the same *direction* as a fast *water stream*, propagated with a *velocity* relative to the pipe (V_A), which was lower than the summation of the veloc-

1 I know all this is very confusing, and the reason is because we are irrationally imposing the notion of the ***absolute***.

ity of light in water (V_B) plus the velocity of the water stream relative to the pipe ($V_{A/B}$) (Chapter 4, Figure 6). The water stream seemed not to fully 'drag' the light beam. The *Galilean Composition*, obviously, was not valid in this case either. So much for piling up *moving* platforms.

But... if the *Galilean Composition of velocities* was wrong, why did it give out so accurate results with macroscopic bodies/radiation moving at *speeds* realizable with the technological possibilities of the time, but failed with subatomic particles like the electron (which moves at much higher velocities) in Kaufmann's experiment, or with *light* (that moves even faster) in Fizeau's experiment? Could the *Galilean Composition*, i.e. the arithmetic sum of *velocities*, be just a limit version of a more elaborated numerical composition which, upon being applied to the former situations, coincided with Galilean's, while differed substantially for much higher velocities?

2.2. 0° Fahrenheit is 'too cold' for a 'Porteño'[2]

My Mother's voice, coming from the Euclidean Necochea, made me return to reality and continue reviewing the details of my imminent visit to the University of Minnesota. While I subconsciously multitasked between my professional preparation for the meeting and remembering the high quality time I had had visiting my natal country and family after 34 years, the video screen was displaying our flight data in real time. Intermittently, two pieces of data attracted my attention while they were being updated: the *velocity* of the plane with respect to *terra firma* (ground speed), and its *velocity* with respect to the air (air speed). The phantom of the *relativity of motion*, once again, distracted me from what was important because of its urgency to what is important because of its transcendence. My Mom knew me very well and would have said: "genio y figura, hasta la sepultura" (genius and character, from cradle to sepulture[3]). I had to rush it; my Mom's sense of duty was her forte, and would soon 'communicate' again, forcing me to return to reality.

Once in the *air*, there were two ways of physical communication between the airplane and *terra firma*: a) *electromagnetic* (radio, radar), i.e. undulatory *radiation* with transmission of *energy* and *mass* without transport of *matter*, critical for taking off, controlling the fight, and landing; and b) *mechanical*, through the *air* interposed between the airplane and solid ground. On the other hand, the *air* was *moving* (wind) with respect to ground and we had three *reference frames*: *terra firma* (SC-A), ***external*** *air* (SC-B), and *airplane* (SC-C). Precisely due to the 'ethereal' mechanical properties of the *air* and the fundamentals of aerodynamics, once in flight, avionic technology could only assure a *speed* relative to the *air* (V_B), without any control whatsoever of its *velocity* with respect to *terra firma* (V_A). I have said the same before, in relation to the *speed* of a bullet relative to the gun. Both technologies define a relative *speed*; fire arms relative to the arm, avionics, relative to the *air*.

It is important to understand that, even when there is no wind in the sense that the expression is normally understood (with respect to ground), the airplane experiences a wind with a *speed* equal and opposite in direction to its

2 In Argentina, people who live in Buenos Aires are called 'porteños' (from the port).

3 There is a powerful rhyme in the original Spanish proverb which, of course, is lost in my literal translation.

speed relative to Earth. We all have perceived this wind when, in our childhood, enjoyed the ride on the open back of a pickup on a day that -according to the weatherman- there was not a scintilla of wind. In that case, the 'ground speed' and the 'air speed' on the video screen of our futuristic pickup would have been the same.

It is obvious that our arrival in Minneapolis depended upon V_A, and employing the *Galilean Composition*, I concluded that $V_A = V_B + V_{A/B}$, with $V_{A/B}$ being the wind velocity relative to *terra firma* and, ergo, a ***tail** wind* would increment the plane *speed* relative to ground reaching Minneapolis ahead of time, while a ***head** wind* would make $V_{A/B}$ negative delaying the arrival -- to the point that had $V_{A/B}$ kept equal in magnitude but opposite to the maximum possible airplane *air speed*, the aircraft would have exhausted its fuel without moving with respect to the planet! I remark once again that this dependence of the aircraft *speed* relative to *terra firma* upon the *wind speed* occurred for two reasons: a) the pilot could only control the *air speed*; and b) the ***external** air* was *moving* with respect to *terra firma*. The flying of an aircraft has many similarities with the propagation of *mechanical* waves.

My imminent meeting with the Director of the Mineral Research Institute at the University of Minnesota (a famous British in the area of computer control of grinding circuits) brought me to reality for a moment and back to my notes... to be ruminating once again on why, given that reading and looking around involved the motion of light from a transmitter to a receiver, these processes were not affected by the airplane motion. Because we were flying at a ***constant** speed* relative to *terra firma*, the plane -according to the *Principle of Newtonian Relativity*- constituted an *Inertial Frame* and, hence, *Newton's laws* were as valid inside the plane as they were on solid ground. But, as we saw in Chapter 1, *light* is not a *mechanical* but an *electromagnetic* phenomenon.

And... what was the inherent difference between 'mechanic' and 'electromagnetic'? It was not easy for me to answer such a question because the atomic nature of *matter* (made of protons, electrons, etc.) implied that *mechanical* phenomena, in their ultimate essence, were really *electromagnetic*: when we say two bodies 'touch', what we mean (without knowing it) is that both *electromagnetic fields* are interacting to avoid physical interpenetration and... that happens well before subatomic particles touch! On the other hand, the physical manifestation of these *electromagnetic fields* is to produce *mechanical forces* (how else, if not so?). Madman's business!

Were the laws of *electromagnetic* propagation the same inside the plane as on solid ground? In other words: *Newton's laws* were **covariant** under *Galileo's Transformation* and that was why all my *mechanical* experiments were the **same** on the plane (SC-C) as at home (SC-A)[4]; what would happen if we applied *Galileo's Transformation* to *Maxwell's equations* which govern *light* propagation? Would these equations **preserve** their ***structure*** with the only difference being that now the *space* and *time* coordinates were those of the *airplane* instead of *terra firma*? Or, on the contrary, would the new equations be ***structurally** different* to the point of, e.g. including the relative *speed* between SC-A and SC-C, i.e. between *airplane* and *ground*? In such a case, reading and

4 I had done all that experimentation while cruising the pampas on the *Giants of the Road*.

looking around would depend upon the particular *ground speed* we had, ergo we could detect it by optical experiments within the aircraft without ***external reference***! Remember that *Newton's laws* only include *acceleration* and that is why the relative *speed* between *Inertial Frames* cannot be detected without ***external*** *reference*. Even though the answer given by Special Relativity Theory to these questions regarding *light* propagation was already known to me, I was in those days far from understanding it with the depth I wanted. Graduating in Electrical Engineering had taken an immense intellectual toll on my part; little time, if any, had been left for philosophical reflection.

Anyway, I quickly realized that, had such an influence of the aircraft *motion* on *light* propagation been real, given that *light speed* was a million times greater than that of the plane, the effect would have been naturally insignificant and impossible to be detected by the passengers. As soon as I had made this conclusion, out of the corner of my eye, I detected a couple having an obvious romantic chat in unison with the captain giving us a brief report about the *time* and the weather in Minneapolis: *time* was noon; temperature was "zero degrees". My English was troglodytic and my mind was metric so that temperature seemed low to me -- though acceptable for somebody like me who had grown up by the Atlantic Ocean with temperatures not very low, but under very humid conditions. Right away, I asked myself whether temperature would also be relative to the *reference frame* and, because it is related to *kinetic energy* of the molecules, I thought that it -most probably- was... but something much simpler and mundane did not occur to me: that the thermal scale the pilot was tacitly using could not be the centigrade scale (Celsius).

While talking, both the pilot and the couple were emitting a *mechanical* wave (sound), the couple in a *direction* ***transversal*** to the plane *motion*, the captain in the same or opposite *direction* (depending on whether the speaker was respectively at the back or at the front). Because I felt much more at home with classical physics, this was an excellent opportunity to investigate the effect, if any, of the plane *motion* on the *speed of sound*. In this case the *speed of sound* is about 1,200 km/h and the airplane *speed* was about 1,000 km/h, both in *air* and comparable to each other so... if there was an effect, it had to be noticeable. However, all passengers' experiences were perfectly 'normal' and identical to those when we were in repose at the gate ready to depart.

And why was that? Because, being sound a *mechanical* process, and being the plane an *Inertial Frame*, the *Principle of Newtonian Relativity* implied that there should not be any difference between our experiences on the plane and on solid ground. Again, *Newton's laws* are ***covariant*** under *Galileo's Transformation* which, in plain English, means they look exactly the same and remain equally valid in any *Inertial Frame* (plane at **constant** *speed* or *terra firma*). If our game of 'truco' was not disturbed while cruising the pampas, the captain's voice could not be disturbed while flying above the Great Lakes; if the butterflies of Galileo did not concentrate aft (tired of catching up with the ship *motion*), the "I love you" from the man to his lover could not be upset by the ***uniform*** *motion* of the aircraft; it was clear to me that the "so do I" from the lady had been issued *ipso facto*.

But it was not as easy as cruising the land of 'Martin Fierro'[5] and 'Don Segundo Sombra'[6], be it on a horse or on a *Giant of the Road*: the experiences in both *Inertial Frames* (*terra firma* and aircraft) would be identical only in identical circumstances; the mere fact that the airplane was flying at a ***constant*** *speed* relative to solid ground was not enough, because sound propagation required a third protagonist, a real *prima donna*: the *propagating medium*, i.e. ***air***.

Was the ***exterior*** *air* the same as the ***interior*** *air*? It certainly was not, because our *air* was trapped inside the plane and was *moving* with us. That had to be the key. What would happen if, hypothetically, we had all windows open so that the relative *speed* between both ***interior*** and ***exterior*** 'airs' became nil and we felt a ***wind*** of 1,000 km/h on our face? And what if our airplane had been a Concorde and we were *galloping* with sound? Even more, what if we 'unsaddled' so as to travel at a ***supersonic*** *speed*? During my experiments on the omnibus, *air* was Mr. nobody; its friction was a phenomenon totally irrelevant to our discussion and I assumed it negligible; objects *motion* did not require the existence of *air*. Sound propagation, instead, needed the *air*: things were not as easy as I had initially surmised. The essence of *motion* for *radiation* seemed to be even more elusive than that of *matter*.

There was no need for my Mom's telepathic intervention; the aircraft began its approximation to the airport of the twin cities Saint Paul and Minneapolis. My meeting was scheduled for 3 pm; I would have just the time to register at the hotel, refresh a little, and take a cab to the University. As soon as I got out of the plane, I was reminded -one more time- of the conventionality of temperature scales: the 'zero degrees' the captain had announced on the plane were degrees Fahrenheit! My native metrical tradition and my provincialism had confused me; temperature in Minneapolis was moderate for the land of 10,000 lakes, but intolerable for a 'porteño': 18° Celsius below zero at noon! My emotions were as extreme as the cold I was feeling.

As a result of that meeting at the University of Minnesota (in which, due to our respective nationalities, the subject of the Falklands War[7] inevitably came up), a few months later (April 1983) the University of Utah offered me to work as a Post-Doctoral Fellow on optimal control of mineral processes. Even though I was leaving one academic post to accept another, this opportunity seemed to be a good intermediate step for me to achieve my objective of finding companies with the big pockets necessary to finance the development of my ideas.

In July 1983, once already settled in Salt Lake City, I started working during my free time on the design of a particle size analyzer based on ultrasonic spectroscopy[8], while looking for companies which could be interested in its development and commercialization. At that moment, even though I worked for the University until 1987, I transformed myself from being an academic into being an entrepreneur, initiating an intense and fruitful process prolonged for almost 30 years, during which I developed four new technologies, while alternating between flourishing economical times and others in which I was on the verge of bankruptcy.

5 'Martín Fierro' is the main character in an epic poem by the Argentinean writer José Hernández (1879).
6 'Don Segundo Sombra' is the main character in a 1926 novel by the Argentinean writer Ricardo Güiraldes.
7 The Falklands War had taken place between April and June of that year (1982).
8 Ultrasonic spectroscopy consists in transmitting/receiving ultrasonic waves with multiple frequencies.

2.3. Mechanical Waves - Galloping with Sound

As I did in Chapter 1, I assume here that the reader has an intuitive notion of what a wave is and, e.g. s/he understands that sound propagates in *air* from our mouth up to the interlocutor's ears as a **local** *disturbance* of its pressure/density (alternating cycles of compression/rarefaction). This *disturbance* around an equilibrium state is initiated by the orator and propagates to the listener without the most minuscule portion of *air* physically being transported from one end to the other. *Air* molecules move only **locally** and relative to the equilibrium position. What propagates is the *disturbance* of local air pressure/density and not the *air matter per se* and, as a result, the *energy* generated by the transmitter, upon arrival, stimulates our ears converting ***mechanical** energy* into ***electrical*** pulses that reach our brain. We say that mechanical *energy* (and *mass*!) transmits through *air* **without** net transport of *matter*.

In the case of sound, given that the perturbation is the **local** pressure of *air*, we say that *air* is the *propagation medium* and it is clear then that sound **cannot** propagate in *vacuum*, as there is nothing to disturb. We also say that a sound wave is ***longitudinal*** because the local *motion* of *air* molecules takes place in the **same** *direction* as that in which the wave propagates. As a counterexample, a sea wave -far from the shore- constitutes a ***transversal*** wave because the local oscillation of water is *vertical* (as easily confirmed by watching a buoy going up and down), while the wave propagates *horizontally*.

Chapter 1 also emphasized that a *wave* of any nature is equivalent to a *field* variable in *time* and *space*. In the case of sound, this field is of *mechanical* pressures so that the presence of a material propagating medium is crucial (air, water, metal, etc.). In short, for *mechanical waves*, the associated *field* simply reflects the physical state in *space* and in *time* of a ***material*** substance which serves as the *medium* for the *wave* to propagate.

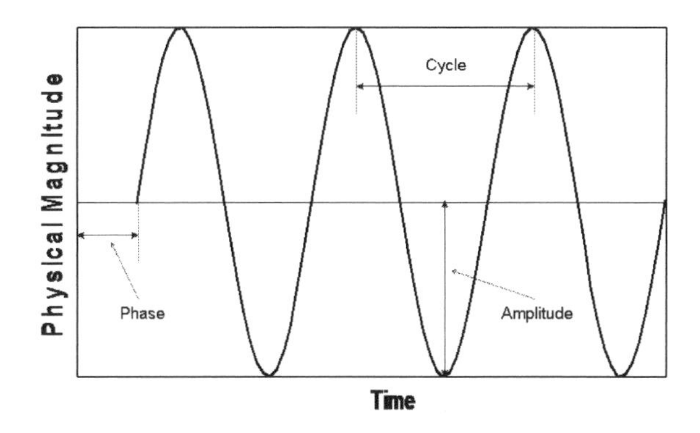

Figure 9. Decomposition into Sinusoidal Components

The concepts of *frequency* and *wavelength* are basic to describe a *wave*. The transmitter produces a *temporal* variation of some physical magnitude, e.g. the amplitude of our vocal cords vibration. This *temporal* variation -or signal as we called it in Chapter 2- can be mathematically decomposed into the sum of many temporal 'pure' variations in the sense that they present a sinusoidal[9] variation that repeats itself indefinitely (Figure 9). This temporal block that repeats itself is called a *cycle*. The *frequency* of this pure signal is the number of *cycles* that occur in the unit of *time* and, hence, is measured in 'cycles per second' or 'Hz' (Hertz)[10]. As an example, if the frequency ***triples***, the number of *cycles* included in the unit of *time* ***triples*** or, equivalently, the *time* a cycle takes to complete is ***three*** times ***less***. Each one of the pure signals into which the complex signal generated by the transmitter can be decomposed, has a different *frequency*, *amplitude*, and *phase* (Figure 9). The numeric inverse of the *frequency* is the *period*, i.e. the *time interval* between the beginning and the end of a *cycle* or, equivalently, between two homologous points of the signal (e.g. two maxima or two minima as indicated in Figure 9).

Human beings can hear signals with pure sinusoidal components which *frequencies* range from 20 Hz to 20 KHz (20,000 cycles/sec), depending on the individual and his/her age. A typical human voice has components with its maximum *energy* around 1 KHz. Professional singers can produce sound with an *energy* maximum around 3 KHz. Sound with *frequencies* below 20 Hz is called infrasound; that with *frequencies* higher than 20 KHz is called ultrasound.

Once the vocal cords modified the *local* pressure in the contiguous *air*, propagation of the mechanical disturbance proceeds with a *speed* and attenuation which are exclusively governed by the physical properties of *air* and, ergo, the transmitter loses all control over what happens to the disturbance. Sound *speed* in *air* is about 1,200 km/h; in water about 5,400 km/h, and in glass around 21,500 km/h. A *field* varying in *space* and *time* is created that covers ***increasing*** regions of *space* as *time* goes by. One more time: it is important to understand that, due to the *mechanical* process of propagation, the *wave speed* is determined by the *mechanical* properties of the *medium*, and only with respect to the *medium*. The *wave speed* with respect to other *reference frames* will depend upon the relative *motion* between the propagating *medium* and those *frames*. Because we are discussing mechanical phenomena, as we know, the relation between *velocities* relative to different *reference frames* is obtained, in Newtonian mechanics, with the *Galilean Composition*.

For the sake of simplicity, imagine the propagation of a pure wave (a single *frequency*). Let us choose the *air* as the *reference frame* (SC-A) and assume that both transmitter and receiver are *in repose* in SC-A. In such a situation, there is no wind for either the transmitter or the receiver. As the initial periodic *temporal* disturbance propagates in the *air*, a *spatial* periodic variation is established in *space* as well: had we measured the local pressure in the space between orator and listener as the *wave* progresses, we would have found that it repeats after having traversed a *distance* which is called the *wavelength* and that it is simply the *distance* the *wave* had traveled during the *time* the transmitter takes for a *cycle* (period). From this, it is easy to understand the relation between frequency (f), wavelength (λ), and velocity (v) which is: $v = \lambda . f$ or in words: if we multiply the number of *cycles* per second of the transmitter by the *distance* traversed during the *time* of each cycle, we

9 Sinusoidal refers to the temporal variation of the trigonometric Sine of an angle varying from 0° a 360°.
10 Unit of frequency in honor of Heinrich Hertz (1857-1894).

obtain the *distance* traversed by the signal in a second, i.e. its *velocity*. More illuminating: the *frequency* is a measure of the *temporal* periodicity, while the *wavelength* is a measure of the *spatial* periodicity. For sound in *air*, the ***minimum*** *frequency* we can perceive (~20 Hz) corresponds to a *wavelength* of 17 meters, while the ***highest*** *frequency* (~20 KHz) corresponds to a wavelength of 17 millimeters (*frequency* a ***thousand*** times ***higher***, wavelength a ***thousand*** times ***shorter***).

2.3.1. The Phenomenon of Interference

When the wave arrives at the receiver, the *temporal* and *spatial* periodicity is converted back into sheer *temporal* periodicity, and the ear receives a *mechanical* pressure varying in *time* which is converted into electrical signals to be interpreted by the brain. A quintessential phenomenon for undulatory *radiation* is that of *interference,* which can be ***constructive*** or ***destructive.*** If our ear receives simultaneously two pure waves of the same *frequency* and *amplitude* but having traveled **different** *distances* (or with **different** propagation *speeds*), the instantaneous arrival pressures can be different and combine arithmetically: if -as it is said- both signals arrive 'in-phase' their crests and troughs will coincide doubling the pressure (Figure 10); if they arrive with opposite phase, maxima (air compression) coincide with minima (air expansion) with the net result that our ear will not hear a thing because there is no net local pressure variation (Figure 11). Intermediate sound sensations will be felt when waves experience intermediate interference phenomena.

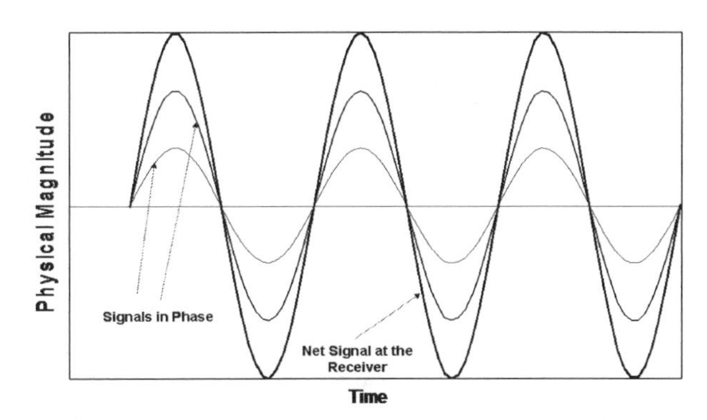

Figure 10. Constructive Interference: net Signal is amplified

Even when the *distance* traversed by each signal is different, if their difference is an **even** multiple of *half the wavelength*, the signals will arrive in-phase, with ***constructive*** interference; if the difference is an **odd** multiple of *half the wavelength*, the signals will arrive with opposite phase experiencing ***destructive*** interference. This is why the measurement of the *interference* between two signals which have traversed **different** *paths*, or have traveled with **different** *speeds*, allows us to measure the traveling *times* for each wave and, if knowing the *distances*, determine their respective *velocities*. This is the operational principle of an extremely precise instrument called *interferometer* and that, for the case of *light* propagation, would play a very important role in the epistemological revolution that was in the making by the end of the 19th century.

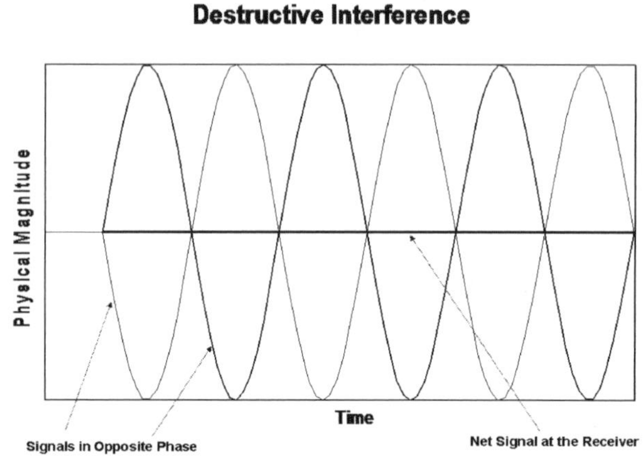

Destructive Interference

Physical Magnitude

Time

Signals in Opposite Phase

Net Signal at the Receiver

Figure 11. Destructive Interference: Net Signal is non-existent

In the supposed case in which *transmitter, medium,* and *receiver* are all in relative *repose*, the *frequency* of the received *temporal* signal is the same as that of the one transmitted. One way of understanding why transmitted and received *frequencies* are identical is as follows: assume that the *distance* between transmitter and receiver is 10 m, the transmitter emits 1 cycle/s (1 Hz), and the wave propagates with a *speed* of 1 m/s. After 1 second, the transmitter has emitted the first *cycle* (which will arrive at the receiver after 9 more seconds) and emits the second *cycle* which, because will propagate with the **same** *speed* as the first one, and the *distance* has **not** changed, will take the **same** *time* to arrive. After 10 seconds, 10 *cycles* will have been transmitted -each one of them separated 1 second from its previous and next *cycles*- while only the first *cycle* will have arrived. After 11 seconds, the second cycle will arrive separated from the first one by again, a second. Following this reasoning, we conclude that after 20 seconds the first 10 *cycles* will have arrived separated by the **same** *time interval* that separated them when transmitted, i.e. the *periods* of *received* and *transmitted* signals are equal and hence *received* and *transmitted frequencies* are equal as well.

In this manner, if the *distance* is sufficiently large and the signal *energy* has not been overly attenuated, we will correctly perceive it though with a *delay*; once the first *cycle* arrives, the others will be received without distortion and the message will be easily understood (sounds familiar to phone conversations using satellites?). Clearly, this ***equality*** of *frequencies* is independent of the particular values for the ***transmitted*** *frequency* and wave propagation *speed*. The only condition we imposed is that *transmitter, medium,* and *receiver* should be all in relative *repose*. Remembering my questions while flying to Minneapolis, regardless of our biological need to breathe, by keeping the *air* trapped inside the aircraft, we were reproducing the same conditions as when we talk on *terra firma*: *orator*, *medium*, and *listener* were all in relative *repose*, and the captain report reached my ear without distortion; besides, given the sound *speed* in *air* and the *distances* involved, the signal *delay* on the plane was as imperceptible as it is at home (around 30 ms to travel 10 m).

Once the conditions on *terra firma* were reproduced, the *Principle of Newtonian Relativity* assured that our experiences on the airplane were identical to those at home -- as long as we were cruising at a ***constant** velocity*. The second hypothetical question I had asked to myself was what would have happen had we -breaking all windows (please do not try it!)- let the trapped air go, experiencing a ***wind*** on our face equal to the 'air speed' displayed on the monitor screen. We will soon be ready to answer this question.

2.3.2. The Doppler Effect in Classical Mechanics

The reader should not despair if s/he cannot follow in technical detail the upcoming discussions. As always, the message (conclusions) is much more important than the messenger (details) and, if so desired, you can simply browse the details, assimilate and reflect conceptually upon the conclusions, and continue with the book.

To evoke familiar images, let us go back to *terra firma* and imagine that the *transmitter* is a train, and that I am at rest on the tracks listening to its whistle while approaching me at ***constant*** *speed* (don't forget to assume the train can run unharmfully over me!). I (*receiver*) and the air (*medium*) are *in repose* in SC-A, but the *transmitter* moves with *speed* V_T in the same direction as the sound wave (whistle traveling with speed v) so that -applying the Galilean Composition- the sound *speed* relative to the train is lower and equal to $v - V_T$. Stretching it, had the *train* moved relative to *air* with the same *speed* as *sound*, the whistle *speed* relative to the *train* would have been ***zero***, and the *train* would have been galloping in unison with its own *sound wave*.

To amplify the effect, let us then assume hypothetically that the train moves with respect to SC-A (tracks, air, and I) approximately at half the speed of sound, i.e. at about 150 m/s, and let us consider two local disturbances of the air caused by the whistle and temporally separated (on the train) by 2 seconds. When the second perturbation leaves the train, it is 300 m ***closer*** to me than when the first one left and, ergo, the second will take 1 second ***less*** than the first (sound covers approximately that *distance* in a second). In this fashion, the *period* in my ear between disturbances will be ***half*** the *period* in the train, and hence the **received** *frequency* will be ***double*** the one **transmitted** by the whistle. During the whole time until the train runs over me, the ***received*** sound will be higher-pitched (***double*** *frequency*), the higher-pitched the higher the train *speed* is. At the instant when the train starts going away from me, the *direction* of *sound* propagation changes abruptly with respect to SC-A (now opposite to the *train motion*) with its *speed* relative to the *train* being higher and equal to $v + V_T$. In this case, given again two air disturbances 2 seconds apart, when the second is emitted, the train is 300 m ***farther*** away than when the first one left and, ergo, will take one ***more*** second to arrive to my ear than the first, so that the *time interval* between them at **reception** will be three seconds instead of two, i.e. the **received** *frequency* will be 2/3 of the one **transmitted** by the train whistle.

In summary, the ***transmitter*** *frequency* is the same when approaching as when going away, but the *wavelength* established in the *medium* (air) is not, and that produces a ***change*** in the *frequency* at the ***receiver***. This ***change*** is the same for a given *train speed, sound* speed in *air*, and a given relative *direction* between transmitter and receiver. Given the relative *velocities*, the **received** *frequency* will increase or decrease, depending upon the **transmitter** approaching or withdrawing from the **receiver**. In our example, when the source approached me, I received ***double*** *frequency* and, when going away from me, I received not half but ***two thirds*** of the **transmitted** *frequency*.

A common folkloric mistake -even in textbooks- consists in asserting that the **received** *frequency* ***increases*** continually until the train gets to our place and then starts ***decreasing*** also continuously when going away. As we saw in our ideal case when standing on the tracks, the **received** *frequency* is ***greater*** (higher pitch) than the **transmitted** one but ***constant*** while coming to me and, suddenly changes to another ***constant*** *frequency* which is ***lower*** (lower pitch) than the **transmitted** one (always the same). The behavior of *light* propagation is similar: a *light* source we see as ***yellow*** when it is in relative *repose* with us, had it approached us sufficiently fast, we would have seen it ***blue*** (***greater*** *frequency*, i.e. ***shorter*** *wavelength*) and, had it withdrawn sufficiently rapid from us, we would have seen it ***red*** (***lower*** *frequency*, i.e. ***longer*** *wavelength*). This so-called redshift of *light* is a purely *kinematic classical* effect; in Chapter 7 (General Relativity) we shall learn about a much more remarkable effect: the ***gravitational*** *redshift of light*.

If we now complicate things a little so as to avoid being killed by the train, I find myself out of the tracks and the *direction* of sound propagation towards my ears is a little different from the *direction* of the train *motion* and, besides, it gradually changes as the train *moves* (from practically ***parallel*** to the tracks when very far away, to ***perpendicular*** to them when passing by, to again ***parallel*** when far away. Because of this, instead of an abrupt change of pitch (from high to low) when passing over me, now there is a gradual *frequency* ***decrease*** consisting in starting with a value higher than that emitted by the whistle (the one calculated in the previous case), passing through a *frequency* identical to the ***transmitted*** one when across me (sound propagates ***perpendicularly*** to train *motion*), and continues ***decreasing*** to the value (lower than the ***transmitted*** by the whistle) calculated in the previous ideal case. Thus, contrary to popular belief, the **received** *frequency* decreases continually, and does so while approaching as well as while going away -- though it is ***higher*** than the **emitted** while approaching and ***lower*** than the **emitted** while withdrawing. I am sure the reader has experienced this phenomenon and probably knows it is called the *Doppler Effect* (or shift) which, together with *interference*, is a fingerprint for undulatory wave propagation.

What happens if the **receiver** is the one that moves at 150 m/s with respect to the *air* and the *tracks*, and the *train* stays *at rest* in SC-A while whistling? The **receiver** *reference frame* (SC-B) is not any more *in repose* with the *air* (SC-A) but moving with a *speed* V_R relative to *air* and *train* so that, in SC-B (the **receiver**), the sound wave travels at a speed $v + V_R$ when the receiver is approaching the train and at a speed $v - V_R$ when going away. Consider again two local air disturbances provoked by the whistle and temporally separated by, this time, only 1 second while the receiver is going ***away***. When the whistle issues its second disturbance, the **receiver** is 150 m farther than when the first one left the train but because, in this case, the disturbance speed relative to the receiver is $v - V_R$ = 300-150=150 m/s, those 150 m are traveled in just one second (150/150). Because of that, the second disturbance will take one more second to get to the receiver than the first one, i.e. the *period* at the **receiver** will be 2 seconds instead of the 1 second at the train or, equivalently, the **received** *frequency* will be ***half*** the **transmitted** one.

If the **receiver** ***approaches*** the train, when the whistle sends the second disturbance, the receiver is 150 m closer than it was when the first left but, in this case, because *sound speed* relative to the **receiver** is $v + V_R$ = 300+150=450 m/s, the 150 m are traveled by the disturbance in one third of a second (150/450). Because of that, the second disturbance will take 1/3 of a second less than the first one, i.e. the *period* at the **receiver** will be 2/3 of a second, instead of the ***transmitted*** one second or, equivalently, the **received** *frequency* will be 50% (3/2=1.5) **greater** than the **transmitted** one.

If we remember that in the case the *train* was approaching while tracks, receiver, and air where in relative *repose*, the **received** *frequency* was 100% ***greater*** (double) instead of the 50% ***greater*** that occurs when it is the *receiver* that approaches the medium and the transmitter (which are now in relative repose), we conclude once again that the *propagating medium* of a *mechanical* wave is the *prima donna* that establishes how the **receiver** will 'see' the wave and not just the mere relative *speed* between **transmitter** and **receiver**. Does this mean that the *Principle of Newtonian Relativity* is invalid? NO; it does not!

In the first case, SC-A (tracks and air) and SC-B (receiver) are in relative *repose* and the ***train*** is *moving* relative to them at 150 m/s, transmitting the sound wave that travels at 300 m/s relative to SC-A (and SC-B); in the second, the train is *at rest* in SC-A with the air, while SC-B (***receiver***) moves at 150 m/s with respect to SC-A. Even though the relative *speed* between ***train*** and ***receiver*** is the same as before (150 m/s) and, of course the *sound speed* with respect to *air* is also the same (300 m/s), the *sonic wave speed* relative to the ***receiver*** in the first case is 300 m/s while in the second case is 450 m/s. In a few words, there was no symmetry between both cases: in order for the second case to be symmetric to the first, the train would have to be at rest in SC-A while the receiver and the medium would have to be in relative *repose* in SC-B which would have to be moving at 150 m/s relative to SC-A. In such a case, it is not difficult to see that the ***receiver*** would detect the ***same*** *frequency* as in the first case. Mechanical undulatory *motion* is as **relative** as the *motion* of *matter* but, for its corroboration, it is necessary to carefully consider the role played by the *propagating medium* in each case.

2.3.3. If you want to hear you are loved... you should recline your seat

Finally, let us consider now a simpler but not less important case which will answer my second question while I was flying over the Great Lakes: both ***transmitter*** and ***receiver*** are in relative *repose* and both in relative *motion* with the *air*. This would have been the case had the train (whistling) and I (receiver) been steady on the tracks while existing a ***wind*** with *speed* V_w; and also had the airplane captain been talking through a speaker located aft or prow, with all windows (hypothetically) open.

Remembering that we call v the *speed of sound* relative to *air*, and seeing things from inside the plane (train), if the ***wind*** runs in the ***opposite*** *direction* to the sound wave (speaker aft), the wave finds a ***head wind*** and, ergo, its *speed* relative to the ***receiver*** (us) is $v - V_w$. If the ***wind*** runs in the ***same direction*** as the sound wave (speaker at bow), the wave experiences a ***tail wind*** and its velocity relative to us is $v + V_w$. In both cases, the romantic couple was emitting sonic waves with a *direction* **perpendicular** to the ***wind***.

In both cases -tail and head winds- given two air disturbances emitted by the whistle/speaker one second apart, by the time the second disturbance is issued, the *distance* between whistle/speaker and us has **not** changed and, even though the wave *speed* relative to us is lower or greater than it is relative to air, it will be the **same** for both disturbances so they will take the **same** *time* to arrive, i.e. **received** and **transmitted** *frequencies* will be the **same**. The ***wind*** accelerates or decelerates the sound wave (relative to us) but it does **not** distort the *frequency* because ***transmitter*** and ***receiver*** are in relative *repose*. Given that aircraft and *terra firma* are *Inertial Frames*, we were able to describe at a stroke our **identical** experiences with the train whistle and the captain report, because the conditions in both *reference frames* were **identical**.

We have come a long way with *mechanical* waves, but we still have to discuss the case in which the man was obviously saying 'I love you' to the lady, while the captain was reporting the temperature in Minneapolis. In this situation, the sound wave was propagating ***transversally*** to the fuselage. Figure 12 is divided in two halves; the one on the right depicts the real situation when all windows are closed so the air transmitting the conversation is trapped and dragged by the plane. External air (wind) and internal air were in relative motion. The external wind had no effect on the conversation; the *propagating medium* was the ***internal*** *air*, in *repose* with the couple. The aircraft was an *Inertial Frame* in which a phenomenon was taking place in identical conditions as on *terra firma* and, consequently, the *speed* and *frequency* of our voice could not depend upon the *direction* and *magnitude* of the airplane *velocity* (which, as we know by now, is unknowable without ***external*** *reference*).

But there were two ways of observing what was going on: the bottom part corresponds to how I perceived the scene, i.e. using the ***plane*** as the *reference frame*. The 'I love you' was emitted ***perpendicularly*** to the fuselage, propagated perpendicularly, arrived at the woman's ear, and her 'so do I' did the same arriving at the man's ear. The top part, instead, corresponds to how the same scene would be assessed using a *reference frame* in which the ***exter-***

nal air is *in repose* and the ***plane*** in *motion*. During the time the 'I love you' took to arrive at the woman's ear, because the plane was moving at a speed comparable to that of sound (and dragged the ***internal** air*), the sound wave trajectory would be seen as ***diagonal*** to the fuselage and the same would happen with the 'so do I'. We confirm one more time that the notion of *trajectory* is relative to the *reference system*. Of course, the objective event of a successful romantic communication should not admit ambiguity and is identically observable in both *systems of reference*.

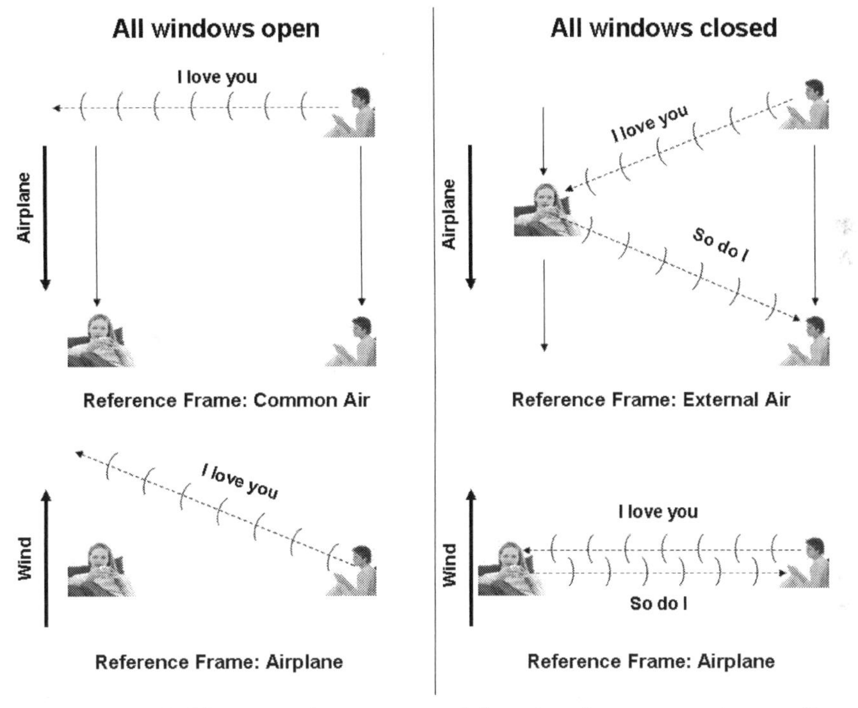

Figure 12. The great importance of dragging the propagating medium

The half on the left depicts the imaginary situation in which all windows are open and, ergo, ***internal*** and ***external*** bodies of *air* are in relative *repose* (they become one body). Using this common air as *reference* (top), the sonic wave propagated ***perpendicularly*** to the plane *direction* of *motion* and, given that it was moving (with the couple and without dragging the air) with a speed comparable to that of sound while the 'I love you' was traveling, the latter expression of love would **miss** the lady's ear and the 'so do I' would never be spoken. Remember that the *speed* of the **source** (lover) does not affect the *speed* of the *wave* with respect to the *propagating medium*, because the latter *speed* is determined by the *medium* ***mechanical*** properties, and only by them.

Using now the ***plane*** as *reference* (bottom part), the *air* would be moving aft dragging the wave with a ***longitudinal*** *speed* comparable to that of the ***transversal*** propagation of sound; the woman would not receive the 'I love you' either (of course!) and the 'so do I' would not be spoken either (of course!). In other words, using the ***aircraft*** as *reference*, the *trajectory* of the sound wave is ***diagonal*** to the plane, while using the ***air*** as *reference*, the *trajectory* would be ***perpendicular*** to the plane. *Trajectories* are **relative** to the *frame*; successful or unsuccessful romantic expressions are **absolute.** In this totally fictitious case, only by reclining her seat, she would know to be loved.

2.3.4. The Doppler Effect when Transmitter and Receiver are in Relative Repose

As we saw in the previous section, if transmitter and receiver are in relative *repose* but in ***uniform*** *motion* with respect to the *air*, the wave will travel ***faster*** (***tail*** wind) or ***slower*** (***head*** wind) relative to the ***receiver*** but **received** and **transmitted** *frequencies* will be the **same.** It is very important for Chapter 7 (General Relativity) that we ask now what would have happened had transmitter and receiver been in relative *repose* but moving ***non-uniformly*** (accelerated) with respect to the *air*.

Given that sound propagates ***uniformly*** with respect to *air*, when the pair source/receiver moves in the same *direction* as the sonic wave (***head*** wind), even though the *distance* between source and receiver does **not** change, because of the ***accelerated*** *motion* of the pair relative to the *air*, the **receiver** goes ***away*** from the air disturbance; the *distance* the second disturbance has to cover is ***greater*** than the *distance* the first disturbance traveled; the *time* between arrivals at the **receiver** is ***greater*** than the *time* between departures at the **source,** and the **received** *frequency* is ***lower*** than the **transmitted** one.

If sound propagates in the opposite direction to that of the pair transmitter/receiver (***tail*** wind) -even though the *distance* between transmitter and receiver does **not** change- due to the ***accelerated*** *motion* of the pair relative to the *air*, the **receiver** gets ***closer*** to the disturbance; the *distance* to cover by the second disturbance is ***shorter*** than the one covered by the first disturbance; the *time* between arrivals is ***shorter*** than the *time* between departures, and the **received** *frequency* is ***greater*** than the **transmitted** one.

2.3.5. A Wave is not a Bullet

All these phenomena occur the way they do preeminently because the mechanic wave *speed* relative to the *propagating medium* does not depend upon the *state of motion* of transmitter or receiver. That is not the case for a bullet coming out of a firearm. Consider the unrealistic case in which sound was an undulatory phenomenon (as it is) but that its *speed* relative to the *air* depended upon the *speed* of the ***transmitter*** with respect to the same *air* (not true). Now imagine that transmitter and receiver are in relative *repose* but move at 150 m/s relative to the *air* in the same *direction* as the sound. Under this assumption, due to the ballistic hypothesis, the *speed of sound* relative to the **gun** is always the same and equal to 300 m/s, while the *speed* relative to *air* would be 300+150=450 m/s. On the other hand, because the *distance* between transmitter and receiver does **not** change, the *speed* of sound relative to the **receiver** is also 300 m/s.

Now, think of our two air disturbances one second apart; when the second disturbance is emitted, given that the *distance* between source and receiver is constant, it takes the *same time* to arrive as the first disturbance took and, hence, **received** and **transmitted** *frequencies* are the same. On the other hand, because the speed in air is 450 m/s, the space-time periodicity of the pressure wave is established per the transmitter *frequency* and this *velocity* (not 300 m/s). However, at the **receiver**, the *space* and *time* periodicity is converted into *temporal* periodicity through $v = \lambda . f$ where v is 300 m/s (not 450 m/s), producing a ***different*** *frequency* than the **transmitted** one! Either transmitted and received frequencies are the same, or they are not. This fact can be experimentally refuted or corroborated and it would be refuted for sound, because we already know that it is a mechanical wave and, ergo, its *speed* relative to the *propagating medium* is determined by the *medium* physical properties, and only by them.

And why did I take the trouble of analyzing a situation that, for sound, we know is unrealistic? Because one of the various theories proposed to explain the negative result of the famous Michelson-Morley experiment with light waves (conceived to measure the ***absolute*** *speed* of our planet in *space*), was based on the ballistic hypothesis, i.e. on postulating that the *speed of light* predicted by Maxwell's equations was not given as **relative** to the *medium* but as **relative** to the *transmitter* (firearm). Had that been the case, the experiment negative results would have not caused such perplexity and frustration (as we shall see in forthcoming sections).

With this brief foretaste, it shall be easy to understand -at the appropriate time- why the *speed of light* predicted by Maxwell's equations is not given as ***relative*** to the *transmitter* -- in the same way as the *speed* predicted by the sound propagation equations is not given with respect to the acoustic *source* (but to the ***medium***). But... is there then a perfect analogy between ***sound*** and ***light***? NO; in the case of ***light***, we shall see that its *velocity* is given neither relative to the **source**, nor to the **medium**! As usual, we have to be patient.

2.3.6. The inexistent Sound Barrier

If we shrank enough ourselves so as to ride on a supersonic bullet, we would never hear the sound produced by the firearm, because we would be always ahead of the sound, and our ear would not detect the pressure wave. Even if the bullet traveled at exactly the sound speed, we would perceive a static pressure field, i.e. with *spatial* but **not** *temporal* variations, and so our ear would not detect any signal to send to the brain. Another way of seeing the same is to notice this situation is the limit of a case we have already studied, in which the ***receiver*** was going ***away*** from the ***source*** (*in repose* with the *air*), so that the wave *speed* relative to the ***receiver*** was obtained subtracting the ***receiver*** *speed* from the *sound speed* in *air*. In this case, the **received** *frequency* was ***lower*** than the **transmitted** one, and the ***lower***, the ***faster*** the withdrawing of the **receiver**, until the latter was moving in *air* at the *speed of sound* -- in which case the **received** *frequency* was ***zero***. Even with an infinite sensitivity, a ***zero*** **received** *frequency* cannot be perceived, as it involves the total absence of pressure variation on the ear.

Let us assume that a supersonic aircraft is nose-diving towards us on *terra firma*. Because the airplane ***decreases*** its *distance* to us ***faster*** than what its own sound does, given two consecutive mechanical air disturbances, when the plane emits the second one, the aircraft has traveled more than what the first disturbance did and, hence, the second disturbance will reach my ear ***before*** the first one. In this case the *sequence* of disturbances ***received*** by my ear is the **inverse** of the ***transmitted*** *sequence*; i.e. -without trying to be mystic- I hear first the future and then the past! But, of course, I am talking about the future and the past **inside** the plane where the events were generated, to be observed in the ***reverse*** *sequence* on solid ground. As long as there exists a signal that travels ***faster*** than the supersonic plane, e.g. *light*, we could use it to communicate from *terra firma* with the aircraft so as to determine the correct *sequence* of *events* in it. Doing so, the mystic flavor of the expression 'hear first the future and then the past' vanishes because -through luminous signals- we shall always receive the events (luminous disturbances sent with each sound disturbance) in the same sequence on *terra firma* as they were emitted on the airplane.

In the real case in which we see the supersonic plane passing by above us, as in the case of the whistling train when I was aside the tracks, the plane *velocity* ***component*** in the *direction* traveled by sound towards us will gradually change, with a ***minimum*** when it is right above us, to ***increase*** again when going away. During its flight, thus, there will be times, when afar, during which the sound is received in **reverse** order, times when the received and transmitted orders **agree**, and others in which **sequentially** generated air disturbances are received, instead, **in unison**. This phenomenon of *temporal* coincidence at a place of *sequentially* generated mechanical signals, can produce the so-called 'sonic boom' that occurs when, as the misleading saying goes, the plane "breaks the sound barrier". Pithily: the 'sonic boom' is the result of the ***constructive*** *interference* of the multiple mechanical waves produced by the supersonic aircraft.

Wrapping up: *undulatory* mechanical *motion*, as *motion* of *matter*, responds to the *Principle of Newtonian Relativity* though, to apply the latter correctly, we have to make sure that the *propagating medium* plays identical roles in both *reference frames* from where we pretend to observe a given phenomenon. Independently of our biological need to breathe, the air pressure was maintained inside the Boeing 727 going to Minneapolis, with the air matter being ***dragged*** with the plane and, ergo, *in repose* with it. With that caveat, there was **no** mechanical experiment inside the plane that could have entitled me to determine -without external reference- that we were moving with respect to *terra firma*.

And, what about non-mechanical, e.g. optical experiments? Could we devise an optical experiment so as to allow us, without external reference, to conclude that the aircraft was moving and so making advocates of ***absolute*** *space* and *motion* happy? Would the presence of *air* be essential for those experiments? Undoubtedly we could conduct the experiments in its presence (I was reading!) but... was it strictly necessary? NO! (otherwise how does the Sun light get here?) Perhaps *light* needed of another *propagating medium* much more 'ethereal' than *air* and that was present even when there was *air* or any other material? No doubt *light* propagated even in solid materials like glass, but... was the *material* strictly necessary? NO! Would that 'other more ethereal medium' have remained in its place had we thrown away the glass and the air? Could we pump that 'other medium' -as we pumped the air inside the plane- to assure it was in repose with us, so as to reproduce the conditions of *terra firma*? But... was that mysterious medium in repose with *terra firma*?

Sound and *light* both behaved as *waves*, but there was something peculiar regarding *light*, as radical as intrinsically different, that was awaiting to be discovered in the first years of the 20th century. We are now prepared to understand the curious and fruitful historical interaction between *Mechanics* and *Electromagnetism*.

2.4. Do I Think in English or Spanish? The Influence of Language on Philosophy

Already settled in Salt Lake City and working hard on the development of a particle size analyzer based on *ultrasonic* spectroscopy, I thought that if I had the fortune of succeeding with this invention, I would continue the research and development thrust using *electromagnetic* waves. My mathematical physics background, combined with the extraordinary experience I would accumulate by working experimentally with *sound* and *light* waves, would constitute an optimal foundation to understand the Theory of Relativity.

After a few months working in the Department of Metallurgy of the University of Utah, I remember my boss asked me whether my thought process was done in English or in Spanish. My instinctive response was "in neither one nor the other". I am partially in agreement with Arthur Schopenhauer in that thoughts die when embodied with words. Einstein himself said in his *Autobiographical Notes*: "I have no doubt but that our thinking goes on for the most part without use of signs (words) and beyond that to a considerable degree unconsciously".

Notwithstanding, the command of a language is crucial to focus the thinking process and, more than anything, to communicate it to others. According to Ludwig Wittgenstein, a good number of the 'great philosophical problems' are inexistent as such, simply because they are ***semantic*** confusions created by the sloppy use of language. As F. David Peat said in [11]:

> *On the one hand, our minds try to probe the ephemeral reality of the quantum world; on the other, we talk, think, and act in a language adapted for discussing trees, rocks, and automobiles -- as well as poetry and emotions.*

There is no better historical example of the negative influence of language on natural philosophy than that of the fruitless search for over a century of a substance called 'ether' that, by its very conception, could not be such. As Bertrand Russell said: "Grammar and ordinary language are bad guides to metaphysics. A great book might be written showing the influence of syntax on philosophy". Paraphrasing Sir James Jeans, scientists could not conceive 'undulations without something to undulate', and needed the mountain of experimental evidence accumulated during the 19th century in order to, after Einstein, start accepting -though not without reluctance- how superfluous and futile the concept of *ether* had been. Still to this day, in the 21st century, the idea of the ***absolute*** and its seduction is irresistible to many people. ***There is***

no problem more difficult to solve than that created by ourselves.

2.5. Light is not somewhere but... goes and comes at 299,792,458 m/s

From time immemorial, *light* was something that *was* somewhere or *was not*, but not something that could *travel*. Even today we colloquially say that 'there is' or 'there is not' *light*. When the notion that *light* 'came' and 'went' was considered, even Kepler thought that it did so instantaneously, i.e. with ***infinite** velocity*. Galileo (what a surprise!) was the first who tried to measure the *speed of light* by arranging to ***measure*** the *time* elapsed for a lantern *light beam* traveling from the apex of one mountain to another 1.5 km away and, upon arrival, a second lamp turned on so as to signal back to the person who had triggered the first *light beam*. Galileo had previously evaluated the human reflexes and coordination for conducting the experiment and, based on that, concluded that *light* either was ***instantaneous*** or must had been astonishingly fast because the ***measured** time* had been indistinguishable from the human reaction and coordination value. Clearly, to measure the ***finite** speed of light*, much ***longer** distances* were needed, and our solar system would be the appropriate laboratory for the endeavor.

Curiously, the eclipses of 'Io', one of Jupiter's moons that Galileo had discovered with his telescope, would be used to prove the finiteness of *light speed*. The moon Io, as ours, shines because it reflects Sun light, and when Jupiter interposes between the Sun and the moon, we stop seeing Io from Earth. Once Io gets out of Jupiter's shadowing cone, we see it again. During the first years of the 1670-1680 decade, Ole Roemer (1644-1710) studied Io eclipses and discovered that when Earth was closer to Jupiter, the eclipses occurred before, and when the planets were farther away, the eclipses occurred later, and that this discrepancy could be up to about 22 minutes.[11]

Roemer realized that, assuming Newton's laws were correct, this temporal discrepancy could have been explained, had it been precisely the *time interval* that *light* took to traverse a *distance* equal to the diameter of the terrestrial orbit around the Sun (D2-D1 in Figure 13).

With the accuracy available for the measurement of *time* and for the estimated *diameter* of the terrestrial orbit, Roemer's idea implied a value for the *speed of light* of about 210,000 km/s. Despite being substantially lower than the currently accepted value (around 300,000 km/s), it was still so incredibly large that few intellectuals were willing to accept it. It is curious that the human mind could blindly accept an ***infinite** speed* but had reservations to accept a ***finite*** one, simply because it was too **large**! The idea of an ***infinite** speed* was inextricably related, and consistent with, the philosophical belief that all events in the Universe occurred *simultaneously* with our perception (vision) of them.

It took more than 50 years for Roemer's idea to be confirmed by another cosmic phenomenon. James Bradley spent several years studying star *motion* and trying to explain the small elliptic trajectory that all of them traversed throughout the course of the terrestrial year. Finally, in 1728, he explained the

11 Roemer overestimated this time by about 5 minutes.

phenomenon and called it 'stellar light aberration'. Ignoring the minor effects due to the daily Earth rotation and our solar system motion, the *speed* of our planet on its solar orbit combined with the ***finite** speed of light* are responsible for the variation of the *incident angle* between the *star light* and our planet.

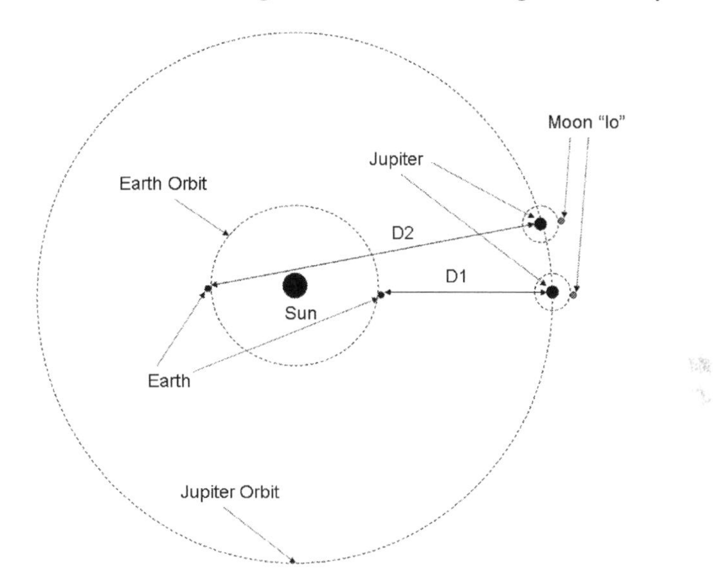

Figure 13. Ole Roemer Explanation for the Finiteness of Light Speed

When there is no wind, if we looked at the rain drops falling vertically from the inside of our car in *motion*, we would see them falling *inclined* towards us; the more ***inclined*** the ***higher*** the car *speed* and the ***lower*** the rain *speed*. Likewise, if we think of *light* as a shower of particles (as Newton did), we will see it coming tilted in the *direction* of our planet *motion* and, hence, we will need to ***change*** the *angle* of our telescopes to track a star as we progress in our annual orbit. If running under the rain, we would tilt our umbrella in the forward direction to minimize our wetting. Given that the needed change in the angle of the telescope depended upon the ratio between the *light speed* and our planet *speed*, and the latter had been estimated with a greater exactitude than in Roemer's time, Bradley was able to calculate the former as 303,000 km/s -- this time very close to the currently accepted value.

More than a century went by until the *speed of light* could be measured with experiments on Earth. Hippolyte Fizeau, in 1849, devised a very ingenious apparatus to measure the *time* it took for *light* to travel to a mirror 8 km away, reflect, and come back to its original place. In 1862, Jean Foucault perfected Fizeau's instrumentation and succeeded in measuring the *speed of light* within the confines of his laboratory. At the same time, he proved that *light speed* in ***water*** was ***lower*** than in ***air***, discrediting Newton's theory for light as a shower of particles (because such a theory demanded that *light* in ***water*** had to propagate ***faster*** than in ***air***). Incidentally, it is worth noting that *sound speed* decreases as we go from ***glass*** (21,500 km/h), through ***water*** (5,400 km/h), to ***air*** (1,200 km/h) but *light*, instead, moves the ***fastest*** in ***vacuum*** (~300,000 km/s), slightly ***slower*** in ***air***, about 226,000 km/s in ***water***, and about 200,000 km/s in ***glass***. Here is a blatantly suggestive difference between the phenomena of *sound* (***mechanical***) and *light* (***electromagnetic***). *Matter* is the *sine qua non* for ***sound*** propagation (the more the better), while *matter* impairs ***light*** propagation.

2.6. Electricity and Magnetism

Thales of Miletus (624-546 BC) discovered that after rubbing a piece of amber[12], it attracted tiny objects. Given that 'electron' is Greek for amber, the word 'electricity' was born to say that the amber had been 'electrified' when rubbed. Later they discovered that two 'electrified' bodies not always attracted each other but that some repelled others, giving birth to two classes of electricity arbitrarily named 'positive' and 'negative'. The phenomena of electric attraction and repulsion were encapsulated in the well-known law saying that charges of the same sign repel and those of opposite sign attract.

It is believed that in the city of Magnesia (Asia Minor), the ancient Greeks also discovered that certain mineral rocks (given the name of 'magnetite') attracted little pieces of iron which, in turn, attracted other pieces of it, and so the tern 'magnetism' was born to refer to a phenomenon which would be studied by the first time also by Thales of Miletus. As with *electricity*, it was soon discovered that the force experimented by the 'magnetized' bodies could be attractive and repulsive, so term 'magnetic pole'was created. Contrary to *electricity*, every *magnetized* body had both classes of pole, and no one of them seemed to exist on its own. In the 12th century, the Chinese had already developed the compass for assistance in navigation, and *electricity* started being studied with a scientific spirit in the 17th century. Nonetheless, *electricity* and *magnetism* were considered as independent phenomena until the 19th century, i.e. before then there was no knowledge of a reciprocal interaction between an *electric* charge and a *magnetic* pole.

Charles-Augustin Coulomb (1736-1806) established the basic quantitative electrostatic laws[13], and conducted many other investigations on *magnetism* and *electricity*. The famous Coulomb Law, as Newton's Gravitation Law, asserts that the ***electrical*** *force* is inversely proportional to the square of the *distance* between the bodies, but now the concept of *mass* for the bodies is replaced by the concept of *electrical charge*, and the *force* can be either attractive or repulsive.

When Alessandro Volta (1745-1827) invented the electric cell, the concept of 'electric current' was conceived and a new discipline was born: *Electrodynamics*, i.e. the study of electric charges in *motion*. Hans Christian Oersted (1777-1851) discovered in 1820 that a compass was affected by a near *electric current*, proving the existence of a relation between *electricity* and *magnetism*. The spontaneous spatial distribution of iron filings around the wire conducting the current showed irrefutably that the *electric current* somehow modified the surrounding *space*, suggesting that -besides the dynamic electric field generated by the charges in motion- it also created a magnetic field similar to that created by a magnet. In the following year, André-Marie Ampère (1775-1836), formulated the quantitative laws which relate the *electric current* with its

12 Hard, light, and brittle yellowish translucent fossil resin.

13 Electrostatics is the discipline which studies phenomena among electrically charged bodies in relative repose.

generated *magnetic field*. The mechanical ***force*** on the compass needle or on the iron filings showed very unique characteristics, radically different from the classical Newtonian *force* between two bodies because: a) besides depending upon the *distance*, it depended on the *intensity* of the *current* (*velocity* of the *charges* in *motion*); and b) it was neither an attraction nor a repulsion, but acted *perpendicularly* to the line joining the needle with the conductor. Later on, Henry A. Rowland (1848-1901) proved that an *electric* charge in *motion* (not necessarily within a conductor) produced a *magnetic field* as well.

The creation of a *magnetic field* by an *electric current* called for pondering whether the reverse would be true: had a *magnet* been near an electric conductor forming a closed circuit, would an *electric current* have been established in the circuit? Joseph Henry (1797-1878) and Michael Faraday (1791-1867) discovered the phenomenon of *magnetic induction*: the mere proximity of the magnet to the circuit was irrelevant but, when *moving* the *magnet* with respect to the circuit, i.e. when subjecting the circuit to a ***variable*** *magnetic field*, then, an *electric current* did appear in the circuit. In summary, a ***variable*** *electric field* (produced by *electric charges* in **motion**) produces a *magnetic field* and, reciprocally, a ***variable*** *magnetic field* (produced by a *magnet* in **motion**) produces an *electric field* (which, if there is a circuit, creates an *electric current*). This interplay between *electricity* and *magnetism* is the phenomenon behind the commercial generation of electric *energy*.

Faraday, who fabricated the first electric motor in 1821, was one of the greatest experimentalists in history and the first scientist who realized the importance of the concept of *electromagnetic field*, as a new physical **reality** at the same level as *matter*. This *field* produced effects on *matter* that were easy to corroborate and, hence, it was hard to accept the *field* as a mere mathematical convenient fiction. For Faraday, when a *current* existed in a circuit, the essence of the phenomenon did not reside in the electrons[14] moving in the wire (tangible), but in the *electromagnetic field* (intangible) existent in the surrounding *space*. If that was true, then -at least in principle- the *electromagnetic field* could exist even if there were **no** *electric charges* or *magnets* in the neighborhood! Before dying, his experimental data and vision were transferred to his colleague and disciple James Clerk Maxwell, who would later present to the scientific community one of the most outstanding theories in history. As you may remember (Chapter 1), due to Faraday's insights and outstanding experimental work, and the astonishing success of Maxwell theoretical framework, we included in the definition of ***physical*** object not only *matter* but also *radiation* (*fields*).

2.7. The Luminiferous Ether -The Culprit is Sound Propagation!

I cannot but warn the reader that many of the upcoming disquisitions may seem -with the benefit of hindsight- illogical or even borderline folly. So as not to fall into that preconception, underestimating the magnitude of the cyclopean epistemological effort undertaken by the best minds of the time, let us reflect -before proceeding- on what William James (1842-1910) said:

A new idea is first condemned as ridiculous and then dismissed as trivial,

14 Regarding tangibility... the wire is certainly tangible but... is the electron tangible?... Let us wait for my next book entitled *What is Reality? Einstein, Quantum Physics, and Folklore*.

until finally, it becomes what everybody knows.

Let us try, then, to temporarily forget the good scientific folklore that we have grown up with by the mere virtue of having been born in the 20th century. As we saw in Chapter 1, Christian Huygens (1629-1695) -contemporary of Newton- had proposed that *light* was an *undulatory* phenomenon with its different colors being *wave components* with different *wavelengths* -- in contraposition to Newton who thought *light* was a beam of material particles. Due to Newton's intellectual stature, Huygens's ideas were not going to be easily accepted.

Even more than a century later, when, in 1803, Thomas Young (1773-1829) experimentally demonstrated that *light **diffracted*** like a *wave*, his ideas were not taken very seriously. Let us remember that wave *diffraction* is a phenomenon whereby the *light* wave, when encountering an object, instead of producing a sharp shadow behind it, surrounds it in the same way as a sea wave proceeds its course surrounding a small boat. The Newtonian theory of a beam of particles could not explain this phenomenon. *Diffraction* and *interference* were the hallmark of *wave* propagation. According to the corpuscular theory, the shadow associated with a coin is easy to understand as *lack of light*, as the coin simply intercepts the *light beam*; in the light of the *wave* theory for *light*, the shadow is instead the result of the ***destructive** interference* between waves surrounding the coin and traveling all the way to the screen where the shadow appears! If it only were for the phenomenon of a shadow, the *undulatory* theory would be just an obviously unnecessary complication. However, even when describing the shadow of a coin, the *wave* theory was superior because it predicted that the dark disk (shadow) would exhibit a luminous point at its center! Such a peculiar prediction obviously could not be matched by the *corpuscular* theory of *light* and, when it was experimentally confirmed, the *undulatory* nature of *light* was solidly established.

Young immensely contributed to scientific knowledge when he experimentally demonstrated the *undulatory* nature of *light* but, ironically, he contributed as well to consolidate the epistemological relevance of the analogy between *light* and *sound* and -given that the latter needed a ***material** medium* for its propagation- to validate the existence of the *luminiferous ether*, omnipresent in vacuum as well as in the substance of all material bodies, as the *propagating medium for light*. Young said (bold type is mine)[15]:

Those who are attached to the Newtonian theory of light, or to the hypotheses of modern opticians, founded on views still less enlarged, would do well to endeavor to imagine anything like an explanation of these experiments, derived from their own doctrines; and, if they fail in the attempt, to refrain at least from idle declamation against a system which is founded on the accuracy of its application to all these facts and to a thousand others of a similar nature.

*From the experiments and calculations which have been premised, we may be allowed to infer, that homogeneous light, at certain equal distances in the direction of its motion, is possessed of opposite qualities, capable of **neutralizing** or **destroying** each other, and of **extinguishing** the light,*

15 As cited in *Great Experiments in Physics*; New York: Dover Publications, 1959.

where they happen to be united; that these qualities succeed each other alternately in successive concentric superficies, at distances which are constant for the same light, passing through the same medium.

*But, since we know that **sound** diverges in concentric superficies and that musical sounds consist of opposite qualities, capable of **neutralizing** each other, and succeeding at certain equal intervals, which are different according to the difference of the note, we are fully authorized to conclude, that **there must be some strong resemblance between the nature of sound and that of light.***

*...And, upon considering the phenomena of the aberration of the stars, I am disposed to believe, that **luminiferous ether pervades the substance of all material bodies** with little or no resistance, as freely perhaps as the wind passes through a grove of trees.*

The *luminiferous ether* was not Young's invention. In Greek mythology, the *ether* (Greek for 'pure air') was breathed by the Gods and even worshipped as a god, with many parallels in other philosophies as Hinduism. According to ancient and medieval science, the *ether* was the substance that pervaded the Universe beyond our terrestrial sphere. Aristotle considered *ether* as the 'fifth element' (quintessence) without physical properties, except for *motion* that had to be in circles. Medieval philosophers, instead, endowed the *ether* with physical properties like density -- that had to be much lower than that of the planets so that the latter could move freely. Robert Fludd (1574-1637) considered the *ether* as something more subtle than *light*, universally penetrating, and immaterial.

Bradley had explained the stellar *light aberration* with Newton's corpuscular theory. My analogy with rain drops falling on my automobile in *motion*, required the assumption that there was no wind, but only to avoid considering an additional effect; in fact, the ***inclination*** of rain fall when using the car as the *reference frame* is a sheer *kinematic* effect, due exclusively to the relative *motion* between rain and car -- and it would have been exactly the same had our planet not had an atmosphere (assuming rain could have existed!). Newton's corpuscular theory was purely ballistic and did **not** require a *propagating medium*. Was the *undulatory* theory -that supposedly did require a *medium*- equally effective to explain stellar *light aberration*?

After our extensive discourse when 'galloping with sound', the reader is very well aware (I hope) of the importance of the *propagating medium* for *mechanical* waves. In addition, the scientific mentality of those days was that every physical phenomenon had to be explicable in *mechanical* terms, so the analogy between *sound* waves and *light* waves proposed by Young made an awful lot of sense. Consequently, the *luminiferous ether* had to exist and be omnipresent, offering absolutely no resistance and, more than anything, had **not** to be *dragged* by *matter* in motion, i.e. to be in ***absolute*** *repose*: if the *ether* was fully *dragged* by our planet in its annual trajectory, there would **not** be any *aberration* of stellar light, for the same reason that the captain voice and the romantic chat inside the airplane on our way to Minneapolis did not experience any aberration -- as long as the *air* was trapped (*dragged* with us) in the aircraft. But there certainly **did** exist *aberration* when we imagined all windows open (air **not** dragged), to the point that the 'I love you' never reached the woman's ear. In the same way astronomers have to tilt their telescopes throughout the year when tracking a star, the lady had to recline her seat to hear she was loved.

Once again: given that the stellar *light aberration* is an indisputable fact, and given the preconception of the *luminiferous ether* as indispensable for *light* propagation, the *ether* has to be in ***absolute*** *repose* and Newton had to be correct in his conception of ***absolute*** *space* and *motion*. At that historical moment, we not only knew our planet *velocity* around the Sun, but we could attempt to measure the *velocity* of our Solar system in the Milky Way... and with it, the *velocity* of the latter in... and with it, the velocity of... in... all the way down to ***absolute*** *space*! Knowing the ***absolute*** *speed* of our planet was equivalent to proving the existence of the ubiquitous *luminiferous ether*. And once we achieved such a herculean feat, would we be content? Or, perhaps we would ask ourselves why in the world was the *ether* the **absolute** *reference*? The science of Physics had been self-inflicting a monstrous migraine that was not going to disappear with a couple of aspirins; only brain surgery would do the job! Hold on... and you'll see.

2.8. Electromagnetic Ether = Luminiferous Ether - The Culprit is Sound Propagation!

As we saw in Chapter 1, Maxwell, in 1865, mathematically encapsulated all the known laws relating *electricity* with *magnetism* in his field equations, and predicted the phenomenon of *electromagnetic radiation*: if a physical system called transmitter produced an *electric field* ***variable*** in *time*, a *magnetic field* would appear in its vicinity ***variable*** not only in *time* but also in *space*, and this *magnetic field*, in a sort of inert symbiosis, would strengthen the *electric field* which, in turn, would do the same with the *magnetic field*, and so forth *ad infinitum* -- with the result that the combined *electromagnetic field*, once generated, would exist independently of the transmitter, propagating with a fixed *speed* that Maxwell was able to calculate from a couple of already-existing laboratory measurements.

But there was something *sui generis* in Maxwell's equations, when compared to the equations for *sound* propagation: in the latter equations, there was no doubt that the mechanical wave *speed* was defined by, and conceived as relative to, the ***material*** propagating *medium*; Maxwell's equations, instead, included a magical number as the wave *speed*, without any hint as to what *reference frame* was supposed to be used for that number to be valid. Even worse: as you may remember, when *Newton's Laws of motion* were expressed in different *Inertial Frames* (applying *Galileo's Transformation*), those laws transformed into themselves, i.e. they were ***covariant***, while Maxwell's equations were **not** ***covariant***, i.e. when transformed from one *Inertial Frame* to another with *Galileo's Transformation*, the *structure* of the equations **changed**!

Given two *Inertial Frames*, if Maxwell's equations were assumed valid in one of them, applying *Galileo's Transformation*, the new equations (supposedly valid in the new frame) contained terms which depended upon the ratio between the relative *speed* of the *Inertial Frames* and the magical number.

In short, Maxwell's equations were not Maxwell's any longer. This mathematical fact suggested that the *laws of electromagnetism*, as opposed to those of *mechanics*, were only valid in a privileged *Inertial Frame*. Would this be, perhaps, an indication that Newton was right? Would this unique *Inertial Frame* -in which Maxwell's equations were valid- be the so wanted but elusive ***absolute*** *space*? If natural laws changed with the *Inertial Frame*, the same experiment conducted in ***different*** *frames* would give ***different*** results and voila! -- we could prove, without ***external*** *reference*, the ***absolute*** *motion* of the *frame*.

On the other hand, in consonance with *sound* propagation, and in opposition to Newton's ***instantaneous*** *action at a distance*, Maxwell's equations clearly indicated that electromagnetic action was **not** *instantaneous*, but propagated through the *electromagnetic ether* with a ***finite*** *speed*. When Maxwell calculated this propagation *speed*, he was astonished to find out that the value was practically identical to the known *speed of light* in its *luminiferous ether* so, at a stroke, he concluded that it was too much of a coincidence, unless *light* was also an *electromagnetic wave* and the *electromagnetic* and *luminiferous ethers* were one and the same thing -- unifying the scientific realms, independent up until then, of *electromagnetism* and *optics*. From now on, in the very few equations that we will present (only as points of epistemological reference), the *speed of light*, and of any other *electromagnetic wave* in *vacuum*, will be represented with the symbol 'C'.

Almost a decade after Maxwell's death, Heinrich Hertz (1857-1894) -a disciple of Helmholtz- experimentally confirmed (in 1887) Maxwell's theory by generating what we call *radio* waves, and proving that they had all the known properties of *light* (reflection, refraction, diffraction, interference, etc.) except -of course- the property of being visible to us. In 1901, Guglielmo Marconi transmitted the letter 'S' in Morse code from England to North America. Later on, it was proved that -reciprocally- *light* had all the properties predicted by Maxwell for *electromagnetic waves*, e.g. the *mechanical* pressure a wave exerts when impinging on *matter*[16]. Beyond doubt, the only difference between Hertz's waves and *light* was their *wavelength*.

Young's *luminiferous ether*, inspired in the necessary *air* for *sound* to propagate, became simply the *ether*, i.e. the ***privileged*** propagating *medium* for *electromagnetic waves* -- from *gamma* rays (wavelength of picometers), through *visible light* (hundreds of nanometers) and *microwaves* (several centimeters), up to *radio* waves (about three meters for FM and 300 meters for AM), and beyond. According to Maxwell, *electromagnetic waves* propagated in the *ether*, which was in ***absolute*** *repose*. But *light* also propagated through *matter* and, in spite of the magnificence of the theory, it had some deficiencies when explaining wave propagation through *matter* and, in particular, when *matter* was in *motion* relative to the *ether*. The discipline of *Electrodynamics* was about to be born.

2.9. The Ether Hunt and the Absolute Motion of our Planet

Using the *Principle of Newtonian Relativity*, I concluded that it was impossible, without ***external*** *reference*, to know the *speed* with which an *Inertial*

16 The tail of a comet moves farther away from the Sun due to the pressure exerted by the Sun light.

Frame moves with respect to another and, hence, assuming that one is in *repose* is as valid as assuming that it moves with any other ***constant** speed*. Consequently, considering the *Inertial Frames* 'Earth' and 'absolute space', there is no mechanical experiment we can undertake on our planet that will allow us to determine the *velocity* with which humanity moves as a whole in ***absolute** space*.

But my reflections during my flight from New York to Minneapolis suggested that when, instead of *matter*, we are dealing with *radiation* as *sound*, assuming that *air* is not the blessing of a *sui generis* planet as ours (that can be trapped inside a Boeing 727), but that -instead- is universally penetrable and penetrating, omnipresent, and the epitome of ***absolute** repose*, then it was clear (by imagining all windows open) that, due to the ***wind*** created by relative *motion*, the *speed of sound* inside the plane would depend upon the aircraft's *motion* in that imaginary *medium*. Now, put the *ether* of the 19th century in place of this imaginary 'air', *light* in place of our *voice waves*, and our *planet* moving at 30 km/s (100 times faster than a Jumbo 747) in place of the Boeing 727, and we have a laboratory called 'Earth' on which we could finally prove our planet ***absolute** motion*.

The airplane *speed* affected the romantic conversation (***transversal***) and the captain voice (***longitudinal***) differently, and the effect on the latter was different when the speaker was at the prow (***tail*** wind) or in the back (***head*** wind) of the aircraft. Similarly, thus, the results of well-thought ***optical*** experiments conducted on our planet would be different when *light* propagated in different directions with respect to our planet *motion* in ***absolute** space*.

In 1818, Francois Arago (1786-1853) tried to measure the ***absolute** velocity* of our planet by measuring the ***refraction*** of *light* through a glass prism. As for the meaning of *refraction*, remember that when *light* passes from a medium (like air) to another (like water or glass), its *direction* changes -- as it is easily confirmed inserting a stick in water. This change of *direction* occurs because *light* propagates with ***different** velocities* in ***different** media*. The ratio between the *speed of light in vacuum* and its *speed* (lower) in a given ***material** medium* is called the *index of refraction* for that material. Because our planet was his laboratory, Arago's prism was *moving* in ***absolute** space* with the same *velocity* as our planet. It is easy to put myself in Arago's shoes because, coincidently, while I was writing the Spanish edition of this book, I had on my work bench -under development- the prototype for a new refractometer which employed a sapphire prism.

It is not difficult to see that, if *ether* is in ***absolute** repose* and the *index of refraction* for glass is 1.5, then the *speed of light* in glass (when in ***absolute** repose*) will be 2/3 (1/1.5) of the *speed of light in ether*. But... what about when the prism is *moving* with the Earth through ***absolute** space*? When *light* travels in the **same** *direction* as the Earth, the *speed of light* relative to Earth will be ***lower*** than in *vacuum* and equal to its ***absolute*** value minus the Earth ***absolute** speed* (both inside and outside the prism). *Mutatis mutandis*, when *light* and Earth travel in **opposite** *directions*, the *light speed* relative to the planet will be ***greater*** because the ***absolute*** Earth *speed* will be added. When planet and *light directions* make ***different** angles*, we will measure ***different** speeds* for *light* and, ergo, ***different*** *indexes of refraction* for the **same** material. In short, my measurements of the *index of refraction* for a given material would change when, capriciously or inadvertently, I would orient my prototype differently on the bench! Furthermore, even if I kept my refractometer anchored to the bench (and the bench anchored to ground), my measurements would vary along the year as we progressed on our trajectory around the Sun!

As we know by now, Arago suffered an astronomical disappointment, and had to conclude that, at least within the accuracy of his experiments, the *motion* of our planet had no effect on the prism *index of refraction* and, perplexed, he told his friend Augustin-Jean Fresnel (1788-1827), who developed a theory as strange as effective to explain the phenomenon. According to Fresnel, *ether* was ***stationary*** (in agreement with Maxwell), *light* propagated as a *transversal* wave in *ether*, and matter ***partially*** *dragged* the *ether*. The *transversal* character of *light* explained the phenomenon of polarization (already known by Newton): interposing two films of a 'polarizing material' *perpendicularly* to the *light beam*, and rotating one with respect to the other, light transmission through them could be gradually varied up to its complete blockage. This fact suggested that the 'undulations' had to take place on a plane *perpendicular* to the *light direction* of propagation.

And why was Fresnel's theory strange? Because, on one side the *ether* was ***stationary***, but on the other it was ***dragged*** by the *matter* in *motion*. Remember our promise of trying to understand the scientists of that time without the benefit of retrospection. The theory postulated that transparent materials trapped different quantities of *ether* which, in turn, was *dragged* by the material when moving without disturbing the ***external*** *ether* (omnipresent and in ***absolute*** *repose*). This was almost perfectly analogous with the ***internal*** and ***external*** *air* during my trip to Minneapolis except that, because the *ether* was omnipresent and the *air* of course is not, the ***external*** *ether* was also ***internal*** (?). The extra quantity (trapped) of *ether* depended on the material *index of refraction*, i.e. there was more *ether* trapped in *glass* than in *air* and a little more in *air* than in *vacuum*. The theory became even weirder because the *index of refraction* is different for different *wavelengths*[17] (colors), so the amount of trapped *ether* had to change with the *light* color.

When *light* entered the *moving glass*, Fresnel predicted that *light* ***absolute*** *velocity* would be the sum of its *velocity* when the *glass* was in ***absolute*** *repose* plus a term which was the result of the coexistence of both ***external*** and ***internal*** *ethers*. This magic term was ***smaller*** than the ***absolute*** *velocity* of the glass (*speed* of Earth), i.e. it violated the *Galilean Composition of Velocities*! From this hypothesis, Fresnel developed a formula that predicted (with good accuracy) the speed of light in transparent bodies in *motion* and, at the same time, explained the negative results of Arago, and the *aberration* of stellar light. The ***lower*** the *index of refraction* of the material in *motion*, the ***less*** extra *ether* was trapped up to the point where, if the body disappeared, only the *ether* remained, the magic term became ***zero***, and the composed *velocity* was simply the known *speed of light in ether*.

The problem was that, with the mechanical mentality of the 19th century, if

17 Otherwise, the rainbow would not appear after the rain.

the 'undulations' were *transversal*, then the *ether* had to have the mechanical *rigidity* of a solid and, in such a case, how could it be so 'ethereal' as to not offer any resistance to the motion of celestial bodies? We know that *air* cannot sustain *transversal* mechanical vibrations precisely because of its lack of *rigidity* (sound is a *longitudinal* wave); on the other hand *air* friction is capable of incinerating a body entering the atmosphere, so the *ether* has to be even more 'ethereal' than *air* but then, it could not sustain *transversal* waves! Perhaps light was not ultimately *mechanical* as people believed. What was it, then?

George Gabriel Stokes (1819-1903), in 1845, developed a theory whereby the *ether* was mechanically schizophrenic, as it had to be *rigid* to transmit a *transversal* wave as *light*, and *fluid* to allow for the free *motion* of celestial bodies; besides, the *ether* was completely *dragged* by *matter* in *motion*. A total *drag* of the *ether* would be equivalent to our pumping the *air* inside the aircraft so as to reproduce the conditions in *terra firma*, in which case the *Principle of Newtonian Relativity* would -supposedly- assure the validity of the *Galilean Composition of Velocities*. In the mean time, the successes of Fresnel's theory had been piling up; the famous experiment of Hippolyte Fizeau en 1851 denied the validity of *Galileo's composition* and added to Fresnel's success. Fizeau's interferometer indicated that the *speed of water* in the pipe did not fully add to the *speed of light in water* and, then, the *water stream* could **not** completely *drag* the *light wave*. This partial *drag* was an effect of the interaction between *matter* and *ether*. Things happened as if the ***internal*** *air* -in which the captain's voice and the 'I love you' waves propagated- *moved* at a *speed* intermediate between that of the airplane and that of the ***external*** *wind*, all with respect to *terra firma* -- instead of being trapped inside the aircraft, i.e. *in repose* with it. But Fresnel himself was conscious of the complexity and limitations of his theory, and indicated that the explication of all these phenomena had a limit, and that sufficiently accurate experiments would not probably be adequately explained.

2.9.1. The Michelson-Morley Experiment

Maxwell had considered measuring the ***absolute*** *velocity* of our planet employing the variations of the eclipses of Io. In a letter to the American astronomer David Peck Todd -published in the scientific magazine *Nature*- Maxwell explained that measurements of *light speed* in a terrestrial laboratory had to be affected by our planet *motion*, but the effect was too small to be easily detected, because it would require measuring an *interval of time* in the order of a ***quadrillionth of a second***! Albert Abraham Michelson (1852-1931) (who would receive the Nobel Prize in 1907) read the letter and, far from being discouraged, he decided in 1881 to measure the 'ether wind', i.e. the Earth *motion* in the ether, using an extremely precise *interferometer* of his own design.

The instrument was based on the tremendous value for the *speed of light* (about 30 cm in a billionth of a second) and the small value of its *wavelength* (400 nm-700 nm). As Maxwell had pointed out, an *interval of time* of a quadrillionth of a second was impossible to measure directly, but the *distance* traveled by *light* in that *time* is in the order of its *wavelength*, and we already learned that the phenomenon of *interference*, i.e. the amplification or annulment of *light* produced by two waves after traveling different *distances* (or at different *speeds*) can be easily measured. Consequently, such a small *time interval* could be ***measured*** through *interference* and all it was needed was to figure out how to force *light* to traverse paths in which, due to the *angle* between those paths and our planet *direction* of *motion*, *light* would travel at ***different*** *speeds* and, ergo, reaching the detector at ***different*** *times*. The instrument sensitivity was so high that during Michelson's first experiment at the Institute of Physics of Berlin, its results were affected by the traffic on the street and had to move his instrumentation to the Astrophysical Institute of Postdam, where he set up his instrumentation underground. Even so, a person hitting the floor a hundred meters away disturbed the instrument readings.

Michelson's results were negative and interpreted as a confirmation of Stokes' theory, i.e. the *ether* was fully dragged by our planet *motion* and, hence, Earth and ether were in relative *repose*. However, as we saw, *ether drag* by the planet did explain neither star *light aberration* nor the violation of the *Galilean Composition of velocities* in Fizeau's experiment. As soon as he read Michelson's results, Mach did not hesitate to suggest the abolition of the notion of *ether*. However, Hendrik Lorentz, in 1886, proved that Michelson's calculations, as far as the conclusion of a full *ether drag* was concerned, was flawed, and concentrated in developing a theory more in consonance with Fresnel's. Meanwhile, in 1887, Michelson and Edward Morley repeated Fizeau's experiment confirming Fizeau's drag coefficient, which compelled Michelson to change mind and start believing in Fresnel's theory and the existence of an 'ether wind'.

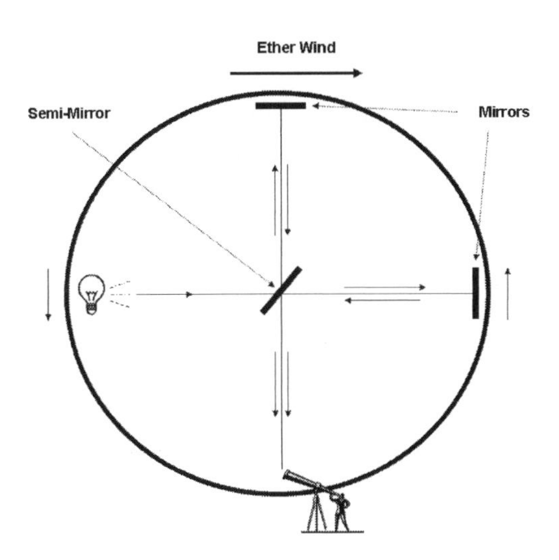

Figure 14. The Experiment of Michelson-Morley

Michelson went back to business and, now together with Edward Morley, conducted the crucial experiment known as the Michelson-Morley's experiment in Cleveland , Ohio, USA (1887), using a very refined version of his *interferometer* with even higher sensitivity and exactitude. Figure 14 depicts a schematic version of the apparatus which floated on a mercury bath so as to minimize the effect of mechanical vibrations and facilitate its rotation (to change its orientation relative to Earth *motion* in the *ether*). The *interferometer* had two ***perpendicular*** arms, one light source, one half-mirror[18] at 45° with the light beam so as to divide it in two beams traveling through the two arms, two mirrors ***perpendicular*** to the respective beams, and a telescope to observe the beam after having been split, traveled eight times through the arms, and recombined at the very end. The bands of *light* and *shadow* typical of *interference* were accurately observed as the instrument was rotated. Given that both arms had the **same** *length*, any variation of the *speed of light* relative to the laboratory due to the *ether wind* would manifest in a *difference* in the *time* that *light* took to traverse both paths and, then, they would arrive at the telescope with a *phase difference* (its crests and troughs not coinciding).

If the orientation of one of the arms coincides with the Earth *direction* of *motion* in the *ether* (horizontal in Figure 14), we have the case of the pilot on the plane to Minneapolis -with all its windows open- giving the weather report. The case where the *light beam* travels *horizontally* from the half-mirror towards the right is analogous to the speaker located at the prow of the aircraft (***tail*** wind); the beam returning to the half-mirror is analogous to the speaker located at the stern of the plane (***head*** wind). The *light beam* traveling through the other arm (vertical), instead, is in a similar situation as the sound wave emitted by the lovers on the plane, and it can be proved (if we accept classical mechanics as valid) that the *round trip time* should be ***shorter*** than when the beam travels *horizontally*. In any other intermediate orientation of the apparatus (with respect to Earth *motion* in *space*), there would exist intermediate effects. If it was true that *light* was a phenomenon entirely analogous to *sound*, the *times* taken by the light beams would be ***different***; besides, this *difference* would ***change*** with the apparatus rotation and, regardless of how small it was, the phenomenon of *light interference* would magnify it, and the instrument would display it. Furthermore, that particular orientation for which the instrument showed the ***maximum*** phase *difference* when the beams recombined, would pinpoint the ***absolute*** *direction* of our planet *motion* in ***absolute*** *space*!

Michelson's disenchantment was again of galactic proportions. Regardless of the apparatus *orientation*, or the *time* of the day, or the *day* in the month, or the *month* in the year, both luminous beams reached the telescope ***in phase***, indicating that the *speed of light* with respect to our planet was independent of the *direction* of *light motion* and the *direction* of Earth *motion* in the supposed *ether*. Due to the astonishing accuracy of the instrument, this time its results were beyond dispute and clearly indicated that the *ether* either did **not** exist or was fully ***dragged*** by the Earth in its orbital trip around the Sun, so that it would never be detected. But, again, upon a complete *drag* of the ether, not only the *aberration* of stellar light would not occur, but the *Galilean Composition of Velocities* had to be valid -- something that Fizeau's experiment had ***disproved***.

Even so... as it usually happens in politics and religion, though very rarely in science, there was a third option for those who were not willing to put up with

18 'Half-mirror' because its silver coating is such that the light beam splits in two, one reflecting as from a regular mirror and the other going through.

brain surgery: YES, *ether* **did** exist; YES, *ether* **was** in *absolute repose*; but... there was a cosmic conspiracy, an exquisite compensation, a Machiavellian plan of Nature, that prohibited us from observing and ***measuring*** the wind that the *ether* produced due to our planet ***absolute*** *motion*. Leibniz's *Principle of the Identity of Indiscernibles* was again at play.

Hendrik Lorentz, a physicist with remarkable mathematical skills, came to the rescue explaining out the negative results of the *ether hunt*, while at the same time keeping the existence of the *ether* in ***absolute*** *repose*. The grand cosmic conspiracy, expressed through his *Principle of Correlation*, was pronounced with the same singular mastery with which it would be demolished by Einstein in 1905.

2.10. The Grand Cosmic Conspiracy - Lorentz's Principle of Correlation

En 1890, Hertz and Oliver Heaviside extended Maxwell's theory establishing the foundation for *Electrodynamics*. Hertz (as Stokes) believed that bodies in *motion* fully ***dragged*** the *ether*, and fervently defended the *mechanical* vision of *electromagnetism*. At the same time, as a result of the increasing frustration with the botched *ether hunt*, the *electromagnetic* vision of *mechanics* was increscent. As we pointed out, Fizeau's experiment and many others were incompatible with a full *ether drag* by our planet. The only successful theory was Fresnel's, but with many inconsistencies and lacking a rationale for the capricious ***partial*** dragging of *ether* in *matter*. At this moment in history, Lorentz concentrated in refining Maxwell's equations and extending them for the case where the *electromagnetic field* existed inside *matter* -- whether this *matter* was *in repose* or *in motion* relative to the *ether*. As a result of this effort, the successful electronic theory of *matter* was born by virtue of which electricity is discrete, with its basic unit being the charge of the *electron*, and the latter being one of the basic components of *matter*.

In this theory, if *matter* was dielectric (non-conductive of electricity), its *electrons* were bonded to the atoms by elastic forces. When *light* traveled through a dielectric material, *electrons* would vibrate because of the electromagnetic field. This interaction between *light* and *matter* caused a delay and attenuation of *light* intensity, which changed with *frequency* (color), explaining *light refraction* (e.g. in water drops when the rainbow forms). The selective *electron* resistance to vibrate in unison with the *electromagnetic field* determined whether *matter* was transparent or opaque.

If *matter* was conductive (as metals), its *electrons* were weakly bonded to its atoms so that the *electromagnetic field* freed the *electrons* from the atoms initiating an *electric current*. Not being elastically attached to the atoms, the *electrons* were incapable of vibrating in unison with the field and *light* could not go through. In this fashion, ***dielectric*** materials were ***transparent*** and poor ***conductors***, while metals were ***opaque*** and very good ***conductors*** of *electricity*.

In opposition to Hertz, Lorentz assumed that the *ether* was in ***absolute*** *repose*, and never disturbed by *matter*. Besides, as in Maxwell's original theory, the ***absolute*** *motion* of a *light source* did not affect the *speed of light in ether* (as is the case with *sound* in *air*). With his electronic theory, Lorentz was able to explain *light* ***refraction***, ***absorption***, and ***reflection***, as well as the ***color*** of *matter*, and many other phenomena. The most refined experiments confirmed Lorentz's theory, and when the *electron* was finally isolated and its charge measured, the *electronic theory of matter* became, and still is, indisputable. Not only did it explain many well-known phenomena but it also predicted many others to be discovered.

In summary, due to the multiple successes of the electronic theory, there had to be a very solid reason for opposing its fundamentals. Fresnel's convection coefficient (*ether* ***partial*** *drag*) was also explained by the theory, though the rationale had nothing to do with a real partial drag of the *ether* (it was, instead, immutable) but with the electronic constitution of *matter* in *motion*. The only problem was that this phenomenon, obvious in Fizeau's experiment and others -like in Fresnel's theory- was explained only approximately -- leaving no doubt that, upon conducting experiments with sufficient accuracy, the ***absolute*** *motion* of our planet would be detected. As Maxwell had predicted, Lorentz acknowledged that it was a matter of accurately measuring extremely short *periods of time*, and the ***absolute*** *space* and *motion* would be detected.

But, for Lorentz's dismay, and that of many others, the experiment of Michelson-Morley had accurately measured the ***hundredth of a quadrillionth of a second*** and the light beams arrived *in unison* (in phase) regardless of the apparatus orientation proving that the *speed of light* was always the **same** with respect to Earth as with respect to the assumed ether... and that could only imply that the *Galilean Composition of Velocities* was miserably failing!

Both George Fitzgerald in 1889, and Lorentz independently in 1892, argued that if all bodies ***contracted*** in the *direction* of *motion* through *ether* in an amount that depended upon the ratio between their *speed* and that of *light* in *ether*, then Michelson-Morley's negative results would be fully 'explained'. Why is that? Because light, having to travel a ***shorter*** *distance* (due to the ***contraction***) in that *direction*, would take a ***shorter*** *time* so that our measurement of its *speed*, though ***different*** 'in reality' (as the *Galilean Composition of Velocities* demanded), would always be the same as in the other arm of the instrument (which did not contract).

And here is the power of the idea and also its vulnerability to epistemological critique: as the *rule* to measure the ***contraction*** was also *moving* with the apparatus (and the whole planet), the *rule* would also ***contract*** in exactly the same proportion and, then, remembering our discussions in Chapter 3, it would be impossible to detect such a change in *length*! Even all of us would ***contract*** in the *direction* of our planet *motion*! Only an individual located in a *reference frame* in *repose* with respect to the *ether* could observe our 'real' physical ***contraction*** (lucky him!). Our fear in Chapter 3 of the Almighty contracting the Universe during the night, without us being able to detect it, would have become a reality. However, instead of doing Her magic during the night, He was doing it permanently -- though only in the *direction* of our planet *motion*! Bizarre theory. If this is not an epistemological crisis, I do not know how to call it.

A much simpler explication would be proposed by Walter Ritz (1878-1909) and others, even after the famous and revolutionary publication of Einstein in

1905. According to them, *light* was something in between a *wave* and a *bullet*: it propagated as a *wave* but its speed was not defined by the *medium*, but by the ***source***. Clearly, this theory would neatly explain Michelson-Morley's results because, being *light speed* relative to the ***source***, and being the latter *in repose* with the planet, all the measurements would provide the same result. However, as with *sound*, there existed similar experiments for *light* that proved such a hypothesis was unsustainable.

Returning to Fitzgerald/Lorentz's ***contraction***, everything seemed to indicate that it was an *ad hoc*[19] (not verifiable) mental makeshift with the only purpose of explaining Michelson-Morley's non-intuitive results and, consequently, it could only be accepted if confirmed by a variety of independent experimental evidence. Experimental work with β *radiation* (electrons) emitted by the chemical element Radio seemed to confirm the ***contraction*** of the *electron* with respect to the experimenter (i.e. to Earth) but, in the Michelson-Morley experiment, the bodies which supposedly contracted were *in repose* with respect to Earth.

But if the ***contraction*** was **real** and due to Earth *motion* in *ether*, then -even though it was undetectable using a rule- there had to be some physical property of bodies *in repose* on Earth (but in *motion* in *ether*) that ***changed*** when their 'real' *length* ***changed***. For instance, when we ***compress*** a transparent body, its *index of refraction* changes in the direction of compression. Various experiments were designed to detect such a ***change*** of optical properties, and all of them miserably failed. The experiment of Trouton-Rankine was conceived to detect the attraction between two electrically charged bodies in repose on Earth, on the assumption that those two *bodies* in relative repose were *moving* (with our planet) through *ether* and, ergo, constituted a couple of *electrical currents* in ***absolute*** *space* which should be attracted to each other. Results: calamitous failure.

Wrapping up: the Fitzgerald/Lorentz ***contraction*** had to be **real** in order to explain the experiment of Michelson-Morley but, no matter how **real** we thought it was, Nature did not allow us to ***measure*** it, not only by *mechanical* means but by *optical* and other *electromagnetic* means. The same had happened with all experiments conceived to detect the planet ***absolute*** *motion*; the Michelson-Morley experiment was *optical* but those of Trouton-Rankine and Trouton-Noble were non-optical electromagnetic, and all of them had tragically failed.

Lorentz then accepted what seemed inevitable, concluding that the Cosmos seemed to conspire so that humanity could never detect our ***absolute*** *velocity*, and expressed his frustration through his *Principle of Correlation*: ***independently of the accuracy and refinement of our experiments, exquisite adjustments occurred, and shall occur in Nature so that the effects of our planet motion through absolute space never were or will be observed.*** Likewise, Poincaré, in 1905, mentioned Michelson-Morley's negative results, and asserted that "this impossibility of proving absolute motion seems to constitute a general law of Nature".

19 If it was *ad hoc*, it was so only -as Adolf Grünbaum points out- from the psychological point of view, because, in 1932, the Kennedy-Thorndike experiment proved that the contraction was, in fact, **not** real.

But Lorentz was not a mediocre physicist and mathematician who would content himself with adopting grandiloquent postures; on the contrary, he was brilliant, and dedicated his efforts to determine -with mathematical precision- the specific structure of those delicate adjustments needed for his Principle to be valid -- so that the ***absolute** motion* of our beautiful planet could remain undetectable by us for eternity.

2.10.1. Lorentz's Transformation vs. Galileo's Transformation

Let us remember that the rationale behind the mechanical equivalence of all *Inertial Frames* was that Newton's laws remained ***structurally* identical** when transformed -using the *Galilean Transformation*- from one *frame* to another. As a reminder, I transcribe the *Galilean Transformation* for the simple case of *motion* along the X-coordinate:

$$X_B = X_A - V_{A/B}.T \qquad Y_B = Y_A \qquad Z_B = Z_A \qquad T_B = T_A = T$$

Lorentz knew that for SC-A and SC-B to be equivalent for *electromagnetic* experiments (i.e. for its Principle of Correlation to be valid), if he applied the *Galilean Transformation* to the *laws of electromagnetism* (Maxwell's equations), they had to transform into themselves; otherwise, experiments in different *Inertial Frames* would give different results and: voila! The ***absolute** motion* of our planet would be finally detected. But the crude reality was that those *laws of electromagnetism* -in contraposition to those of *mechanics*- were **not *covariant*** under the *Galilean Transformation*, but ***changed*** in their *structure*, giving a new *speed of light* in the new *reference frame*. It was precisely this disturbing fact that had catapulted the infamous and fruitless *ether hunt*.

Only two circumstances could explain the devastating failure of the *ether hunt*: either a) Maxwell's equations were wrong; or b) the *Galilean Transformation* was wrong. The former had been experimentally confirmed with extreme accuracy; it only remained the sacrosanct *Galilean Transformation*, whose validity had been already challenged by the experiments of Arago and Fizeau. Lorentz, without doubting the validity of the *Galilean Transformation* in the realm of *mechanics*, decided to investigate how he had to change it so that Maxwell's equations could remain ***invariant in structure*** after transforming them from one *Inertial Frame* to another. The result of his research was the *transformation* that carries his name in Relativity Theory. The *Lorentz's Transformation* for the simple case of *motion* along the X-coordinate is as follows:

$$X_B = \left(\frac{1}{\beta}\right)(X_A - V_{A/B}.T_A); \qquad Y_B = Y_A; \qquad Z_B = Z_A;$$

$$T_B = \left(\frac{1}{\beta}\right)(T_A - V_{A/B}X_A / c^2) \qquad \beta = \sqrt{1 - V_{A/B}^2 / c^2}$$

The reader should not be disheartened by the apparent complexity of this transformation when compared to Galileo's. Let us start by understanding their differences from the intuitive standpoint. First, notice that the *speed of light in vacuum* (c) appears in the formulae regardless of whether *light* intervenes or not in the phenomena we are studying. Regarding the relation between the X-coordinates (X_A and X_B), there is a coefficient (β) that multiplies the term that was previously interpreted as the *distance* to the origin in SC-B and calculated as the *distance* to the origin in SC-A minus the *distance* between SC-A and SC-B after the common *time* $T_A = T_B = T$. This equation was valid because, for Galileo and Newton, the *distance* between two *points* at a given *instant* was ***invariant***, i.e. the same whether measured in SC-A or in SC-B.

In *Lorentz's Transformation*, instead, the *distance* between two *points* did **not** seem to be ***invariant***, unless β was equal to 1. But, analyzing a little the values that can take, we see that, if the *relative speed* between SC-A and SC-B ($V_{A/B}$) is much lower than c, $V_{A/B}^2/c^2$ is practically zero, the radicand is practically equal to 1, and so is β, obtaining Galileo's relation. Figure 15 shows that β decreases from 1 to zero as the ratio between the *frames relative speed* and that of *light in vacuum* increases ($V_{A/B}/c$). It is also remarkable that the latter ratio for a car, an airplane, the space shuttle (using Earth as reference), and even for our planet (using the Sun as reference), is so small that, for all practical effects β is equal to 1. Only when the ratio approaches 10% of the *speed of light in vacuum,* is when β becomes appreciably less than 1 and, $1/\beta$ begins to increase.

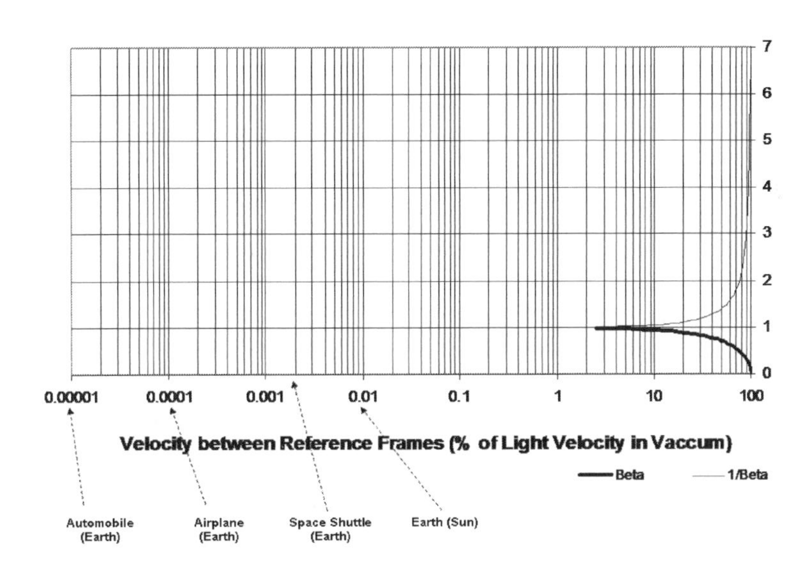

Figure 15. Values of β for different relative speeds (in % of light speed in vacuum)

Fascinating! All those cases in which *Galileo's Transformation* had been confirmed, corresponded to *velocities* much ***lower*** than that of *light in vacuum,* and *Lorentz's Transformation* reduced to Galileo's for those cases. At the same time, those cases in which *Galileo's Transformation* had been found invalid, corresponded to the world of subatomic particles (as the electron) moving at speeds comparable to that of *light in vacuum*, or to situations where *light* itself participated (as in Fizeau's experiment with water and light). Now it was not difficult for Lorentz to prove that a *segment* measured in SC-A (*ether* in ***absolute*** *repose*) would be ***contracted*** with respect to the same *segment* measured in SC-B (Earth) by exactly the amount that was needed to explain Michelson-Morley's experiment. Given that ***absolute*** *space* (ether) was supposedly the privileged *reference frame* from which the ***absolute*** **reality** was to be observed, the FitzGerald/Lorentz ***contraction*** was for Lorentz **real** and measurable in SC-A but impossible to be so in SC-B (where we human beings inhabit). The factor β was precisely the fraction of that 'real' length by which every object/subject on Earth (including itself) would contract by virtue of our motion throughout absolute *space*. If we assumed that the Sun was fixed in ***absolute*** *space*, then our planet ***absolute*** *speed* would be 30 km/s, i.e. 0.01% of the light speed in vacuum (100x30/300,000) and β would be 0.999999995 ($\sqrt{1-0{,}0001^2}$) and because our planet has a diameter of about 12,000 km, a privileged observer *in repose* with ***absolute*** *space* would see the Earth ***contracted*** in about 6 cm ((12.000x(1-0,999999995)=0,00006 km=6 cm).

However, if the ***contraction*** was **real**, it was difficult to understand why it only depended upon the body *speed in ether* and not on mechanical properties such as its *rigidity*. The *ether* seemed to 'decide' it had to apply a bigger *force* the more *rigid* the body was; otherwise, the latter would not ***contract*** always by the same factor for a given *speed*. A quarter of a century later, the Kennedy-Thorndike experiment, even though it was conceived to prove the *dilation of time* (Chapter 6), would irrefutably demonstrate that the ***contraction*** could **not** be **real**, by essentially repeating the experiment of Michelson-Morley, but with one arm much ***longer*** than the other.

In respect to the temporal coordinates in *Lorentz's Transformation*, we see that, besides β appearing again, both *temporal* and *spatial* coordinates in SC-A play an important role to determine the *temporal* coordinate in SC-B. Given an *event* occurring in SC-A at time T_A, had we wanted to know the *time* at which the same *event* happened in SC-B, we would have needed to know -besides T_A and β- the place in SC-A where the *event* occurred (i.e. X_A). As with the *spatial* coordinate, if $V_{A/B}$ is very small with respect to c, then β is practically equal to 1, the term subtracted from T_A is practically zero and the relation transforms into Newton's absolute time: $T_A = T_B$. It was crystal clear then that *Lorentz's Transformation*, necessary to justify *Lorentz's Principle of Correlation*, was reduced to *Galileo's Transformation* when the *speed* of the *reference frame* relative to *ether* was much lower than the *speed of light in vacuum*.

But what Lorentz discovered about the transformation of the *temporal* coordinate was a pill much harder to swallow for a physicist at the end of the 19th century, than what he discovered about the *spatial* coordinates. For his *Principle of Correlation* to be valid, there had to exist two *times* (T_A and T_B), one for the immutable ***in-absolute-repose*** *ether* and another for us, miserable human beings! Nothing closer to scientific heresy than to question the ***absoluteness*** of *time*. How could it be that, once the clocks in both *frames* were synchronized, the *time* in SC-B would not be the same as the *time* in SC-A? And not only that: that *difference* between both *times* changed with the *speed* with which SC-B *moved* relative to SC-A! Lorentz could not accept that both T_A and T_B could be real in both SC-A and SC-B, so he defined the concept of 'local time' for a *reference frame* in *motion* with respect to *ether* as a mathematical artifact only valid for *electromagnetic* experiments -- but he never accepted that his 'local time' could be the *time* ***objectively measured*** and psychologically perceived by the human beings living in such world in ***absolute*** *motion* that he thought our planet was. Joseph Larmor had arrived at the same notion of 'local time', and he did not attribute to it any physical **reality** either. Admitting that the *time* we experienced on Earth was as **real** as, though **different** from, the ***absolute*** *time*, was equivalent to admitting that the *ether*, besides ***contracting*** bodies, ***retarded*** clocks!

Finally, in the same way as the *Galilean Transformation* implied *its Galilean Composition of Velocities* between SC-A and SC-B, the *Lorentz Transformation* implied its own *composition of velocities* that, most surely, would be reduced to the Galilean one for very low speeds. For now, the only important aspect of this new *composition of speeds* that I would like to disclose is that when the velocity that Lorentz and Larmor wanted to measure in SC-B (Earth) was that of *light* moving in the **same** *direction* as our planet in *ether* (***head*** wind) -and assuming that its *speed* in *ether* was 300,000 km/s- then the new *Lorentz's composition* did **not** give 299,970 (300,000-30) as *Galileo's Transformation* would do, but -instead- exactly 300,000 km/s! Likewise, when *light* displaced in the **opposite** *direction* to that of our planet in *ether* (***tail*** wind), the new composition did **not** give 300,030 km/s (300,000+30) as Galileo's would, but again exactly 300,000 km/s! In words:

The new composition of light speed with the planet's speed always gave the same light speed, as demanded by the dreadful failure of the ether hunt.

Another way of expressing this singular result is that given a luminous beam traveling in vacuum, and our intention of measuring its speed from different *Inertial Frames*, our measurement would always be 299,792,458 m/s for any of the infinite number of possible *Inertial Frames*. For lovers of ***absoluteness***, here was a good -though unexpected- candidate.

But, by the end of the 19th century, our *time* had to be ***absolute*** and only the ***time*** that an observer *in repose* with the King 'ether' would measure and psychologically perceive could be considered **real**. All the other *times* had to be **distortions**, mere perspectives accomplices of a gigantic conspiracy to prevent us from seizing the supposed **reality**. So strong was the preconception of ***absolute*** *space* and *time* in the scientific mentality of those days, that Lorentz did not realize the grand transcendence of what he had discovered, and contented himself with remodeling the edifice of Physics -- instead of rebuilding it with a new foundation. Despite his great intellectual caliber, being surrounded by trees, Lorentz was incapable of seeing the forest.

3. What we have learned

We have, at last, discussed -in five long chapters- all the concepts and history necessary to understand Relativity Theory. Let us summarize what we have learned in this chapter:

• The *velocity* of a *bullet* is defined with respect to its gun; the *velocity* of an *airplane*, as that of *mechanical* waves, is defined with respect to the *propagating medium*. The *bullet* does **not** require a *medium*; *sound* and *aircrafts* **do**;

• According to Galileo and Newton, there is **no limit** for *velocity* in this Universe. By the end of the 19th century, experiments with subatomic particles and Fizeau's experiment with *light* traveling in a water stream, seemed to indicate that the electron *inertia* increased with *speed* and that the *Galilean Composition of Velocities* was **not** valid;

• *Undulatory mechanical motion*, as the *motion* of *matter*, responds to the *Principle of Newtonian Relativity* though, in order to apply it correctly, we have to make sure that the critical role played by the *propagating medium* in one *Inertial Frame*, is identical to the role played in the other *Inertial Frame*;

• *Interference* between two undulatory signals that have traveled different *paths*, or at different *speeds*, allows us to measure the *times* for each wave and, if *distances* are known, determine their respective *velocities*. This is the operational principle of an extremely precise instrument known as *interferometer* which, in the case of *light*, played a preeminent role in the epistemological revolution that took place at the end of the 19th century;

• Ole Roemer in the 17th century, observing the eclipses of one of Jupiter's moons, showed that the *speed of light* had to be ***finite***, and James Bradley in the 18th century determined the *speed of light* with good accuracy through the phenomenon of *stellar light aberration*. Fizeau and Jean Foucault, in the 19th century, measured the *speed of light* within the confines of their laboratories;

• Thomas Young experimentally demonstrated the *undulatory* nature of *light* but, ironically, contributed to consolidate the epistemological relevancy of the ***analogy*** between *light* and *sound* and -given that the latter needed a *propagating medium*- to solidly establish the notion of a *luminiferous ether* (omnipresent in vacuum and in the substance of all matter) as the *propagating medium* for *light*;

• According with Maxwell's theory, Young's *luminiferous ether* -inspired in the *air* needed for *sound*- became to be simply *ether*, i.e. the *propagating medium* for *electromagnetic waves*. *Light* is an *electromagnetic wave*;

• If *Galileo's Transformation* was applied to *Maxwell's equations*, the latter changed structurally! This fact suggested that *electromagnetic laws*, as opposed to *mechanical* ones, were valid only in a ***privileged*** *reference frame*: the *ether* in ***absolute*** *repose*. Earth ***absolute*** *motion* could be detectable through *optical* experiments, and thus the ***ether hunt*** started;

• ***The ether hunt***, through Michelson-Morley's experiment and many others, was a major failure. In any orientation of the Michelson-Morley's apparatus, the two light beams arrived *in unison*, indicating that the *speed of light* relative to Earth was **independent** of the *magnitude* and *direction* of Earth *motion* in the supposed ***absolute*** *space*;

• Hendrik Lorentz attempted to explain the negative results of the ***ether hunt***, while keeping the existence of *ether* as the ***absolute*** standard of *repose*. The grand cosmic conspiracy was pronounced by Lorentz with the same singular mastery with which it would be demolished by Einstein in 1905;

• Lorentz discovered his transformation whereby every object in *motion* through *ether* had to mysteriously ***contract*** and experience a 'local time'. The composition of *light speed* with any other *speed*, instead of being their arithmetic sum (as Galileo's), gave again the *speed of light* as the result, precisely as the failure of the ***ether hunt*** demanded;

• The preconception of ***absolute*** *time* and *space* was so strong by the end of the 19th century, that Lorentz did not realize the grand transcendence of what he had discovered, contenting himself with ***remodeling*** the edifice of Physics -- instead of ***rebuilding*** it with a new foundation. Despite his great intellectual caliber, being surrounded by trees, he did not see the forest.

Additional Recommended Reading

[1] d'Abro, A., *The Evolution of Scientific Thought - From Newton to Einstein.* New York: Dover Publications, 1950.

[2] Einstein, Albert and Infeld, Leopoldo, *The Evolution of Physics.* New York: Simon & Schuster, 1966.

[3] Russell, Bertrand, *ABC of Relativity.* London: Routledge, Taylor and Francis Group, 1997.

[4] Jeans, James, *Physics and Philosophy.* New York: Dover Publications, 1943.

[5] Hoffman, Banesh, *Relativity and its Roots.* New York: Dover Science Books, 1983.

[6] Poincaré, Henri, *Science and Hypothesis.* New York: Dover Publications, 1952.

[7] Carnap, Rudolf, *An Introduction to the Philosophy of Science.* Edited by Martin Garner, New York: Dover Publications, 1996.

[8] Reichenbach, Hans, *From Copernicus to Einstein.* New York: Dover Publications, 1970.

[9] Kamenov, K.G., *Space, Time, and Matter, and the Falsity of Einstein's Theory of Relativity.* Vantage Press, Inc., 2008.

[10] Crotti, Marcelo A., *La Relatividad Conceptual.* En http://www.crotti.com.ar /Relatividad/.

[11] Peat, David, *Einstein's Moon - Bell Theorem and the Curious Quest for Quantum Reality.* Chicago: Contemporary Books, 1990.

[12] Goldsmith, Donald, *The Ultimate EINSTEIN.* Byron Press Multimedia Books, 1997.

Chapter 6

Galloping with Light

If I pursue a beam of light with the velocity c (velocity of light in a vacuum), I should observe such a beam of light as an electromagnetic field at rest though spatially oscillating. There seems to be no such thing, however, neither on the basis of experience nor according to Maxwell equations...

The Special Theory of Relativity

...From the very beginning it appeared to me intuitively clear that, judged from the stand point of such an observer, everything would have to happen according to the same laws as for an observer who, relative to the earth, was at rest. For how should the first observer know, or be able to determine, that he is in a state of fast uniform motion? Albert Einstein at 16 (*Autobiographical Notes*).

At last! After five long chapters interleaving semantic and epistemological discussions, history and philosophy of science, personal conundrums and experiences that enriched my life, while debunking popular (and not so popular) folklore, I would like to believe that we are prepared to conceptually understand the Special Theory of Relativity. One more time: my objective is not to convert the reader into an expert on the subject; to the contrary, that would require much more effort than reading this book. The objective is to show that Einstein's theory, experimentally corroborated for the last hundred years, regardless of how outlandish and opposed it may seem to our prejudices (disguised as they are with the 'common sense' costume), is rational, consistent, and intelligible to the layperson -- if s/he has the audacity of accepting the unfounded nature of those prejudices. It is the desire and the honor of stimulating that necessary boldness in the non-scientific person that motivated my writing of this book. And this indispensable fresh intellectual strength I thrust to instill has nothing to do with our academic or professional credentials; furthermore: had I believed those credentials were strictly necessary to seize a concrete and positive message out of this book, I would have not gone through the trouble of writing it.

Chapters 1, 2, and 3 proved that both *space* and *time* are notions much more complex than what we are used to believe, and that neither *time* nor *space* have the absolute character we subconsciously assign to them. In Chapter 4, I went to great lengths in trying to convince the reader that, despite the apparent absoluteness of ***non-uniform*** (accelerated) *motion*, this type of *motion* was as ***relative*** as the ***uniform*** type. I also anticipated that the General Theory of Relativity (Chapter 7), presented by Einstein in 1916, would finally integrate ***non-uniform*** *motion* to the family of relative concepts. In Chapter 5, instead, I -for the most part- ignored *accelerated motion* to concentrate on *mechanic* and *electromagnetic* ***radiation***, their similarities and disparities, and how those waves manifest themselves when observed from different *Inertial Frames of Reference*.

Let us start by understanding the name of what we want to comprehend. Even though the Special Theory was published by Einstein in 1905 through two scientific articles[1], the label 'Theory of Lorentz-Einstein' was initially assigned to it, and Max Planck also used the term 'Relative Theory' until, in 1906, Alfred Bucherer (1863-1927) coined the locution 'Theory of Relativity'. Finally, when Einstein published his masterpiece on gravity in 1916, so as to distinguish his old creation from the new one, the epithet 'special' was chosen for the former and 'general' for the latter. As we already pointed out, 'special' means 'restricted to Inertial Frames of Reference', i.e. this theory cannot describe the Universe from an arbitrary *reference frame*. This restricted scope is not to be ignored, though not overemphasized either, because there is a great variety of practical phenomena than can be accurately treated and predicted by the Special Theory.

Ironically, and precisely because of the great success of this simple version of Relativity Theory, most of its detractors have chosen to ignore (out of ignorance or malice -- you judge case by case) its philosophical foundation and restriction to *Inertial Frames*, so as to declare it invalid. The famous twins' paradox is regularly explained within Special Relativity in all popular science books, when the mere conception of a twin traveling through *space* at *speeds* close to that of *light* and coming back to Earth to meet his twin brother, involves relative *accelerations* between the two (i.e. non-Inertial Frames) and, hence, situations which are excluded from the Special Theory. If when I was waiting for the kids to leave the *Carrousel by the Church* (in 1968), the idea of finding myself at that place with shorts and without stubble (even though I had been born in 1948), stimulated the reader's imagination and curiosity, this chapter will prepare us to comprehend the rationale behind my speculation that day, the fallacy behind those who consider it a real paradox when trying to explain it within Special Relativity, and its complete discussion and demystification in the following chapter when presenting the General Theory of Relativity.

The relativity that the Special Theory refers to is that manifested if we compare the different descriptions of **reality** we obtain when employing any two of the infinite *Inertial Frames of Reference* available to us. In sum: when we affirm that a given physical magnitude is ***relative***, we mean that its value is **different** in different *Inertial Frames* and, conversely, when we assert that a magnitude is ***absolute***, it means that its value is the **same** in **all** *Inertial Frames*.

The objective and merit of Einstein's theory is to identify those physical magnitudes which are ***absolute***, i.e. common for all *Inertial Frames*, distinguishing

1 A. Einstein, *On the Electrodynamics of Moving Bodies*, Annalen der Physik, 17:891-921, June 30, 1905; *Does the Inertia of a Body depend on its Energetic Content?* Annalen der Physik, 18:639-641, Sept. 27, 1905.

them from those which are a mere perspective, only shared by those observers *in repose* within a given *Inertial Frame*. At the end of this chapter, I hope I will have convinced the reader that both *space* and *time* are meager perspectives, while *space-time* is ***absolute***, i.e. shared by all inertial spectators. It was about time to take the bull by the horns!

1. The Principles of Relativity throughout History

A. d'Abro, a physicist-philosopher as brilliant as unknown, in his great book *The Evolution of Scientific Thought - From Newton to Einstein*[2], enumerates six different *Principles of Relativity* that have been enunciated along history:

- The primordial Relativity of space and time;
- The Kinematic or Visual Principle of Relativity;
- The Dynamic or Classical Galilean and Newtonian Principle of Relativity;
- Einstein's Principle of Special Relativity;
- Einstein's Principle of General Relativity;
- The Radical Mach-Einstein Principle of Relativity.

The Primordial Relativity of *space* and *time* alluded by d'Abro refers to the conventionalism of any ***measurement*** technique that we treated in detail along Chapters 2 and 3. We empathized that the definitions of *congruence*, *uniformity*, and *standard* body/process all have a degree of ***arbitrariness*** and, ergo, the ***measurements*** of *space* and *time* depend upon those conventions. In short: both *space* and *time* are metrically amorphous, i.e. they do not have -despite how strongly we believe so- an inherent *metric* which would allow us to ***measure*** them without any definitions. In this sense, thus, neither *space* nor *time* is ***absolute***.

I insist: both the ***measurement*** of *space* and *time* can only quantitatively inform us of the relation between the Universe and the *standard* body/process: if all *matter* of the Universe changed its size in the same proportion, and all processes of the Universe speeded up or down in the same proportion, we would never know it -- as the corresponding *standard* body/process would ***change*** accordingly. For instance, to say that the average *lifetime* of a human being on our planet is 70 years, is an assertion that only has meaning when is interpreted in terms of the particular physical ***process*** chosen to define the physical unit called 'year', but it does not mean at all that our organism 'has' a biological ***absolute*** notion of that *period of time*, and therefore *ages* statistically in accordance with it. *Aging* of our organism is a physical *process* and, as such, constitutes a natural *clock*, made up of the trillions of atomic *clocks* that make what we call the material or corporal component of a human being -- even though this process is not particularly adequate to be selected as the standard (for the many reasons articulated in Chapter 2).

The Kinematic or Visual Principle of Relativity was treated in detail in Chapter 4 and patently appeared during my experiences with the *Giants of the Road* and the *Carrousel by the Church*. Visually, there was no difference between my

2 D'Abro had to pay the original publication with his own money, as no publisher considered him because of his lack of professional degree or affiliation. Note that he still does not appear in Wikipedia.

accelerated *motion* while the carrousel was *rotating* with respect to the landscaping in repose, and the opposite assumption (carrousel and I *in repose* while the landscaping was *rotating* in the opposite direction). Nevertheless, there existed obvious dynamical effects (*inertial forces*) that sharply differentiated ***non-uniform*** *motion* (carrousel rotating or bus braking) from ***uniform*** *motion* (carrousel stationary or bus at constant speed), so that it was not obvious that ***rotation***, and in general any ***accelerated*** *motion*, could be as **relative** as ***uniform*** *motion*.

The Dynamical Principle of Relativity of Galileo and Newton was presented in depth in Chapters 4 and 5. Employing both *kinematic* and *dynamic* criteria, it was impossible to distinguish between two *Inertial Frames*, (i.e. two frames whose relative *velocity* is ***constant***) and, hence, *velocity* was **relative** while *acceleration* was **absolute** -- at least from the *mechanical* point of view. Nevertheless, the calamitous failure of the ***ether hunt***, depicted in Chapter 5, seemed to indicate the same impossibility of discriminating between *Inertial Frames* by means of *electromagnetic* experiments so that Einstein -as we shall see immediately- pronounced the *Principle of Special Relativity* as one of the pillars (First Postulate) of his Special Relativity Theory.

The *Principle of General Relativity* and what d'Abro calls the Mach-Einstein Radical Principle of Relativity were anticipated in Chapter 4 during my discussions on ***accelerated*** *motion*, and constitute the foundation for the General Theory of Relativity (Chapter 7).

1.1. The Principle of Special Relativity

Let us remember that our first definition of *Inertial Frame of Reference* was a frame in which the *Galilean Principle of Inertia* was valid and that, if we found in Nature one *Inertial Frame*, we had *-ipso facto-* an infinite number of them: all those *moving* ***uniformly*** with respect to the one we had found. This fact was expressed as the *Galilean Principle of Relativity*. After Newton, we extended the definition of *Inertial Frame* to that in which *Newton's Laws of Motion* were valid and concluded, again, that once we found one *Inertial Frame*, we had in infinitude of them: all those *moving* ***uniformly*** with the former. This fact was expressed as the *Newtonian Principle of Relativity*. In essence, this means that, given that the *Laws of Mechanics* are the **same** in all *Inertial Frames*, there is **no** *mechanical* experiment performed in a given *Inertial Frame* that -without external reference- would tell us that we are *moving* relative to other *Inertial Frames*. Besides, with external reference, we can only detect and measure our *motion* ***relative*** to other *Inertial Frame*.

We should remember as well that, even though they are 'principles', i.e. postulates ***inductively*** obtained by means of experimental observation (e.g. Galileo's ship), I demonstrated their validity ***deductively*** -- as if they were *theorems* instead of *axioms*. My ability to prove them *deductively* was based on two subliminal hypotheses so inextricably established in our philosophical marrow that not even Lorentz dared to touch them -- even after having discovered his eponymous *transformation*. These hypotheses are: a) the *length* of a body (*distance* between two points in *space* at a given *time*) is an immanent property of that body, i.e. the **same** regardless of our *reference frame*; and

b) *time* is ***absolute***, i.e. the **same** for all *reference frames*. Once these two assumptions are accepted as truthful, the relation between *space* and *time coordinates* in any two *Inertial Frames* is governed by the *Galilean Transformation*, and the relation between the respective *velocities* of a given object in those two *Inertial Frames* is given by the *Galilean Composition of Velocities*.

But Nature is much more than the set of all *mechanical* phenomena, and the ***ether hunt*** had proven that not only *mechanical* experiments but also *electromagnetic* ones had failed to detect the *motion* of our planet. Poincaré, in a lecture given in Saint Louis (Missouri, USA) in 1904, and published in *The Monist*, Vol. XV, No. 1 (1905), was prophetic:

> *Perhaps, too, we shall have to construct an entirely new mechanics that we only succeed in catching a glimpse of, where, inertia increasing with the velocity, the velocity of light would become an impassable limit. The ordinary mechanics, more simple, would remain a first approximation, since it would be true for velocities not too great, so that the old dynamics would still be found under the new.*

In the same publication, Poincaré refers to:

> *...The principle of relativity, according to which the laws of physical phenomena must be the same for a stationary observer as for an observer carried along in a uniform motion of translation; so that we have not and can not have any means of discerning whether or not we are carried along in such a motion.*

Leaving aside the epistemological amphibology of expressions like 'stationary observer' and 'observer in uniform motion', Poincaré's last excerpt is what we call the *Principle of Special Relativity*. Undoubtedly, Poincaré's philosophical depth was remarkable, and the prophetic nature of his writings has prompted many historians to assert that, had Einstein not appeared completely out of the blue in 1905, the great French philosopher would have eventually conceived Special Relativity Theory.

2. Annus mirabilis 1905: Few Words and plenty of Epistemology

For the delight of a unifying mind like Einstein's, everything seemed to indicate that, given any two *Inertial Frames*, not only the *Laws of Mechanics*, but also those of *Electromagnetism* were valid in both frames; otherwise, the ***ether hunt***, conducted by the best minds of the 19th century to detect our planet *motion*, would not have ended in such a major disappointment.

The pernicious full analogy between *sound* and *light* had to be discarded: *sound* is a ***macroscopic*** phenomenon intimately bonded to the macroscopic concept of ***substance***, and that is why there is no *sound* in interatomic or intersidereal spaces; *light*, instead, travels as easily in the former as in the latter. Furthermore, *matter* impairs *light* propagation! This innate difference between the two classes of *undulatory radiation* indicated that something quite *sui generis* separated *light* from *sound*, and that new mental schemes were necessary. It appeared mandatory to accept that there could be *undulatory radiation* without an *undulating* ***material*** *medium*. The *electromagnetic field* had to be the *prima donna*, not the *propagating medium* as it was for *sound*.

At the early age of sixteen, Einstein had imagined that he was galloping on a luminous *wave*, i.e. he was travelling at the **same** *speed* as *light*. In such hypothetical circumstances, 'horse' and 'rider' would be in relative *repose* and, consequently, Einstein could only observe the ***spatial*** variations of the *electromagnetic field* (not its *temporal* variations), i.e. the wave would be ***stationary*** for him. But Einstein was not a typical adolescent as we were: he immediately realized that Maxwell's equations that governed propagation of the wave when observed from Earth, would fail in a *reference frame* traveling *in unison* with the wave itself. To Einstein, it was instinctively inconceivable that the laws of Nature would change with the point of view of the observer (frame of reference), and dedicated his life to prove he was right.

In his article of 1905 entitled *On the Electrodynamics of Moving Bodies*, after a brief introduction highlighting that "Maxwell's electrodynamics leads to asymmetries which do not appear to be inherent in the phenomena", Einstein said:

Examples of this sort, together with the unsuccessful attempts to discover any motion relatively to the "light medium" suggest that the phenomena of electrodynamics as well as of mechanics possess no properties corresponding to the idea of absolute rest. They suggest rather that, ... the same laws of electrodynamics and optics will be valid for all frames of reference for which the equations of mechanics hold good.

Einstein elevated the last conjecture (the one that starts after the ellipsis) to the status of ***postulate*** and, hence, extended *Newton's Principle of Relativity* to *electromagnetic* phenomena. In the same way as Newton extended *Galileo's Principle of Relativity* by extending the definition of *Inertial Frame*, Einstein extended it one more time as a *reference frame* where both the laws of *mechanics* and of *electromagnetism* are valid. Ten years after imagining he was *galloping with light* to see it *in repose* and, at the same time, intuitively realizing that such a feat would be incompatible with *Maxwell's equations*, Einstein initiated the resolution of his adolescence paradox proclaiming: *Maxwell's equations* are to be ***valid*** in any *Inertial Frame of Reference*, because experimental facts so dictate it, and because the unity and beauty of the Universe so demand it. This ***First Postulate*** is known as the *Principle of Special Relativity* and, with the style I chose for this book, it would be expressed as follows:

If we found in Nature an Inertial Frame of Reference, i.e. a frame in which both the laws of mechanics and electromagnetism are valid, then, we have -ipso facto- an infinite number of them: all those frames which move uniformly with respect to the former, independently of their velocities.

Immediately, Einstein enunciated a ***Second Postulate*** (known as the *Principle of the Constant Velocity of Light in Vacuum*), and he articulated it as:

We will introduce another postulate, which is only apparently irreconcilable with the former, namely, that light is always propagated in empty space with a definite velocity c which is independent of the estate of motion of the emitting body.

Even though -due to the particular way he expressed this postulate- it is not clear what Einstein had in mind, ten years after imagining he was *galloping with light* to see it *in repose* and, at the same time, intuitively realizing that such a feat would be incompatible with *Maxwell's equations*, Einstein completely resolved his adolescence paradox with this second decree which I would express thus: no matter how fast I may be *moving* relative to a given *reference frame*, the *velocity of light* relative to me will be always the same and, ergo, I will never see it *in repose*.

And... why did Einstein think that, for the physicist of that time, this ***Second Postulate*** would seem *prima facie* 'irreconcilable' with the first one? After all the warming up we went through in previous chapters, we know that the *Principle of Newtonian Relativity* implies that *Newton's laws* must be structurally invariant (***covariant***) in any two *Inertial Frames* so that, with our new definition of *Inertial Frame*, the *Principle of Special Relativity* now implies that not only *Newton's laws* but also *Maxwell's equations* must be structurally invariant when transformed from one *Inertial Frame* to another. We also learned that this ***covariance*** of Newton's laws is achieved when the *transformation* to convert the *space* and *time coordinates* from one *Inertial Frame* to another is the *Galilean Transformation*. Finally, we discovered that the failure of the ***ether hunt*** demonstrated that the *Galilean Composition of Velocities* was not valid with *electromagnetic* experiments, i.e. that *Maxwell's equations* were **not** *covariant* under the *Galilean Transformation*. It is the latter that is irreconcilable with the ***Second Postulate*** of Einstein, because the *Galilean Composition of Velocities*, being a simple arithmetic sum, would never produce a "definite velocity c which is independent of the state of motion of the emitting body" -- as Einstein proclaimed.

The *Galilean Transformation* was the *prima donna* of a scientific folklore that, due to its outstanding success, had acquired sacrosanct status. If Einstein dethroned it, introducing another transformation for which Maxwell's equations transformed into themselves (and so both of his postulates could be true), then, unless that new transformation (miraculously) also transformed Newton's laws into themselves, he would dethrone the whole science of *Mechanics*! Einstein would solve one crisis and create another one. Lorentz, though with an incorrect epistemological interpretation, had found such a *transformation* -- but he never considered it as replacing Galileo's. Einstein had to have a drastic and revolutionary card under his sleeve, so as to not throw away thousands of years of scientific work supported by mountains of experimental data on mechanical phenomena. Relativist Mechanics would soon be born.

2.1. Relativity of Simultaneity as a Result of its Conventionality

After such a succinct but revolutionary introduction of his two postulates, Einstein started the first section of his article with an outrageous title borderline offensive for the scientific mentality of those days: "Definition of Simultaneity". If we forget for a moment all we have learned in previous chapters, we could imagine the erudite spontaneous reaction (in 1905) as something like this:

What? What an insolent! Does he really think we need to define something so obvious? What can an unknown patent office clerk tell us about

> ***something so solidly established? What else is there to say about simultaneity beyond saying that two events are simultaneous when they occur at the same time?***

Einstein starts irritating the expert even more by reminding him/her that "If, for instance, I say "That train arrives here at 7 o'clock," I mean something like this: "The pointing of the small hand of my watch to 7 and the arrival of the train are simultaneous events." -- but also asserting that this conception of *time* has only unequivocal meaning when clock and train are *spatially* ***contiguous***. He proceeds showing that two identical clocks, located at distant places A and B, allow us to measure ***local*** *times* of events occurring at A and B but not a ***common*** *time* useful to order those *events*. Only a mind so perspicacious could have made such an elementary but profound observation: that the objective notion of ***global*** *time* could only make sense if we accepted the intuitive idea of ***local*** *simultaneity*, but abstained from believing that such intuition was extendable to ***distant*** places without further ado. I emphasize once again (Chapter 2) that we are dealing with *events* occurring at A and B which are ***distant*** but they are in relative *repose* and, hence, the conventionality of *simultaneity* is not -as the folklore states- an exclusive consequence of relative *motion*, but -instead- of something much more fundamental. Here is Einstein's definition:

> *We have so far defined only an "A time" and a "B time". We have not defined a common "time" for A and B, for the latter cannot be defined at all unless we establish by definition that the "time" required by light to travel from A to B equals the "time" it requires to travel from B to A.*

Pretending we do not know what we learnt in Chapter 2 regarding the vicious cycle that exists between *velocity* and *simultaneity*, let us imagine again the reaction of the erudite in 1905:

> ***What? What sort of idiocy is this? Do we need to define that light travels with the same velocity from A to B as from B to A? Is there any doubt about it? And, if there is, why do we not measure it so as to experimentally confirm whether it is so or not?***

As I said, with the benefit of historical retrospection and after having read Chapter 2, we know that the notion of *velocity* needs the notion of *simultaneity* and the latter -for ***distant*** *events*- needs the notion of *velocity*, i.e. we have a vicious circle to open before we can proceed. What Einstein was precisely doing was breaking that cycle by defining *simultaneity* in one of the many possible ways. Einstein's definition is the one that provides the maximum descriptive simplicity (Figure 3, Chapter 2). Using his definition, he shows how to *synchronize* two distant ***stationary*** *clocks* and also that this synchronization is symmetric and transitive. In this fashion, with an imaginary ***stationary*** *clock* at each place of an *Inertial Frame* -all of them synchronized with a master stationary clock- Einstein defined the *time* of an *event* as that given by the ***local*** *clock* which is *simultaneous* with the *event* occurrence. Please, observe that the latter 'simultaneous' corresponds to the intuitive notion of ***local*** *simultaneity*, i.e. to the irritating comment on the *simultaneity* of 'the pointing of the small hand of my watch to 7 and the arrival of the train".

In this manner, *time* is defined within an *Inertial Frame* and is common for all observers *at rest* in that *frame*: if two *events* are *simultaneous* for an observer at a given arbitrary *place* and *at rest* in the *frame*, they are *simultaneous* for any other observer at any other *place*, as long as she/he/it is *at rest* in the *frame*. Snappishly: so defined, *simultaneity* is ***absolute*** within a given *Inertial Frame*. So far so good: that is our 'common-sense' understanding of how *time* works. Will *simultaneity* also be ***absolute*** between different *Inertial Frames of Reference*?, i.e. given two *events simultaneous* in an *Inertial Frame*, will they also be *simultaneous* in any other *Inertial Frame*?

In an extremely compact manner, Einstein proves that if two *Inertial Frames* in relative *motion* use his definition of *simultaneity* to define *time* in each frame, and its ***Second Postulate*** is to be valid, then the *length* of a body *at rest* in the *frame* is different from its *length* when it *moves* in the *frame* (i.e. when it is *at rest* in the other *Inertial Frame*), and also that two *events simultaneous* in one of the *frames*, do **not** have to be *simultaneous* in the other *frame*. In short: *length* and *simultaneity* (and hence *time*) are ***relative*** magnitudes.

Figure 16[3] . Two lightning bolts simultaneous on the tracks but not on the train (or vice versa).

Despite the arduous discussions in Chapter 2 regarding ***topological*** and ***metric*** *simultaneities* that led us to discover the necessity of a definition to convert the ***multiplicity*** of the former into the indispensable ***uniqueness*** of the latter, and the disquisitions in Chapter 3 on the need for another definition for the *length* of an object in *motion*, I suspect the reader is still vulnerable to a potential anaphylactic reaction when considering the idea of various observers in relative *motion* not agreeing in the *simultaneity* of two events A and B or, equivalently, one of them saying that A and B were ***simultaneous***, another assuring (besides assuring that he is not drunk and that his clock is not malfunctioning) that A occurred ***after*** B, and a third observer swearing (besides

3 Composite image per license given in http://commons.wikimedia.org/wiki/File:VIRM_Echten.JPG.

pledging he is mentally sane and that his clock is perfectly accurate) that A occurred ***before*** B.

I will make a last attempt to get rid of our prejudices and, for doing so, I go back to the Master. Einstein wrote his first popular science book on Relativity Theory in 1916 [5], immortalizing the usage of a high speed train and two lightning strokes observed from two *Inertial Frames*: the ***train*** and the ***tracks***. The upper part in Figure 16 shows that the bolt on the left -seen from inside the train- falls on A' while -seen from the tracks- falls on A; likewise, the bolt on the right -seen from inside the train- falls on B' and, in the frame of the tracks, falls on B. Besides, M' is the middle point between A' and B' in the *reference frame* defined by the ***train***, and M is the middle point in the *frame* defined by the ***tracks***.

Let us assume that we are at M and perceive both bolts *in unison*. Given that both *light signals* arrived at my eyes *simultaneously* (**local sensorial** *simultaneity*), that we are at the middle position between A and B, and that -by Einstein's definition- *light speed* is the same in both directions, we conclude that the two strokes were *simultaneous* (***distant*** **inferred** *simultaneity*). More precisely: if we *synchronize* identical clocks at A and B according to Einstein's definition, when the two bolts fall on A and B, those clocks will indicate the **same** *time* and, ergo, they were *simultaneous*.

Points A', M', and B' are defined inside the ***train*** by simple *spatial* coincidence at the moment of the *simultaneous* electrical discharges on the ***tracks*** but, due to the train *motion*, A', M', and B' -fixed on the ***train*** *reference frame*- move towards the right in the *reference frame* in which the tracks are *at rest*. Imagine now that a passenger is at M' and pretends to determine the *simultaneity* of the two bolts in his natural *reference frame* (the ***train***). The passenger's criterion of *simultaneity* will be (as it was for us) that, because M' is the middle point between A' and B', and because the *speed of light* is -by definition- the same when it travels from B' towards M' as when travels from A' towards M', if s/he sees both luminous signals *simultaneously* (***local*** **sensorial** *simultaneity*), both bolts will have fallen *simultaneously* (***distant*** **inferred** *simultaneity*). Contrariwise, that signal (left or right) seen first will be the one corresponding to the bolt which occurred the first. More precisely: if the passenger *synchronizes* two identical clocks at A' and B' according to Einstein's definition, when the two bolts fall on A' and B', simply reading those two clocks will indicate whether the *events* were *simultaneous* or not, and if not, which one was the first.

Because we assumed that ***train*** and ***terra firma*** are *Inertial Frames*, *light speed* is the same inside the ***train*** as on ***solid ground***. Besides, as the passenger at M' is approaching B and parting from A (lower part in Figure 16), the *light signal* from the bolt at B will be seen by the passenger at M' ***before*** than the signal from the bolt at A and, ergo, the passenger will conclude that the bolt on the ***right*** struck **before** the bolt on the left! If we now assume that the ***train*** is the one *in repose* and the tracks (and the rest of the universe) are moving towards the left, given that -on the train- the bolt on the ***right*** fell before the bolt on the left, for us -on the tracks at M- because we are approaching A', the ***shorter*** propagation *time* for the signal coming from A' compensates with its

later departure and, vice versa, because we are separating from B', the ***longer*** propagation *time* for the signal coming from B' compensates with its ***earlier*** departure -- both signals reaching our eyes *in unison*, concluding they were *simultaneous* (as we assumed in the first place). *Mutatis mutandis*, had we assumed that the bolts struck *simultaneously* for the passengers on the ***train***, we would have concluded that there were **not** *simultaneous* when observed from the ***tracks***.

The great majority of the popular science books on Special Relativity somehow reproduce the above explanation given by Einstein in 1916, but they present the relativity of *simultaneity* as an inevitable logical consequence of the relative *motion* between the ***train*** and the ***tracks***. This is epistemologically incorrect and, of course, it is not Einstein to blame but the uncritical reader and writer who forgot (or missed) that Einstein's popular explanation tacitly conveyed his own definition of *simultaneity*, as explained in his original 1905 article. In other words, the fact that the passenger sees as *non-simultaneous* two events which the person by the tracks sees as *simultaneous* is a consequence of not only the relative *motion* but also the particular definition of *simultaneity* which is common for both *Inertial Frames* (train and tracks).

As Adolf Grünbaum demonstrates in his monumental work *Philosophical Problems of Space and Time*, if we *synchronized* the clocks at A and B with Einstein's definition and, with it, we concluded that the two bolts were *simultaneous*, because the passenger had the liberty to *synchronize* his clocks at A' and B' utilizing another of the many possible definitions of *simultaneity* (Figure 3, Chapter 2), among all those, there would be one which would entitle the passenger to infer that the two bolts struck also *simultaneously* in his *reference frame* (***train***)! Naturally, common sense and the objective of maximal simplicity compel us to adopt Einstein's definition for all *Inertial Frames*. For the last time: Einstein's definition, used for all *Inertial Frames* provides the descriptive simplicity that is so important for Science but, at the same time, tends to hide the epistemological foundation for the concept of *simultaneity* and the basic reasons for the inherent relativity of the notion of *time*. As we have reiteratively found, a good part of what appears to us *-prima facie-* as objective **reality** is, instead, just a consequence of our conventions to discover it.

2.2. The Lorentz's Transformation a consequence of Einstein's Postulates

Let us remember that Lorentz discovered his eponymous transformation when trying to explain the negative results of Michelson-Morley's experiment, and that his interpretation was that the arm of the interferometer which was ***aligned*** with our planet *motion* in *ether* (***absolute*** *space*) experienced a **real** *contraction* (the so-called Fitzgerald/Lorentz's contraction) but that, despite being **real**, was undetectable because the instrument, we, and the planet were *contracted* by the same proportion due to the *ether wind*. Only somebody located *at rest* with the *ether* could measure the 'real' *length*; what we, on Earth, measured was a *distortion* (an 'apparent' *length*). The *contraction factor* (ratio between the 'real' *length* measurable in the supposed ***absolute*** *space* and the *length* measured on Earth) was $\beta = \sqrt{1 - V_{A/B}^2 / c^2}$ where $V_{A/B}$ was the ***absolute*** *velocity* of our planet, i.e. the *velocity* of Earth (SC-B) relative to *ether* (SC-A). We also saw that this causal explication of the experiment through a physical *contraction* was very difficult to accept because, for a given *speed*, the *contraction* was the same regardless of the body physical properties and, besides, all attempts to detect it through changes of other physical properties associated with *length* (like electrical conductivity and index of refraction) had been unsuccessful.

Likewise, in accordance with the same interpretation, *Lorentz's Transformation* affected the measurement of *time* on Earth in comparison with the 'real' *time* measured in ***absolute*** *space*, so that the *clocks* on Earth ran behind with respect to the ***absolute*** *time* indicated by a *clock at rest* with the *ether*. Only that fortunate individual *at rest* in ***absolute*** *space* would perceive the 'real' *time*; our *time* measured and lived on Earth was a ***distortion*** (apparent). The *dilation factor* (ratio between the 'real' *time interval* measurable in ***absolute*** *space* and that measured on Earth) was the same as for the *space interval* (*length*), i.e. $\beta = \sqrt{1 - V_{A/B}^2 / c^2}$.

Einstein, in his 1905 article, tossed away all this scientific folklore about *ether*, 'real' and 'apparent' *length*, 'real' and 'apparent' *time*, etc. by simply asserting the following about his two postulates and the *ether* assumption:

> *These two postulates suffice for the attainment of a simple and consistent theory... The introduction of the "luminiferous ether" will prove to be superfluous inasmuch as the view here to be developed will not require an "absolutely stationary space" provided with special properties...*

With his two postulates, Einstein mathematically deduced *Lorentz's Transformation* as a transformation between *space* and *time coordinates* of **any** two *Inertial Frames* -- instead of between any *Inertial Frame* and ***absolute*** *space* in ***absolute*** *repose*, as it had been interpreted till then. The difference between *Lorentz's Transformation* in ***Lorentz's theory*** and *Lorentz's Transformation* in ***Einstein's Special Relativity*** is not mathematical but *ontological* and *epistemological* and, being so, it was to be expected the emergence of historians, scientists, and philosophers that, not having understood in depth the philosophical content and transcendence of the theory, would minimize Einstein's contribution.

Figure 7 in Chapter 4 has been reproduced here as Figure 17, as a reference for the *Galilean Transformation* and its associated *Composition of Velocities*, so we can compare them with the *Lorentz's Transformation* and its corresponding new *Composition of Velocities* (Figure 18). Please remember that the ***absoluteness*** of *time* ($T_A = T_B = T$) was the essence of classical science and, still to this day, it seems natural and viscerally correct for most of humanity. Likewise, the *distance* between two *points* in *space* at a given *time* (*length* of a body) was considered by classical science as an invariant, i.e. an intrinsic property of an object independent of the *reference frame* in which it was measured.

In both figures, X_B is the object coordinate measured in SC-B (*distance* between the body and the Y_B axis) and $X_{A/B}$ is the same *distance* measured in SC-A. It is the hypotheses of ***absolute*** *time* and the validity of $X_B = X_{A/B}$ (***absolute*** *space*) that justify the equalities (geometrically 'obvious') stated in Figure 17 and deliver the *Transformation of Galileo* and its simple arithmetic *composition of velocities.*

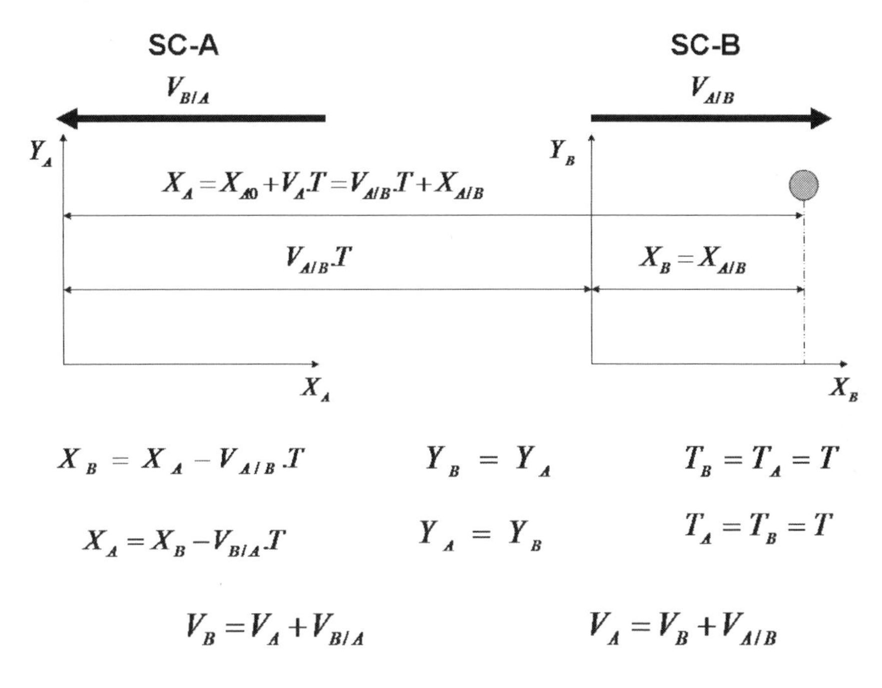

Figure 17. Galilean Transformation and its Composition of Velocities

It is crucial to understand the ***reciprocal*** character of the notion of relativity. In Figure 17, the first line below the graph shows the transformation to obtain the coordinates in SC-B (X_B, Y_B, T_B) when we know them in SC-A (X_A, Y_A, T_A); the second line shows how to obtain the latter when knowing the former; the third line (left) depicts the Galilean Composition of Velocities giving the object velocity in SC-B (V_B) as the sum of its velocity in SC-A (V_A) and the velocity $V_{B/A}$ of SC-A measured in SC-B; and -on the right- the object velocity in SC-A (V_A) as the sum of its velocity in SC-B (V_B), and the velocity $V_{A/B}$ of SC-B measured in SC-A.

It is the sheer fact that all these equations in Figure 17 (including Newton's laws not shown[4]) are ***structurally identical*** in both *reference frames* (only the variable names are different to refer to different frames), that guarantees the ***full equivalence*** of both *Inertial Frames*, i.e. the impossibility of finding out -even with ***external*** *reference*- which one of them is *moving*. Given that $V_{A/B} = -V_{B/A}$, looking through the window of a *Giant of the Road* only allowed me to detect our ***relative*** *motion*. At the same time, the ***covariance*** of *Newton's laws* ensured that, **without** external *reference*, there was no *mechanical* experiment I could conduct so as to detect not even our relative *motion*. This ***total equivalence*** is what the *Principle of Newtonian Relativity* asserts.

4 Figure 8 in Chapter 4.

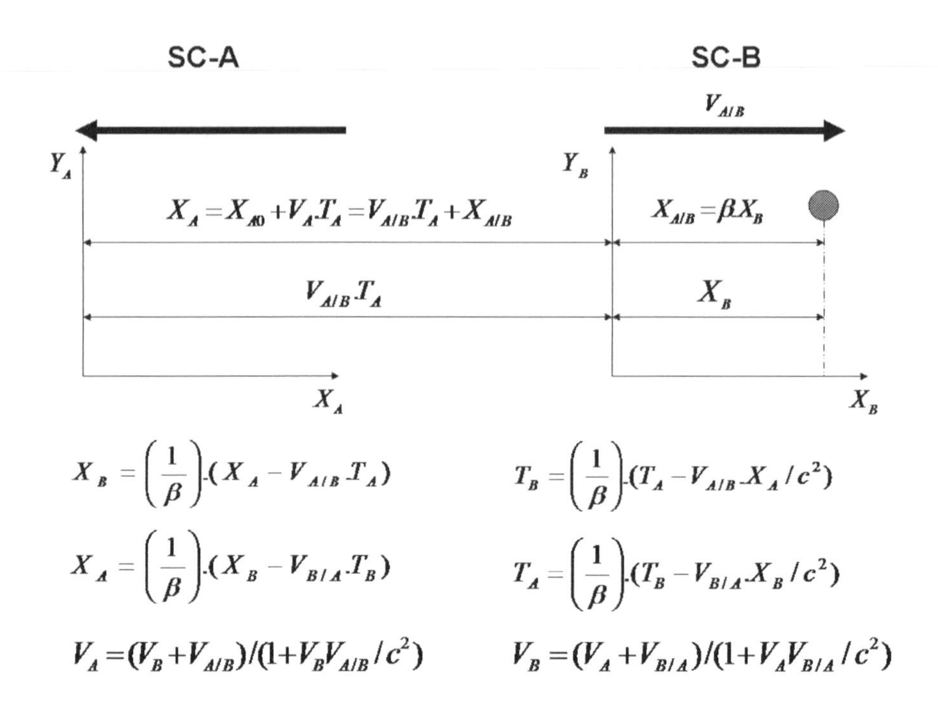

$$X_B = \left(\frac{1}{\beta}\right).(X_A - V_{A/B}.T_A) \qquad T_B = \left(\frac{1}{\beta}\right).(T_A - V_{A/B}.X_A / c^2)$$

$$X_A = \left(\frac{1}{\beta}\right).(X_B - V_{B/A}.T_B) \qquad T_A = \left(\frac{1}{\beta}\right).(T_B - V_{B/A}.X_B / c^2)$$

$$V_A = (V_B + V_{A/B})/(1 + V_B V_{A/B} / c^2) \qquad V_B = (V_A + V_{B/A})/(1 + V_A V_{B/A} / c^2)$$

Figure 18. Lorentz's Transformation and its Composition of Velocities

As for the *Lorentz's Transformation* (Figure 18), as we saw in Chapter 5, $X_{A/B} = \beta.X_B$ but, with the new ontological interpretation, the Fitzgerald/Lorentz *contraction* is **not** a phenomenon that requires a ***causal*** explanation; it is simply a ***reciprocal effect*** between any two *Inertial Frames* due to the definition of *length* for a *moving* object that we introduced in Chapter 3. This *contraction* is **not real** in the classical sense because disappears if the observer is *in repose* with the observed object. Also, please note that the *contraction* occurs only in the direction of *motion*; the object dimension in the *direction* ***perpendicular*** to that of *motion* does not change. As with the *Galilean Transformation*, all equations are ***structurally identical*** for all *Inertial Frames* -- so that it is impossible to ascertain which frame is the one *moving*.

An observer in SC-A obtains as the *length* for an object stationary in SC-B (in relative *motion* with SC-A) a value which is β times ***shorter*** than the one s/he would have obtained had the object been *at rest* in SC-A and, reciprocally, an observer in SC-B obtains as the *length* of an object stationary in SC-A (in relative *motion* with SC-B) a value which is β times ***shorter*** (**not** longer!) than the one s/he would have obtained had the object been *at rest* in SC-B. If the length of an object were β times ***shorter*** when measured from one frame and β times ***longer*** when measured from the other, the contraction would be **real** in the classical sense, and we could distinguish one *frame* from the other and... we could prove the existence of ***absolute*** *repose*! Only when SC-A and SC-B are in relative *repose* ($V_{A/B} = -V_{B/A} = 0$) is when the *length* of an object will be the same in both frames and is then called the object ***proper*** *length* -- corresponding to our classical and provincial notion of absolute *length*. The new notion of *length* is ***relative*** to the *Inertial Frame of Reference*.

Likewise, with the new interpretation, the *dilation of time* is not a phenomenon that needs a **causal** explanation; it is simply a ***reciprocal effect*** between any two *Inertial Frames* due to the conventionality of *simultaneity* and its *definition* by Einstein. This *dilation* is **not real** in the classical sense because disappears if both *Inertial Frames* are in relative *repose*. Using SC-A as reference, an *interval of time* between two *events* measured with a *clock* ***stationary*** in SC-B (in relative *motion* with SC-A) will be β times ***shorter*** than the one measured with a *clock* ***stationary*** in SC-A and, ***reciprocally***, using SC-B as reference, the *interval of time* between the same *events* but measured with a *clock* ***stationary*** in SC-A (in relative *motion* with SC-B) will be β times ***shorter*** (**not** *longer*!) than the one measured with a *clock* ***stationary*** in SC-B. If the *interval of time* were β times ***shorter*** when measured from one frame and β times ***longer*** when measured from the other, then the *dilation* would be **real** in the classical sense, and we could distinguish one frame from the other and... we could prove the existence of ***absolute*** *repose*! Only when SC-A and SC-B are in relative *repose* ($V_{A/B} = -V_{B/A} = 0$), *time* is **common** for both frames and is called the ***proper*** *time* which corresponds to our classical and provincial notion of ***absolute*** *time*. The new notion of *time* is ***relative*** to the *Inertial Frame of Reference*.

The *Relativistic Composition of Velocities*, shown on the last line of equations in Figure 18, replaces the *Galilean Composition* of classical physics, is easily deducible from the *Lorentz's Transformation* and, when $V_{A/B}$ is much lower than c, we obtain $V_A = V_B + V_{A/B}$, i.e. it becomes the *Galilean Composition*, as we expected.

As with the *Galilean Transformation*, it is the sheer fact that all these equations in Figure 18 (including Maxwell's equations not shown) are ***structurally identical*** in both *reference frames* (only the variable names are different to refer to different *frames*), that guarantees the ***full equivalence*** of both *Inertial Frames*, i.e. the impossibility of finding out -even with ***external*** *reference*- which one of the them is *moving*. At the same time, the ***covariance*** of *Maxwell's equations* ensured that, neither the Michelson-Morley experiment nor any other electromagnetic experiment, will allow us to detect the *motion* of our planet on its solar orbit. This ***full equivalence*** is simply what the *Principle of Special Relativity* asserts.

But... what about mechanical phenomena? We shall soon see that Newton's laws had to be extended per the relativistic vision of the Universe and, once done, the *Lorentz's Transformation* is valid for all natural phenomena when observed from any *Inertial Frame*, it reduces to the *Galileo's Transformation* for relative velocities which are insignificant with respect to the *speed of light in vacuum*, and, hence, classical Newtonian mechanics retains its successful reign, but only for those cases where that condition is met.

2.3. The Imagination of an Adolescent reduced to... an Impossible Reality

Now it is not that difficult to understand how Einstein resolved his adolescence paradox. In Figure 19, the wild horse represents *light* traveling in *vacuum* and replaces, this time as *radiation*, the object being observed in Figure 18. Einstein rides on *Babieca*[5] (which constitutes the *Inertial Frame* SC-B) and attempts to reach the *light beam* which was generated by a luminous source *at rest* in *terra firma* (SC-A). The noble and miraculous horse gradually increments its *velocity* relative to SC-A and, hence, the *distance* between horses ***decreases*** with the *time* in SC-A. If *Babieca* could get side by side with the *light beam*, Einstein would jump and continue *galloping with light*. Could he?

Imagine that Einstein has reached a *velocity* $V_{A/B}$ with respect to SC-A. Given that SC-A is inertial, Maxwell's equations are valid in it and, ergo, the speed of light in SC-A (V_A) is equal to c; what will the speed of light measured by Einstein, i.e. in SC-B (V_B) be? The answer to this question is obtained through the *Relativistic Composition of Velocities* in its version appearing at the bottom of Figure 19 and, when doing so, something algebraically quite curious happens after replacing c for V_A. Let us proceed and see what it is:

$$V_B = (c+V_{B/A})/(1+c.V_{B/A}/c^2) = c^2.(c+V_{B/A})/(c^2+c.V_{B/A}) = c^2.(c+V_{B/A})/(c.(c+V_{B/A})) = \left(\frac{c^2}{c}\right).\left(\frac{c+V_{B/A}}{c+V_{B/A}}\right)$$

I am sure the reader knows that dividing by zero is prohibited because it does not produce a unique result so, given that $c \neq 0$, we are entitled to cancel c^2 on the numerator with c on the denominator, obtaining simply c, so as to reach the equation:

$$V_B = c.\left(\frac{c+V_{B/A}}{c+V_{B/A}}\right)$$

Now, may we again cancel numerator with denominator inside the brackets? After all, they are equal and its quotient should be one... well, YES... as long as $c+V_{B/A}$ is not zero, i.e. as long as $V_{B/A} \neq -c$; and, when would that happen? When Einstein reached the *speed of light* with respect to *terra firma*, because -in that case- the *velocity* of SC-A measured in SC-B would be $V_{B/A} = -V_{A/B} = -c$ and we would be dividing by zero. This means that Special Relativity does not permit a material object to reach the *velocity of light in vacuum* though, in principle, may have a *velocity* as close (from below) to c as desired. As long as the denominator is not zero, it cancels out with the numerator and $V_B = c$, i.e. as long as Einstein's *velocity* relative to *terra firma* (SC-A) is ***lower*** than c, Einstein shall always measure a *speed* for the *light beam* equal to c and, hence, he will never see the *electromagnetic field at rest*! This is no less and no more than what the *Principle of the Constant Velocity of Light in Vacuum* demands, which -in conjunction with the *Principle of Special Relativity*- was used by Einstein to deduce the *Lorentz's Transformation* and its associated *compositions of velocities*.

Besides, the *Principle of Special Relativity* implies that *Maxwell's equations* transform into themselves when transferred from *terra firma* to Einstein's

5 'Babieca' was the name of the Cid Campeador's horse that "ran more than wind and jumped more than deer".

world (SC-B), so that **no** *electromagnetic* experiment conceived while galloping will reveal whether he, *terra firma*, or both is/are moving with ***uniform*** speed. Of course, due to the ***reciprocity*** inherent in relativity, the other version of the *composition of velocities* shown in Figure 19 is ***structurally identical*** to the one we used and, hence, if we assume that the *speed of light* relative to Einstein is c, then it will be also c with respect to *terra firma*, independently of how fast the latter could be *moving* with respect to Einstein.

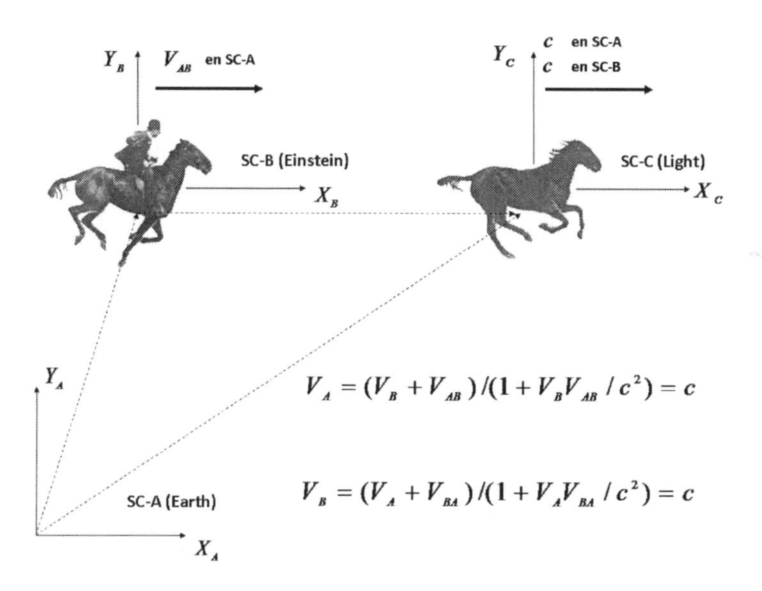

Figure 19. The imagination of an adolescent

Finally, let us imagine that Einstein -a singular adolescent- still distrusted the solution to his paradox that Einstein -a special adult- would give 10 years later with his Special Relativity Theory, and decides to deploy the wings of his *Pegasus*[6] so as to travel at 99.999999% of c with respect to SC-A, picks a lantern out of his pocket, and measures the *speed of light*; what value will he obtain? Exactly the value c, even though the ***source*** is moving at a *speed* relative to SC-A practically equal to that of *light in vacuum* (but **not** equal!). And why? Because, given that Einstein *gallops* at a ***constant*** *speed* relative to Earth, and given than the latter is supposedly *inertial*, Einstein is also *inertial* and, then, his lantern produces a *light beam* traveling -relative to him- with the *velocity* c.

Einstein's solution to his own paradox consisted, thus, in asserting the physical impossibility of ***galloping with a light beam traveling in vacuum*** and, hence, there is no body, radiation, or physical process capable of reaching or surpassing that *speed.* It should be clear, however, that the *speed of light* in *matter* is always ***less*** than c, so that Einstein could *gallop* with *light* as he imagined, or even leave it behind. For instance, the *index of refraction* for ***water*** is 1.333 which means that ***light speed in water*** is about 3000,000/1.333=225,056 km/s, so that Einstein's even more miraculous (now amphibious!) horse could catch up with the *light beam,* he could transfer to the wild horse and see the

6 Pegasus was -in Greek Mythology- a winged divine horse.

electromagnetic field at rest! We only need to apply the relativistic *composition of velocities* replacing 225,056 km/s for V_A and -225,056 km/s for $V_{B/A}$ obtaining the value zero for V_B (the denominator is **not** zero), i.e. *light* and Einstein would be in relative *repose*, though vertiginously traveling in ***water***.

Further stretching the horse herculean abilities, Einstein breaks the *sound barrier* with his whip[7], *accelerating* so as to break the *light barrier in water* and... what happens? Nothing! Because neither the *sound* nor the *light barriers* in *matter* are real physical barriers; he leaves *light* behind and continues *accelerating* to break the *light barrier in vacuum* and... what happens? As we anticipated in Chapter 1, the titanic *energy* of the horse had been and is being used ***more*** and ***more*** to *increase* its own *inertia* and ***less*** and ***less*** to *increment* its *speed* so that, now yes, the *speed of light in vacuum* constitutes a **real** and **universal** physical barrier.

A pithy way of summarizing the aforesaid is as follows: a) for any body (*matter*) or *radiation* in *matter*, there is always an *Inertial Frame of Reference* in which its ***proper*** *mass* (mass at rest) is **not *zero*** and, ergo, given that its relative *mass* (inertia) increases indefinitely with its *velocity*, it cannot ever reach the ***limit*** *speed* c ; and b) for *electromagnetic radiation in vacuum*, its *speed* is equal to the limit c for **all** *Inertial Frames*, ergo, it has **no** *proper mass* because there is **no** *Inertial Frame* in which it is *at rest*.

Besides being a **universal** barrier, we saw that the *speed of light in vacuum* is **absolute**, i.e. the same for all *Inertial Frames*. But then... is Einstein asserting on one hand that -as Galileo and Newton already knew- all *velocities* are **relative** but, on the other hand, that there is **one** *velocity* (that of *light in vacuum*) which is **absolute**? YES, that is what Einstein affirms, but -as I already pointed it out- in classical mechanics there was also an ***absolute*** *speed*, except that it was **infinite** --and that is why *time* and *space* were ***absolute***. According to Special Relativity, this ***absolute*** *speed* is **finite** and equal to the *speed of light in vacuum* -- and that is why *space* and *time* are ***relative***.

3. Relativistic Mechanics - Mass and Energy are relative and related by $E = m.c^2$

Are *Newton's laws* ***covariant*** under *Lorentz's Transformation*? The answer to this question is a categorical NO! But... then... are we confronting a new crisis? The answer is another categorical NO!

We know that for relative *speeds* between *frames* much ***lower*** than c , the *contraction* and *dilation factor* β is practically equal to 1, so that *Lorentz's* and *Galileo's Transformations* are indistinguishable, and the *relativistic composition of velocities* becomes the *Galilean composition*. Because -in Chapter 4- we proved that the *Galilean Transformation* left *Newton's laws* ***structurally invariant***, it is clear that for ***low*** *velocities*, classical mechanics remains valid, explaining its solid success up until the 19th century. But, from the above considerations, we also infer that *Newton's laws* cannot be valid for ***high*** *speeds* -- even though for ***low*** *speeds* they have to be an excellent approximation to more general laws which must be ***covariant*** under the *Lorentz's Transforma-*

7 The whip was the first man-made device which broke the sound barrier.

tion. Einstein showed that the key for *Newton's Second Law* to be ***covariant*** under *Lorentz's Transformation* resided in accepting that *mass* was **not** an intrinsic property of an object, but **relative** to the *reference frame* within which was measured, i.e. that it depended upon the relative *speed* between the object and the observer, with the dependence law simply being:

$$m = m_0 / \beta = m_0 / \sqrt{(1 - V_{A/B}^2 / c^2}$$

In this equation, where the *contraction/dilation factor* β is again present, the symbol m_0 stands for the classical *mass*, now interpreted as the *mass* when measured in a *reference frame* within which the body is *at rest*, and referred to as its ***proper*** *mass*.

From Figure 15 (Chapter 5), we see that β tends to 1 for values of $V_{A/B}$ approaching zero and, hence, when the body moves at low speeds, the relativistic *mass* is indistinguishable from the classical *mass*. From the same Figure 15, we see that $1/\beta$ grows beyond limit when $V_{A/B}$ approaches the limit c (without ever reaching it) so that the relativistic *mass* departs from the classical one and ***increases*** without limit.

In this fashion, the experiments of Kaufmann and Bucherer on the variable *mass* of the electron were explained at once; equally, the physical impossibility of reaching the *speed* c was also explicated -alternatively- by the limitless ***increase*** of a body's *inertia* when *accelerated*. Once this step was accomplished, it was straightforward for Einstein to write his article entitled *Does the Inertia of a Body depend on its Energetic Content?*, giving birth to the most famous equation in history, and to all its astonishing consequences treated in Chapter 1 with considerable detail.

Incidentally, now the reader -if willing to put up with a little algebra- can understand why the classical expression we all learned in college for the *kinetic energy* of an object is $(1/2).m.v^2$, where m and v were the *mass* and the *speed* of the object. What the teacher most probably did not tell you was that those two physical attributes were measured in ***different*** *reference frames*! In fact, m is actually the *proper mass* (i.e. m_0 in the previous equation), while v is measured from a *reference frame* whose relative speed with the former *frame* is precisely v (i.e. $V_{A/B}$ in the previous equation). Now, applying the binomial series expansion[8] to $\{1/\beta\} = \{(1 - V_{A/B}^2 / c^2)^{-1/2}\}$ we get:

$$E = m.c^2 = (m_0 / \beta)c^2 = m_0 c^2 \{1 + (1/2)V_{A/B}^2 / c^2 - (3/8)V_{A/B}^4 / c^2 + ...\}$$

And, finally, we obtain:

$$E = m_0 c^2 + (1/2)m_0 V_{A/B}^2 - (3/8)m_0 V_{A/B}^4 / c^2 + ...$$

In English: the total energy of the object (E) is equal to its ***intrinsic*** *energy* $m_0 c^2$ (i.e. as measured from a *reference frame* in which the object is *at rest*)

8 It was Newton who generalized the well-known finite binomial expansion of $(a+b)^n$ to the case in which the exponent n is not an integer (in our case it is -1/2) and the expansion becomes an infinite series.

plus its ***extrinsic*** (*kinetic*) *energy* that results from its *motion*. This much we already knew from Chapter 1. What we now learned is that, for *speeds* ***small*** with respect to c (so that all the terms in the series except the quadratic one can be neglected), the *kinetic energy* can be very accurately approximated by half the product of its *proper mass* and the square of its *speed*, i.e. $(1/2)m_0V_{A/B}^2$. And this is precisely the formula we knew from classical physics.

3.1. And... What about Newton's Universal Gravitation?

Einstein soon realized that *Newton's Law of Gravitation* imposed a serious problem, not easy to resolve within Special Relativity. As we know, the *force of gravitation* between two bodies is proportional to the product of their *masses* and inversely proportional to the square of their *distances*. But... Einstein had proved that both *mass* and *distance* were ***relative*** notions! Which of the infinite *masses* and *distances* for the infinite possible *Inertial Frames* were the appropriate values for the *gravitation* law to be valid? Once again we conclude that the notion of *force* is relative to the *reference frame*. Was it enough to simply apply the law in the selected *Inertial Frame*? But... was the *Law of Gravitation* ***covariant*** under the *Lorentz's Transformation*? Or was it only valid in a privileged frame? For Einstein, the idea of a privileged *frame of reference* was aesthetically unacceptable -- besides, it would violate his *Principle of Special Relativity*, the validity of which he was hoping to extend from only *Inertial Frames* to any fully arbitrary frames.

Furthermore, Newton had conceived *gravitation* as an *instant action at a distance*, while Special Relativity imposed an ***absolute*** *speed* **limit** for *motion* irrespective of the *reference frame*, so *gravitation* had now to be conceived as a *field* variable in *space* and *time* propagating at a **finite** *speed* resembling *electromagnetic radiation*. The only way for *gravitation* to obey the *Principle of Special Relativity* -inspired by *electromagnetic* phenomena- was that, instead of constituting a mysterious *instantaneous action at a distance*, it was governed by **local** interactions in *space* and *time* that, eventually, would cover sidereal *distances*. Only so, the ***relativity*** of *distance* and *mass* would not impose an insurmountable difficulty.

Based on that rationale, Einstein thought that the problem consisted in finding a sort of 'Maxwell's equations' for *gravitation* phenomena. Notwithstanding, he soon realized that the idea of a *gravitational field* acting on, and totally independent of the properties of, *space* was not viable. He had to develop a new theory of universal *gravitation* compatible with his Special Relativity Theory. On the other hand, as I said, accepting that the *Principle of Relativity* was only valid for *Inertial Frames* was instinctively and aesthetically repugnant for Einstein's mind. In agreement with Alfonso X 'The Wise', he could not believe that the Deity had missed the opportunity of creating a Universe which was uniform, aesthetic, and governed by the same laws regardless of our vantage point. His General Theory of Relativity would solve both problems with a 'single stroke' that took him ten years to execute.

4. Year 1908: From 'Space' and 'Time' to Minkowski 'Space-Time'

In classical physics, the world of *events* is of course also tetra-dimensional, but *space* (tridimensional) and *time* (one-dimensional) can only be assigned with a *metric* independently: both *interval of time* and *interval of space* (at a given *time*) are supposed to be ***absolute***, i.e. the **same** for all *Inertial Frames*, or -in common parlance- for all observers regardless of their relative *motion*. Hermann Minkowski (1864-1909), who had been Einstein's professor, discovered in 1908 that, according to Special Relativity, there existed a combination of *spatial* and *temporal coordinates* that remained ***invariant*** within all *Inertial Frames* and, consequently, could be used to define a tetra-dimensional *metric* which would provide *space* and *time* (as a whole) with a ***geometry***: *space* and *time* of classical physics fused into the *space-time* of relativistic physics. The ***objectivity*** that seemed to be lost when realizing with disenchantment that *space* and *time* were ***relative***, was now recovered when discovering that *space-time* -with such a *metric*- was ***absolute***.

4.1. Space-Time Diagrams in Classical Physics

For the sake of simplicity, and so as to be able to graphically expressing the ideas, we will assume that *space* has only one dimension (instead of three) and, in this manner, we can represent *space* and *time* in a two-dimensional *Cartesian Coordinate System* -- *space* horizontally and *time* vertically. When representing *space* and *time* jointly, the *trajectory* of an object is called its *world line*. In Figure 20 (upper left), the time axis 'OT' is the *world line* of an object which is *in repose* at X=0; the vertical line 'aA' is the *world line* of an object *in repose* at X=a; the oblique line 'OB' is the *world line* of an object that *moves* with ***constant*** *velocity*; 'OC' is the *world line* of an object that *moves* ***uniformly*** but ***faster*** than the previous one (the *slope* is ***lower*** so it covers ***more*** *space* in the ***same*** *time*); the curve 'OD' corresponds to a body/radiation that *moves* with ***variable*** *velocity* (very ***fast*** at the beginning while gradually ***slower*** as *time* goes by); the *space* axis 'OX' is the *world line* of an object that -hypothetically- *moves* with ***infinite*** *velocity* (it is ***everywhere*** in ***zero*** *time*).

It is interesting to realize that, if we see the *trajectory* of an object as *time* passes in the classical way, i.e. as a *succession* of *points* in *space*, *motion* can be *accelerated* and still the *trajectory* be ***rectilinear*** (autobus braking); if, instead, we see its *trajectory* as a *world line*, i.e. in a *space-time* diagram, ***uniform*** *motion* (including *repose*) has always a ***straight*** *line* as its *world line*, while ***non-uniform*** (accelerated) *motion* (even when the object *moves* in a *straight line* in *space*) has always a ***curved*** *line* as its *world line*.

In order for the graph to have a quantitative meaning, we have to choose a representation scale for each axis, i.e. to decide with what length we will represent the unit of space on the X-axis, and with what length we will represent the unit of time on the T-axis. Due to the preponderant role we know the speed of light in vacuum plays in Special Relativity, we agree in representing the unit of time (second) on the T-axis and the distance that light covers in such time (~300,000,000 m) on the X-axis with the same length. With this convention, the world line for light is a diagonal at 45° when traveling towards right, and at 135° (with the X-axis) when traveling towards left (Figure 20, upper right). Please note that, had we chosen the same length on both axis to represent 1 second and 1 meter, then -due to the tremendous value of the speed of light- its world line would have been graphically indistinguishable from the X-axis.

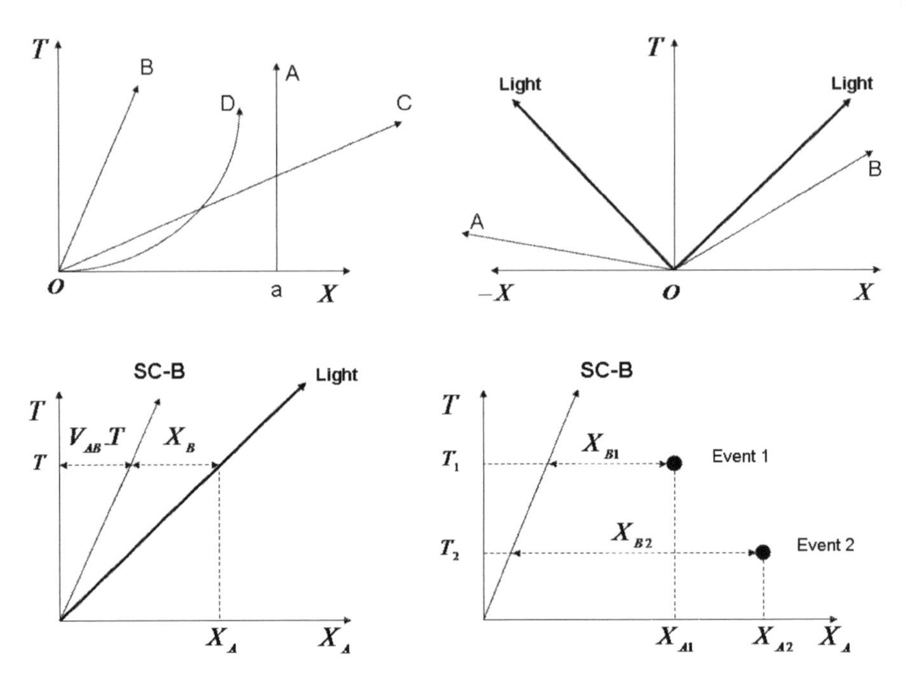

Figure 20. Classical Physics in Space-Time Diagrams

Equivalently, we could select the **same** *scale* for both axes, but change the ***unit of space*** to 'light-second' (the *distance* light covers in a *second*) and, doing so, the *speed of light* would naturally be 1 light-second/second and its *world line* would again be a diagonal at 45° or at 135°. We could also choose light-year for *space* and year for *time* as **units**, with the same graphical depiction. The two diagonals at 45° and at 135° are simply the result of a convenient choice for the ***scales*** of the graph and for the ***units*** of *space* and *time*.

In classical physics there is no *speed limit* in the Universe: just keep a ***constant** force* applied to a body and, because its *mass* (inertia) does not depend upon its *speed*, the *acceleration* is ***constant*** and, therefore, its *speed **increases*** beyond limit. It is clear then that, according to Newton, both 'OA' and 'OB' on the upper right graph are possible *world lines*. In Special Relativity, there is a limit for speed and, ergo, 'OA' and 'OB' are not physically realizable (their *speeds* are ***higher*** than the *speed of light in vacuum*).

So as to deepen our conceptual understanding before leaving classical physics, let us see how to represent the *Galilean Transformation* on a *space-time* diagram. In Figure 17 we drew two *Cartesian Coordinate Systems* SC-A and SC-B in relative *motion* with velocity $V_{A/B}$. In Figure 20 (bottom left), SC-B is represented inside SC-A (X_A, T_A) by means of the world line that SC-B (its origin 'O') traverses in SC-A, i.e. with an oblique whose slope corresponds to the speed $V_{A/B}$ between both Inertial Frames. The *world line* of a light beam is also shown, and we want to calculate the *distance* covered by light in SC-B, if we know the distance covered in SC-A. Given that *time* is assumed ***absolute***, i.e. $T_A = T_B = T$, we graphically infer that $X_B = X_A - V_{A/B}.T$, and, hence, that the *speed of light* in SC-B is the ***lower*** the ***greater*** is $V_{A/B}$ -- as demanded by the *Galilean Composition of Velocities*. Another way of saying graphically the same is that, because the *light world line* does not bifurcate the angle between the *world line* of SC-B and the *spatial* axis of SC-A (and SC-B) in two ***equal*** parts, the *speed of light* in SC-B is ***lower*** than in SC-A. The adolescent Einstein would not have had any problem in *galloping with light in vacuum*, not even in leaving it behind -- with all the inconsistencies that we know now would that bring about.

In this *space-time* classical world, we cannot define a tetra-dimensional *metric*, i.e. given any two *events*, there is no mathematical expression -made out of combining their *spatial* and *temporal* coordinates- that is ***invariant*** within the set of all *Inertial Frames*. All we can do is to ***independently*** define a *metric* for the classical *Euclidean space* (Pythagorean Theorem) and a *metric* for the classical *time*. For the inquisitive reader who wonders why is that so, let us take a look at Figure 20 (bottom right). What we are looking for is an expression including the *space* and *time coordinates* of both Event 1 and Event 2 in such a fashion that it takes the same value in SC-A as in SC-B. If we found it, we would have a 'distance' or *space-time interval* between any two *events* that would not depend upon the relative *motion* of *inertial* observers. Given that *time* is assumed ***absolute***, the *interval of time* between Event 1 and Event 2 is the same in SC-A as in SC-B; contrariwise, the *distance* between them in SC-A is $X_{A2} - X_{A1}$ while their *distance* in SC-B is $X_{B2} - X_{B1} = X_{A2} - X_{A1} - V_{A/B}(T_2 - T_1)$ which is always different, except when Event 1 and Event 2 are simultaneous ($T_2 = T_1$). As we know, in classical physics, the *distance* between two ***simultaneous*** *events* is ***absolute*** while the *distance* between two ***non-simultaneous*** *events* is ***relative*** to the *reference frame*[9] but, given that the *interval of time* is ***absolute***, there is no ***absolute*** *metric* in the classical *space-time*. In order for such a *metric* to exist, *time* and *distance* for a given *time* would have to be ***relative***! -- as we learned they are in chapters 1 through 5 and the failed ***ether hunt*** so required it.

4.2. Space-Time Diagrams in the Special Theory

In Special Relativity, the *spatial* part of an *Inertial Frame* is treated as a *Euclidean space* and, ergo, a *metric* for *space* can be established in which the *distance* between two *points* is calculated with the Pythagorean Theorem, and a *metric* for *time* is achieved with the *synchronization* technique of Einstein (Chapter 2). As long as we limit ourselves to observe the Universe from a given *Inertial Frame*, without pretending to interchange our experiences and measurements with other *frame* in relative *motion* with ours, our classical notions of ***absolute*** *space* and *time* are not affected. The *Principle of Special Relativity* (first Einstein's postulate) demands that Nature laws have to be exactly the same in every *Inertial Frame*, regardless of its *state of motion*. However, ***rela-***

9 Remember the multiple distances between Salt Lake City and Las Vegas.

tive** motion* is a constant in the Universe and, as soon as we desire to communicate with other worlds in relative *motion*, concepts as *length*, *simultaneity*, and *interval of time* become relative, changing of value from world to world. We have lost *space* and *time* as ***absolutes, but -in the process- we gained an ***absolute*** we did not know: the *speed of light in vacuum*.

It is interesting to note that, given that nothing can *move* with respect to itself, the *time* axis of a *space-time* coordinate system is, also, its own *world line*! As we saw in Figure 20 (bottom left), due to the scale/units adopted, the *world line* for *light* bisected the angle between the *world line* of SC-A (its *time* axis) and the *space* axis of SC-A but, as soon as we wanted to observe the *light beam* from SC-B (with *velocity* $V_{A/B}$ relative to SC-A), the *world line* of *light* did not bisect the angle between the *world line* of SC-B and the ***space*** axis (common for SC-A and SC-B) any longer, and that was why the *speed of light* was ***lower*** in SC-B than in SC-A.

But Special Relativity requires that the *velocity of light in vacuum* has to be the same in both SC-A and SC-B so, for that to happen, i.e. for the *light world line* to bisect the angle between the *space* and *time* axes of SC-B, obviously the *space* axis of SC-B has to move towards the *light world line*, as indicated in Figure 21 (left). The greater $V_{A/B}$, the closer the SC-B axes have to be so that -independently of how fast Einstein's Pegasus flied- he would always see the wild horse (light) going away with the same velocity c and, hence, he would never see the electromagnetic field *in repose*. It is also graphically clear that the *spatial* axis X_B in SC-B contains all the events that occur at *time* ***zero*** in SC-B (that is the definition of *spatial* axis), but the same axis considered as a *world line* in SC-A appears as a ***succession*** of *events*!, i.e. we rediscovered the relativity of *simultaneity*: two events *simultaneous* in SC-B (all those on its X-axis are) are seen *sequentially* in SC-A.

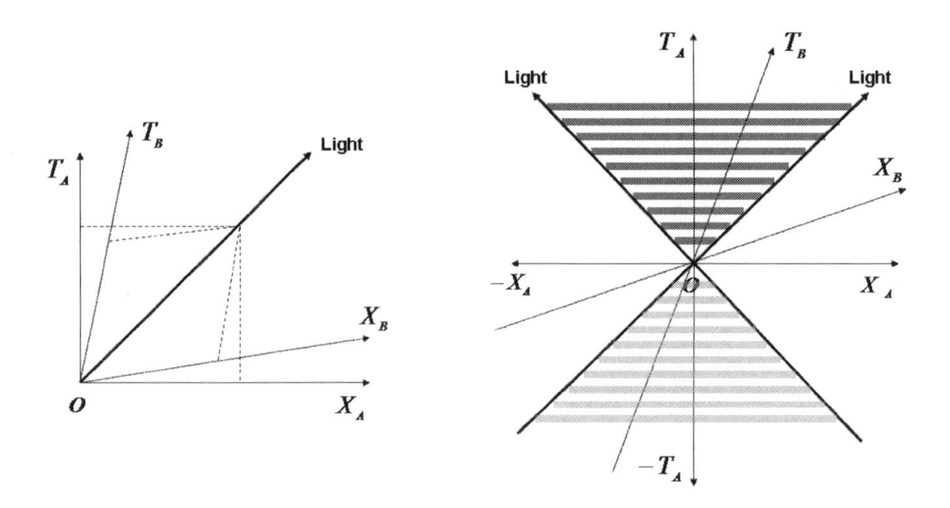

Figure 21. Relativistic Physics in Space-Time Diagrams

We conclude that, seen from a given *Inertial Frame*, there are infinite *directions of time*, one for each inertial observer, as well as infinite *directions of space*, one for each inertial observer. Given that no object (body or radiation) can travel faster than *light in vacuum*, the stripy cone between the two *world lines* for *light* defines the locus of all possible *directions* for the *time* axis, and its complement (not striped) defines the place of all possible *directions* for the *space* axis (Figure 21, right). The fact that the new axes for SC-B are not Cartesian (perpendicular) is simply due to our representing SC-B by its *world line* in SC-A but, of course, we can always use a Cartesian system for any *Inertial Frame* -- as long as we define it in such a way that the *speed of light* is always the same for all of them.

4.3. The Space-Time Interval: The Grand Absolute in the Special Theory

Minkowski discovered an ***invariant*** between all *Inertial Frames*, which he called the *space-time interval* (better: *event interval*), i.e. an *interval* that combines the *interval of space* (*distance*) and the *interval of time*. Given two events A1 and A2 with *coordinates* (X_1, Y_1, Z_1, T_1) and (X_2, Y_2, Z_2, T_2), and defining $\Delta X = X_2 - X_1$, $\Delta Y = Y_2 - Y_1$, $\Delta Z = Z_2 - Z_1$, and $\Delta T = T_2 - T_1$, it is not difficult to mathematically prove that the *Lorentz's Transformation*, besides leaving unchanged the *speed of light in vacuum*, it also leaves invariant the following expression:

$$s^2 = c^2.\Delta T^2 - (\Delta X^2 + \Delta Y^2 + \Delta Z^2) = c^2.\Delta T^2 - D^2$$

In this equation 's' is the *space-time interval*, 'D' is the *distance* between A1 and A2 (*space interval* using Pythagorean Theorem), and '$c.\Delta T$' is the *distance* that *light* covers in the *time interval* 'ΔT' between A1 and A2. We see then that the *event interval* is expressed in **units** of *spatial distance*, but is a combination of *space* and *time intervals*, i.e. it is a *distance* inside *space-time*. Observe also that, except for the negative sign, the formula is similar to that of a *distance* in a four-dimensional *Euclidean space* (Pythagorean Theorem extended to four dimensions). As in the *metric* for 'Necochea space', though in opposition to the *metric* for 'Madrid space', there are no terms containing products between the increments of different coordinates (only squares), and the coefficients multiplying each one of the squares for the increments of *distance* and *time* (i.e. the *metric* tensor) are ***constant*** (c^2,-1,-1,-1). Because of that, even though *space-time* is clearly ***non-Euclidean*** due to the intrinsic difference between *space* and *time*, its *curvature tensor* is ***zero*** as it is for a *Euclidean space* and that is why it is said that *space-time* in Special Relativity is ***semi-Euclidean*** and sometimes people use the confusing terms 'plane' or 'flat'.

Naturally, if A1 and A2 coincide in *space-time*, then both D and ΔT are zero and the *event interval* between A1 and A2 is ***zero***. But because this *interval* is an ***invariant***, it will be ***zero*** in all *Inertial Frames*. This simply reflects the well-known fact that a *space-time* coincidence is ***absolute***: if two cars collide, a *space-time* coincidence occurs with potentially fatal consequences. Imagine then that an observer on *terra firma* lamented the death of the driver and passengers, while a military pilot -that happened to fly low in the neighborhood at ***high*** speed- objectively assured that the two cars never crashed and their passengers are alive and on their way home!. ***The intersection of two world lines has to be an absolute, and so is indicated by the metric of space-time***.

Let us now assume that A1 is the event of *departure* of a *light beam* and A2 is the event of *arrival*; the *distance* between A1 and A2 is the *velocity of light* multiplied by the *time interval* between them and, ergo, the *event interval* has to be again equal to ***zero*** for all *inertial* observers (second Einstein's postulate). The *distances* and *times* will be different for different *Inertial Frames*, but the *space-time interval* will be ***zero*** for them all. In this case, we say that the *event interval* between A1 and A2 is ***light-like***. We then infer that the ***geometry*** of *space-time* is such that all pairs of events on the two 45° and 135° world lines in Figure 21 (right) have their mutual *event interval* equal to ***zero***! While in a *Euclidean space* the only way for the *distance* between two *points* to be ***zero*** is for them to ***coincide***, in *space-time* two *events* can be as far away as we want in terms of their *spatial distance* but be still 'together' because their *event interval* is ***zero***. The ***intersection*** of two *world lines* (previous case) is a special case of ***light-like*** *interval* where both *space* and *time intervals* are ***zero***. A ***light-like*** *event interval* is another ***absolute*** that derives directly from *Einstein's Second Postulate*, and so is indicated by the *metric* of *space-time*.

If, instead, the *distance* D and the *time interval* ΔT between events A1 and A1 are such that *light* covers D in ***less*** *time* than ΔT or, equivalently, the *distance* that *light* would cover in the *time interval* ΔT is greater than D, then: s^2 is positive. The *distance* and *time* will be different for different *Inertial Frames*, but the *event interval* will be identical for all of them. In this case, we say that the *event interval* is ***time-like***. The reason behind the name is that, when s^2 is positive, there always exists an *Inertial Frame* in which A1 and A2 coincide in *space* ($D = 0$) and, hence, the *event interval* is purely *temporal*. Remembering Chapter 2, it is clear that in this case A1 and A2 can be causally related so that its *temporal* order is univocal. If we supposed that A1 is the coordinate origin, Figure 22 shows A2 on the solid hyperbola with a value for $s = 5$ light-years. In fact, the two branches of the solid hyperbola define all the possible events in space-time for which s^2 =25, both in SC-A and SC-B, as well as in any other *Inertial Frame*.

As an instance of a ***time-like*** *event interval*, remember my trip to Las Vegas from Salt Lake City: given that -of course- the speed of my automobile relative to *terra firma* is ridiculously small with respect to that of light, departure (A1) and arrival (A2) are separated by a *time interval* much larger than what takes light to go from Salt Lake City to Vegas, $c.\Delta T$ is then much larger than the 672 km separating both cities and, hence, s^2 is positive. It is clear, thus, that there is an *Inertial Reference Frame* for which D is ***zero*** and, as we already saw, it is my own car! In this frame, I did not move at all during the whole trip so that the *event interval* that I experienced was strictly *temporal* -- leaving no doubt that A2 (arrival) occurred after A1 (departure). Also, my frame *temporal* axis is (Figure 22) the T_B axis of SC-B and the *time* measured on this axis is the so-called ***proper** time* (in repose). Using the standard *terra firma* as reference (SC-A), the *event interval* is a different combination of *space* and *time* (clearly with ***non-zero** time* and *distance*), but its value is exactly the same as when using the automobile (SC-B) as reference.

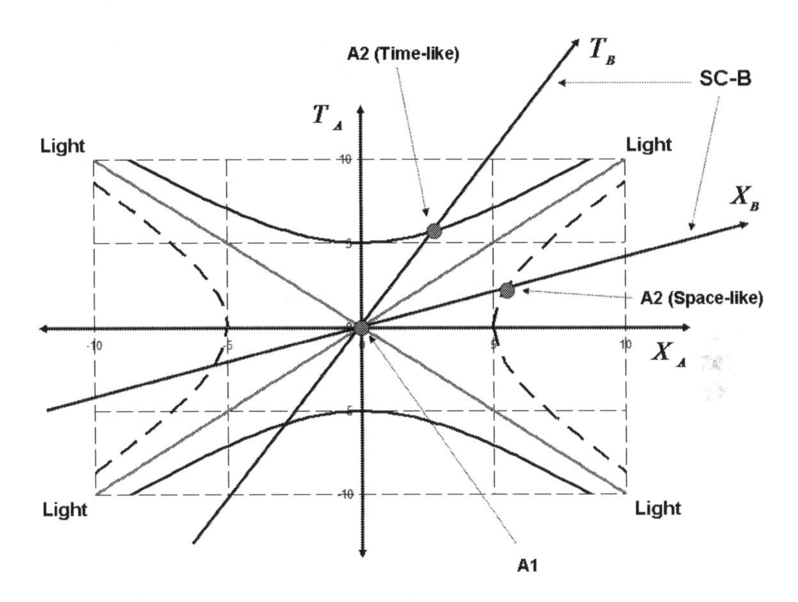

Figure 22. Event Intervals which are 'light-like, 'time-like' and 'space-like'

The *dilation of time* can be graphically confirmed: In SC-B, the time $T_B = s/c$ is 5 light-years divided by the *speed of light* (1 light-year/year), i.e. 5 years, while in SC-A, it is clear that the horizontal line drawn from the event (A2) toward the T_A axis intersects it above the solid-line hyperbola, so that it corresponds to another hyperbola with a larger *event interval* (larger s), i.e. T_A is larger than T_B . ***Using SC-A as reference, the clock -stationary in SC-B- runs more slowly (behind) than the clock stationary in SC-A.***

Finally, if the *distance* D and the *time interval* ΔT between events A1 and A2 are such that *light* takes from one to the other longer than ΔT or, equivalently, *light* cannot get from one place to the other in the time interval between the events, then s^2 is negative[10]. Not even *light* can establish a ***causal*** relation between both *events. Distance* and *time* will be different for different *Inertial Frames*, but the *event interval* will be identical for all of them. In this case, we say the *event interval* is ***space-like***. The reason behind the name is that, when s^2 is negative, there is always an *Inertial Frame* we can choose in which A1 and A2 coincide in time ($\Delta T = 0$) and, ergo, their *event interval* is purely ***spatial***. Figure 22 shows an *event* on the broken-line hyperbola whose interval is ***space-like*** with a value for $s^2 = -25$. In fact, the two branches of the broken-line hyperbola define the locus of all possible events in space-time for which $s^2 = -25$, both in SC-B, SC-A, and any other *Inertial Frame*.

10 The square of a number can be negative if we allow it to be complex (instead of real).

It should be clear (Chapter 2) that A1 and A2 -in this case- cannot be ***causally*** related because the fastest signal in the Universe (light) cannot travel from one place to the other in the *time interval* that separates the *events* and, therefore, the *temporal* ***order*** between them is ambiguous so that -besides the *Inertial Frame* in which A1 and A2 are ***simultaneous*** ($\Delta T = 0$), there are other *Inertial Frames* in which A1 occurs ***before*** A2, and others in which A1 occurs ***after*** A2. In the language of Chapter 2, A1 and A2 are *topologically* ***simultaneous*** and, then, their ***metrical*** *temporal* ***order*** changes from one *Inertial Frame* to another.

The contraction of space can be graphically confirmed: In SC-B the spatial coordinate X_B is equal to $D = \sqrt{-s^2} = 5$ light-years, while in SC-A, it is visually clear that the vertical line drawn from A2 towards the X_A axis cuts the latter to the right of the broken-line hyperbola, i.e. the intersection corresponds to another hyperbola with a greater value for $-s^2$ and, ergo, X_A is greater than X_B. ***Using SC-A as reference, the distance between events A1 and A2 -stationary in SC-B- is shorter than the distance between A1 and A2 in SC-A.***

It is the ***invariance*** of the *event interval* that gives the *space-time* of Minkowski a *metric* different to the *Euclidean metric* of classical *space* -- even though, in the latter, nothing prevented us from combining *spatial* and *time* coordinates. The classical *space-time* was simply a *temporal* ***succession*** of the classical *space*. Each term in the ***succession*** was a snapshot of the ***tridimensional*** *space*, simply because we thought that all places in the Universe 'lived' the same *instant*.

In Minkowski *space-time*, there is **no** snapshot common for all *space*; *space* and *time* cannot be separated in a ***unique*** way; each *inertial* observer experiences a ***different*** combination of them. Notwithstanding, the *event interval* is ***unique*** for all of them. The fact that the ***temporal*** *coordinate* in this *metric* appears with the ***opposite*** sign to the ***space*** *coordinates*, patently indicates the essential ***difference*** between *space* and *time* and, ergo, that *time* is not, as it is common to hear, "the fourth dimension of space".

4.3.1. Other Absolutes

In the same way that the *event interval* is ***absolute*** and resulted from combining two ***relatives*** (*space* and *time*), it is possible to combine *energy* and *momentum* in a **tetra**-dimensional expression that is ***invariant*** for all *Inertial Frames*. The *energy* is the ***temporal*** part, and the *momentum* is the ***spatial*** part for this new ***absolute***. *Different* inertial observers will measure *different* values for *energy* and *momentum* of an object, but all of them will measure the same value for the tetra-dimensional magnitude 'momentum-energy'.

It is also possible to combine *electric charge* with *electric current* in a single **tetra**-dimensional magnitude, the ***temporal*** part being related to *charge*, and the ***spatial*** component being related with *current*. If we are *in repose* with an *electric charge*, we only see its *electric field*; if we *move* relative to the *charge*, the latter becomes an *electric current* and, besides its *electric field* we see a *magnetic field*. *Different* observers will see *different electric* and *magnetic fields*, but there is an underlying **reality** which is unique for all of them: the *electromagnetic field*.

It is now apparent why Einstein never liked the name "Theory of Relativity" for his intellectual creature -- as the latter is simply the result of his passionate and rational quest for the ***absolute***.

4.4. Past, Present, and Future in Classical and Relativistic Physics

Let us choose the *place* where we are at this *moment* as the origin for all *Inertial Frames*. In Figure 23, then, my *space-time* ***present*** is represented by $X_A = 0; T_A = 0$. However, in classical physics (left), all the spatial Universe, i.e. all points on the X_A axis, share my ***present***, regardless of the *place* and their relative *motion*. The term 'present' does not need of the ideas of *place* or *motion*. The ***present*** in classical physics is thus the whole X-axis and not just the origin.

Similarly, in classical physics, *time* elapses -to use Newton's jargon- "equably" so that, after having passed 10 minutes for me, 10 minutes will have passed for the whole Universe, and my ***future*** then coincides with everybody else's ***future*** till the end of *time*. The same can be said regarding the ***past***. In this fashion, ***past***, ***present***, and ***future*** are ***absolutely*** *independent* of *space* and the relative *motion* between observers. I can only be conscious of ***present*** and ***past*** *events*, and ***past*** *events* may have occurred a few seconds ago though infinitely ***far*** away and still are in my ***past*** and everybody's ***past***. That is why the ***past*** is represented by the **bottom half** and the ***future*** by the **upper half** of the *space-time* diagram in Figure 23 (left).

Suppose that 99 years ago a supernova occurred at a place which is 100 light-years away from Earth. Can we say that this *event* is in our ***past***? According with classical physics, even when Newton would have accepted that *light* still did not have *time* to reach us and, then, we will not be aware of its occurrence till next year, his categorical response would have been: YES, it is in our ***past***. The vertical arrowed line in Figure 23 (left) corresponds to the *world line* for the star which -we assume- stays at a fixed distance from us (we are at the origin). The explosion could have happened just a nanosecond before our present ($T_A = 0$) and it would still -per our 'common sense' thinking- be in our past (even though only our grandchildren could not about it). The same can be said with respect to the classical ***future***. For Newton, the flow of *time* had nothing to do either with the ***causality*** of the Universe or with the impossibility (he never imagined) of univocally establishing -without a convention- the *simultaneity* of two *events*.

Contrariwise, for relativistic physics, the ***present*** is **not** an *instant* shared by **all** *space*, but an ***event***, i.e. an ***instant*** at a ***place*** in *space*. Likewise, the ***past*** and the ***future*** are **not** a group of *instants* shared by **all** *space*, but a ***group*** of ***events*** responding to an ***order*** relation with the ***present event***. The ***past*** for the ***present*** *event* is the ***set*** of all those *events* for which there is a physical signal that could have propagated from their *place* and *time* up to our ***present*** *place* and *time* and, given that *light* is the fastest signal in the Universe, the ***set*** of *events* called ***past*** is defined by the bottom cone in Figure 23 (right).

In other words: if an *event* **cannot** somehow influence our ***present*** *event*, it is **not** in my ***past***. I insist: the ***present*** is an ***event***; the ***past*** is a ***collection*** of ***events***. This ***set*** is a subset of what Newton called (and, still today, we do in our provincial world) the ***past***.

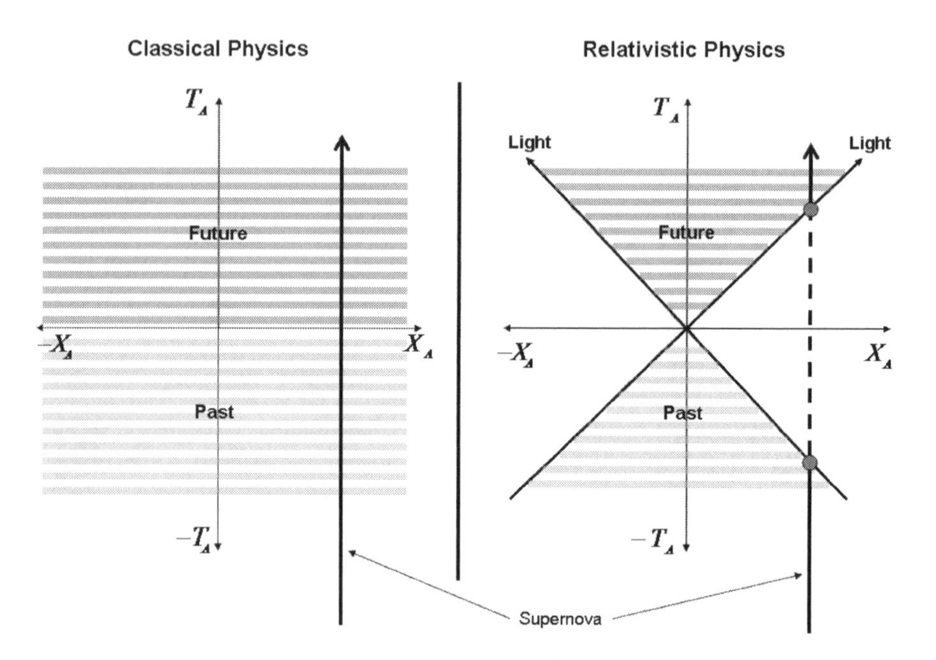

Figure 23. Past and Future in Classical and Relativistic Physics.

It is paramount for the reader to understand that, being the *event interval* ***absolute***, if all inertial observers in relative *motion* (in ***different*** *Inertial Frames*) share my ***present***, they will also share my ***past*** irrespective of the relative *motion*. Given that the definition of ***past*** is based on the possibility or impossibility of a ***causal*** relation existing between two *events*, the ***past*** for two observers that share the ***present*** has to be ***absolute***.

Could I be writing this book ***before*** I was born? Of course not! Unmistakably, the *event* of my birth is prior to the *event* of being writing this book, so it has to be part of my ***past***. Now, imagine an astronaut that flies by my side and *synchronizes* his/her *clock* with mine at the *moment* of our ***spatial*** coincidence so that we share the ***present***. Given that we know by now that *time* is **relative**, would it be possible for the astronaut to affirm that my birth is **not** in his past? Of course not! Otherwise, our sacrosanct *Principle of Causality* (so essential for our rational vision of the Universe) would be violated. The astronaut will **differ** on the amount of *time* that has elapsed since I was born, but s/he **never** could infer that our shared 'now' and 'here' is, for her/him, or ***simultaneous*** with, or ***prior*** to my birth. Both for her/him and for me, as for the rest of the Universe, I am writing this book **after** I was born.

The ***future*** of our ***present event*** is the ***collection*** of all those ***events*** for which there exists a physical signal that departing from *here* and *now* would arrive at those *places* within the *time interval* that separates them and, given that *light in vacuum* is the fastest signal in the Universe, the ***future*** of our ***present*** is defined by the upper cone in Figure 22 (right). In other words: if an *event* cannot be influenced by my ***present event***, it is not in my ***future***. I insist: the ***present*** is an ***event***, the ***future*** is a ***collection of events***. This **set** is a subset of what Newton called (and, still today, we do in our provincial world) the ***future***. If the supernova *event* is inside the upper cone or on it, we -at least in principle- could do something to prevent the explosion.

It is again supremely important that the reader realizes that, being the *event interval* ***absolute***, if all inertial observers in relative *motion* share my ***present***, they share my ***future*** as well, irrespective of our relative *motion*. Given that the definition of ***future*** is based on the possibility or impossibility of existing a ***causal*** relation between two *events*, the ***future*** for two observers sharing the ***present*** has to be ***absolute***.

Could the reader be reading this book ***before*** I wrote it? Of course not! Doubtlessly, the *event* of reading this book is posterior to my writing it and, ergo, it has to be part of my ***future*** (I am writing it now). Now imagine an astronaut passing by my side that *synchronizes* her/his *clock* with mine at the *moment* of ***spatial*** coincidence so that we share the ***present***. Given that we know that *time* is **relative**, would it be possible for the astronaut to objectively assert that the purchase and reading of this book is not in his ***future***? Of course not! Otherwise, the *Principle of Causality*, so essential to our rational vision of the Universe, would be violated. The astronaut could **differ** on the amount of *time* that will elapse between our shared *here* and *now* and the purchase and reading of my book, but s/he could **never** infer that our *here* and *now* is for her/him either ***simultaneous*** or ***posterior*** to his reading. For him/her, me, and for the rest of the Universe, I am writing this book **before** it is purchased and read (I hope).

Finally, in relativistic physics, there is a kind of 'limbo' associated with our ***present*** *event*: as we saw in Chapter 2, given our ***present*** *event*, all those *events topologically* ***simultaneous*** with it are neither *absolutely* in our ***past*** nor *absolutely* in our ***future***. Had the supernova been an *event* on the broken-line in Figure 23 (right), there would not have been a physical signal that could have reached our ***present*** to let us know of its occurrence. The *event* would have been too **far** away for the *time interval* between the *event* and our ***present***; its *temporal* ***order*** with us is **undetermined** and we have the liberty to establish it by *definition*. Employing Einstein's definition of ***metric*** *simultaneity*, the *temporal* ***order*** will become ***unique*** but, now, according to the *Lorentz's Transformation*, this unique order will be ***different*** for every *inertial observer*! For one observer in an *Inertial Frame*, the supernova is in the ***past***; for another in relative *motion* with the first, it is in the ***future***; for a third, it is ***simultaneous***. This 'limbo' is relative to the *frame*. When it is physically impossible for a ***causal*** relation between two *events* to exist, there is no inconsistency in obtaining **opposite** *temporal orders* for them, when measured from ***different*** *Inertial Frames*.

5. The Optical Clock as a Consequence of Einstein's Postulates

The *dilation of time* can be deduced with elemental algebra (Pythagorean Theorem), if we analyze the behavior of two ***optical** clocks*, each one stationary in a different *Inertial Frame*. As we learned in Chapter 2, the *Principle of Causality* demands that if a dynamic physical process that is fully ***isolated*** from the exterior is found on the same state twice, then the *succession* of states between the first state and its return to it, constitutes by definition a *cycle* which, being the process completely isolated, will continue recurring in *time* and, ergo, could be employing as a *clock* by assuming that each *cycle* has the same *duration* (***uniformity*** of the *time* ***unit***).

Einstein's Second Postulate states that *light speed in vacuum* measured by ***different*** observers, each ***stationary*** in a ***different*** *Inertial Frame*, will be always the ***same***. In short, the process of *light propagation in vacuum* is ***universal*** and fully ***insensitive*** to external influences of any nature, i.e. it is perfectly ***isolated***. Being that the case, an obvious way of defining the ***unit*** of *time* and assuring its ***uniformity*** in an arbitrary *Inertial Frame*, is to make *light* travel a ***fixed** distance* forth and back between two mirrors (Figure 24, left) -- conceiving an ***optical** clock*. For instance, if the mirrors were separated 150,000 km[11], and the *clock* emitted one tick every time that the *light beam* completed a round trip, then we would have defined the unit called 'second'; if we choose a more practical *distance* so we can fabricate the *clock*, say 15 cm, then we would have defined the unit called 'nanosecond', and so forth. Given that the *light beam* always travels the **same** *distance* between ticktacks, and that its *speed* is ***constant***, we have the assurance of ***uniformity***. Likewise, every other *Inertial Frame* can define the flow of *time* with their corresponding ***optical** clocks*. Please, do not forget that, had the ***ether hunt*** been successful, i.e. had the ***ether wind*** exerted an action on *moving* objects, the *speed of light in vacuum* would have been different in different *Inertial Frames*, so that the ***optical** clock* would not have been an ***isolated*** process, and we could detect *motion* by conducting ***optical*** experiments in our *reference frame*. It is the *Principle of Special Relativity* that guarantees an ***optical** clock* as a perfectly ***isolated*** system, so as to provide a ***uniform unit*** of *time*.

If we have side by side an ***optical** clock*, an ***atomic** clock*, and a piece of the element ***radium*** disintegrating -all in relative repose- the *time* defined by all of them will be ***identical*** because they are perfectly ***isolated*** processes. The disintegration pace of ***radium*** measured with a ***stationary optical** clock* will be always the same independently of the *Inertial Frame* state of *motion*. What we are measuring is the ***proper** disintegration time* of ***radium***, i.e. a *time* ***measured*** with a *clock at rest* in the *frame*. Had it not been this way, the *difference* in the ***proper** disintegration time* of ***radium*** between two different *Inertial Frames* would have allowed us to distinguish one frame from another, and we would have abandoned the *Principle of Special Relativity*. The ***optical** clock* is thus as valid as the ***atomic** clock* or a ***radioactive*** process. All of them are virtually ***isolated***.

Let us build two ***identical** optical clocks*, i.e. the *distances* between their respective mirrors are congruent when put side by side. Each clock has an inte-

11 For the sake of simplicity, we assume light speed is exactly 300,000 km/s.

grated microcomputer which simply counts the number of times the *light beam* completes a round trip, and we corroborate that they run ***in unison***. Now, let us use one of the *clocks* to define *time* in the *Inertial Frame* SC-A, and the other to define *time* in another *Inertial Frame SC-B*. Both an observer in SC-A and another in SC-B, when measuring -with their respective rules- the *distance* between mirrors (***proper** length*), they obtain 15 cm. Consequently, from *Einstein's Second Postulate*, each tick of the clocks indicates the passing of one nanosecond in their respective *reference frames*. In this way, the *time* elapsed in nanoseconds between two events will simply be the number of *cycles* accumulated by the respective microcomputer. Let us call the *light beam* in SC-A the A-beam, and the *light beam* in SC-B the B-beam.

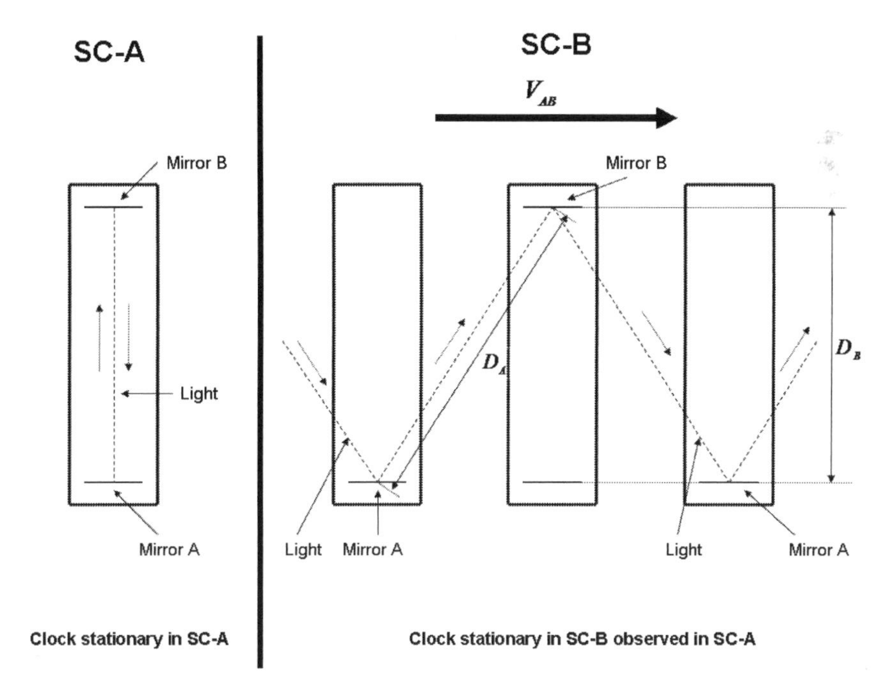

Figure 24. The Optical Clock is a perfectly isolated System

We now assume that SC-B is ***uniformly** moving* relative to SC-A with *velocity* $V_{A/B}$ *perpendicularly* to both *light beams* in both frames (Figure 24). *Einstein's Second Postulate* still is valid, so each tick still indicates the passing of one nanosecond in each frame. On the other hand, the *First Postulate* (*Principle of Special Relativity*) assures that there is no experiment (mechanical or electromagnetic) that, without ***external** reference*, will entitle the observers to determine that their *reference frames* are in *motion* and, if with ***external** reference*, they can only infer their ***relative** motion* (all they can aspire to). Will the two clocks -now that they are in relative *motion*- still run ***in unison***? If we answer this question with our millenary prejudices, we will respond with a categorical 'of course they will, unless one of the clocks is faulty'.

Let us first deal with the observer in SC-A who, using his/her *clock*, observes the clock in SC-B. Given that SC-B and its *clock* are *moving* relative to SC-A with *velocity* $V_{A/B}$, in the *time* that the B-beam travels from mirror A to mirror B, the very *clock* has ***moved*** to the right in SC-A; likewise, in the *time* that the B-beam returns from mirror B to mirror A to complete a tick, the *clock* moved farther to the right in SC-A (though ***stationary*** in SC-B). It is clear then that the B-beam travels ***more** distance* when observed from SC-A than when observed from SC-B. ***Reciprocally***, the A-beam travels ***more** distance* between ticks when observed from SC-B than when observed from its own *reference frame*. By now this should not surprise us, as we know that the *trajectory* of an object is ***relative*** to the *reference frame*.

But... then -using SC-A as reference- because the *speed* of the A-beam is the same as the *speed* of the B-beam, when the *clock* in SC-A has completed a *cycle*, emits its click, and the microprocessor increments the count by one, the *clock* in SC-B still has not completed its *cycle* because it has to cover ***more** distance* at the ***same** speed* and, ergo, from the vantage point of SC-A, the *clock* in SC-B runs ***behind*** (runs ***more** slowly*) and a *time interval* $\Delta T_{A/B}$ ***measured*** by the *clock* in SC-B (the number of cycles completed) but ***observed*** in SC-A, will be ***shorter*** than the *interval* $\Delta T_{A/A}$ ***measured*** by the clock in SC-A (number of cycles completed) and ***observed*** in SC-A.

In a few words: even though, without external reference, the observers in each *Inertial Frame* do not notice anything unusual, when they interact, they find that their respective *clocks* do **not** run ***in unison*** as they did when they were in relative *repose*. It is not difficult to prove, applying Pythagoras' theorem to the triangle defined by the B-beam when observed from SC-A, that the relation between $\Delta T_{A/B}$ and $\Delta T_{A/A}$ is:

$$\Delta T_{A/B} = (\sqrt{1 - V_{A/B}^2 / c^2})\Delta T_{A/A} = \beta.\Delta T_{A/A}$$

Mutatis mutandis, if the observer stationary in SC-B compares her/his *clock* with the *clock* in SC-A, the latter is the one with a saw-tooth *trajectory*, while the *trajectory* for the former is vertical. When the *clock* in SC-B completed a *cycle*, the A-beam is still on its way, so that it is ***behind*** the *clock* in SC-B. As expected, the relation between the interval provided by both clocks is fully reciprocal, i.e. ***structurally identical*** to the previous one:

$$\Delta T_{B/A} = (\sqrt{1 - V_{B/A}^2 / c^2})\Delta T_{B/B} = \beta.\Delta T_{B/B}$$

Given that it is an experimental fact that all ***isolated*** periodical processes -when used as *clocks*- define the same *time* as long as they are in relative *repose* ($\Delta T_{A/A} = \Delta T_{B/B}$), it is concluded that the *dilation of time* we deduced using an ***optical** clock* is valid in general, i.e. we have proved -through a different route- that the two *Einstein's postulates* imply the validity of the *Lorentz's Transformation* (with the ontological interpretation of Einstein). In order to conceptually understand the practical consequences of these strange phenomena of ***spatial** contraction* and ***time** dilation*, as well as to accept its physical **reality**, we shall now study in detail an ***imaginary*** case of human *space* travel, and a **real** case in the subatomic world.

6. Space Contraction and Time Dilation in Sidereal Space

Imagine a spaceship traveling with a speed 'v' (relative to Earth) equal to 80% of the speed of light in vacuum ($v = 0.8c$), and passes by our planet heading to Gliese 581 -- a star about 20 light-years away from us, and rotating around which there are several planets seemingly meeting the conditions for the existence of life. The fraction 0.8 and the distance 20 light-years are very convenient as they produce a factor $\beta = \sqrt{1-0.8^2} = 0.6$ and all distances and times are integers. Besides, we assume that Gliese 581 can be considered in relative *repose* with our planet, so that both bodies define one of the *Inertial Frames* we will employ. The spacecraft is the other *frame*.

It is essential to understand the relevance of the expression '...passes by our planet...', i.e. we do not allow the spaceship to take off from Earth or land on a planet around Gliese 581: both events would involve *non-Inertial Frames* and, hence, out of the validity field for Special Relativity. The conclusion we shall attain are only valid if Earth and Gliese 581 can be considered as an *Inertial Frame*, and the spacecraft *moves* with ***uniform** velocity* with respect to Earth/ Gliese 581 since it passes by Earth until it passes by Gliese 581. Under those conditions, the *spaceship* constitutes another *Inertial Frame*.

We ***synchronize*** the *clock* on Earth with the *clock* on the star with Einstein's technique and in such a way that they display ***zero** time* at the *instant* when the spaceship passes by our planet; the astronaut sets his/her *clock* to ***zero*** as well. In Figure 25 (top half), *events* are shown as they would be observed in the Earth/Gliese 581 *frame*. Its bottom half depicts the *events* as seen when using the spaceship as *reference frame*. Notice that the spaceship *length* measured by the astronaut (bottom) is ***longer*** than when measured in Earth/Gliese 581 (top) because the former is the ***proper** length* and the latter is its ***moving** length*. Similarly, the Earth is ***smaller*** in the *direction* of *motion* when ***measured*** from the spaceship (bottom) than when ***measured*** by us on Earth (top) because the former is its ***moving** size* while the latter is its ***proper** size*. In both halves, at the top, the *spatial **coincidence*** of our planet with the spaceship is depicted; at the bottom, the *spatial **coincidence*** of the ship with the star is shown. We also see the indication of the *clocks* on Earth, star, and ship.

Given that the *distance* between Earth and Gliese 581 -measured in their own Inertial Frame- is 20 light-years and the ship travels at 0.8c, it will take 20/0.8=25 years for the ship to pass by the star. Because the *clocks* on Earth and star displayed ***zero*** at the start and we assumed both are in relative *repose*, both will indicate ***25 years*** when ship and star spatially coincide. This is our usual thinking process. Again: in that time ***measured*** with a *clock **stationary*** with us (our ***proper** time* on Earth), all of us will have aged (within individual variations) the equivalent of ***25 years***. Now, let us investigate the evolution of the *clock* inside the spacecraft. We know that when the ship passed by us on Earth, its clock indicated ***zero*** (because the astronaut so set it). What *time* will it indicate when passing by the star? 25 years? NO!

The ship is an *Inertial Frame* in relative *motion* with the *Inertial Frame* Earth/Gliese 581 within which, according with the *Lorentz's Transformation*, the *time* indicated by a *clock **stationary*** on the ship, will be the fraction β

of the *time* elapsed on Earth, i.e. 25x0.6=***15 years***. When the ship and star are ***contiguous*** in *space*, the two clocks can be compared and confirm without ambiguity that one show ***25 years*** while the other indicates ***15 years***! Does it mean that the astronaut aged ***10 years*** (40%) ***less*** than Earthlings and any aliens living on a planet around Gliese 581 did? Not so fast! Let us investigate the *events* from the vantage point of the astronaut.

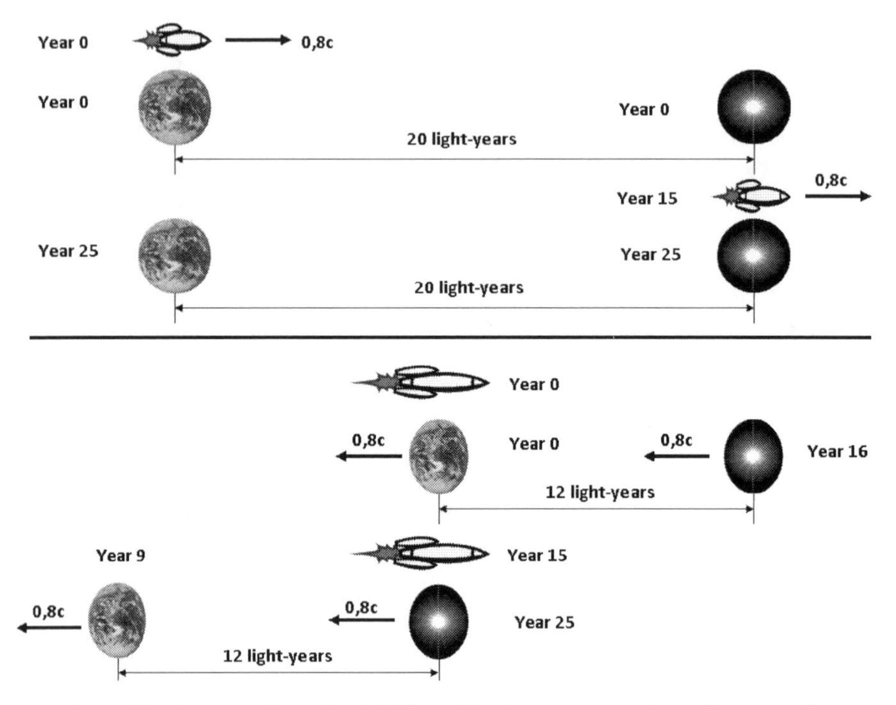

Figure 25. Top: seen from Earth/Gliese 581; Bottom: seen from the spaceship

We can now imagine the spaceship *in repose*, and the Earth/star system approaching the astronaut with a speed of 0.8c[12] so that his ***measurements*** of *space* and *time* are entirely ***reciprocal*** to ours. The ***fixed*** *distance* between Earth and the star is now a ***moving segment*** and, when the astronaut measures it, obtains a value which is the fraction β of the value we obtain on earth, i.e. 20x0.6=***12 light-years***! Likewise, the Earth and star diameters are -for the astronaut- 40% (100-60) ***smaller*** in the *motion direction* than the values we measure *in repose* with those celestial bodies. Naturally, when the *spatial* ***coincidence*** between ship and Earth occur, both *clocks* indicate ***zero*** because they so agreed on. Naturally as well, given that the results of the *clocks* when checked side by side cannot depend of the *reference frame*, when the ship is ***contiguous*** to the star, both ***clocks*** must indicate the *time* they did when we observed everything from the earth/star reference system: ***15 years*** on the ***ship*** and ***25 years*** on the ***star***.

Being the *distance* between Earth and star -measured in the spaceship- only ***12 light-years***, and because the earth/star system travels at 0.8c, the star will

12 To be precise, from the discussions in Chapter 4 regarding the symmetry needed for motion to be relative, not only the Earth/star system but the whole rest of the Universe is to be assumed moving relative to the spaceship.

take 12/0.8=***15 years*** to arrive at the ship, and that is why the *clock* on the ship indicates ***15 years*** when they pass by each other. It is remarkable that the *time* elapsed on the ship when we used the earth/star reference system was also ***15 years*** and it is so because, as we know, *space-time* ***coincidences*** are necessarily ***absolute***. Now, how does the *clock* on Earth evolve if seen from the spaceship? We know that when Earth passed by the ship, both *clocks* displayed ***zero*** by convention. What *time* will the *clock* on Earth indicate when the star passes by the spacecraft? 15 years? NO! Oh, I know: 25 years? NO!

The Earth/Gliese 581 is an *Inertial Frame* in *motion* relative to another *Inertial Frame* (the ship) within which, according to the *Lorentz's Transformation*, the *time* shown by a *clock* ***stationary*** on Earth will be the fraction β of the *time* elapsed on the spaceship, i.e. 15x0.6=***9 years*** so that the *clock* on Earth will indicate 'year 9' when ship and star are ***contiguous***. Using, thus, the ***earth/star*** system as reference we concluded that, on Earth (and on the star) had elapsed ***25 years***, but using the ***spaceship*** as reference, we now conclude that on Earth (and on the star) only ***9 years*** elapsed. Surprised? You betcha! We still resist accepting that both the *time interval* between two *events* and the *distance* between them are ***relative*** to the *reference frame*. So... **not** 15 years, **not** 25 years, but ***9 years*** is the right answer!

On the other hand, when ***ship*** and ***star*** are contiguous, their *clocks* can be contrasted and, again, the *clock* on the star must indicate ***25 years*** -- as its ***local*** reading cannot admit ambiguity. But, because Earth and star are in relative *repose*, their *clocks* run ***in unison*** and, therefore, if ***9 years*** elapsed on Earth, the same happened on the star, so that the *clock* on the star must have indicated the ***year 16*** (25=16+9) when the Earth *spatially* ***coincided*** with the ship.

Now, here is the big question which baffles to this day all detractors of the theory: given that, seen from the spaceship, ***9 years*** passed on Earth/Gliese 581 while ***15 years*** did on the ship, does it mean that Earhlings and aliens on Gliese 581 aged ***6 years*** (40%) ***less*** than the astronaut? Did we not conclude -when seen from the earth/star reference- that the astronaut had aged ***10 years*** (40%) less than Terrestrials and aliens? It is impossible for an individual to age 40% ***less*** than another and the latter to age 40% ***less*** than the former, simply because we changed our vantage point. Once they are face to face, only one of them may have more gray hair and wrinkles! There must be an explanation for this apparent paradox. Hold on till the upcoming chapter.

The *time* elapsed on Earth according to the astronaut is ***9 years***, but the ***proper*** *time* for Terrestrians, i.e. the *time* ***lived*** and ***measured*** in a frame *at rest* with us is ***25 years***; the *time* elapsed on the spaceship according to the *clocks* here on Earth and the *time* measured by the astronaut (his ***proper*** *time*) are, instead, the ***same*** and equal to ***15 years***. The first two *time intervals* ***differ***; the second two *time intervals* agree. The reason for the agreement in the second case is that the spacecraft is the only one of the three protagonists which is ***present*** in both *events* and, ergo, constitutes an *Inertial Frame* in which its *event interval* is purely ***temporal***. From the vantage point of the Earth/star, the spaceship traverses 20 light-years in 25 years (20/0.8) and the *time* in the spaceship dilates to 60% of 25 years (15 years); from the vantage point of the spaceship, it is the *distance* between Earth and Gliese 581 that

-fully equivalently- contracts to 60% of 20 light-years (12 light-years) so that -at the speed of 0.8c- the spaceship experiences 12/0.8=15 years for the trip. Obviously, the time elapsed inside the ship has to be the same from both *frames of reference*.

Regardless of all these discrepancies and agreements between different vantage points, the time lived by Terrestrials cannot be other than their ***proper** time*, i.e. ***25 years***, and the *time* lived by the astronaut cannot be other that his/her ***proper*** time, i.e. ***15 years***; leaving room for individual differences, it is people on Earth that have more gray hair and more wrinkles! Notwithstanding, the apparent paradox described two paragraphs back (due to the ***reciprocity*** of relative *motion*) has yet to be explained. The perspective from Earth/ Gliese 581 frame agrees with the conclusion based on the ***proper** times*, but the perspective from the spaceship imposes a paradoxical conclusion.

I finish this imaginary case reminding the reader that even though we talked about ***contrasting** clocks*, gray hair, and wrinkles, under no circumstance we allowed the astronaut and Earthlings to be in relative *repose* for such a comparison, as that would have implied that at least either the former or the latter had changed of *Inertial Frame*, with the necessary *acceleration* and, in such a case, Special Relativity would not have applied. The so-called 'twin paradox', in which one of the twins leaves Earth (of course *accelerating*) and returns (of course reversing his course and *decelerating*) to find his twin much ***older*** than s/he is, can only be discussed and understood after presenting the General Theory of Relativity in the next chapter.

7. You better believe it! Space Contraction and Time Dilation in the Subatomic World

It is then obvious that for the effects of the relativity of *space* and *time* to be ***measurable***, the relative *speed* between two *frames* has to be a significant fraction of the *speed of light in vacuum*. In our macroscopic world, and particularly if one of us pretends to be the traveler as in the previous section, we still do not command the appropriate technology to develop such relative *velocities*. However, without realizing it, we are daily moving at *speeds* comparable to that of *light* with respect to, e.g. the electrons hitting the screen on a TV tube, or to the ions traveling in each cell of a plasma TV screen so we can watch our favorite movie, or to *cosmic radiation*. Cosmic radiation is a well-known phenomenon and was utilized in 1963[13] to prove that the effects of *space contraction* and *time dilation* really occur between the *frames* of a ***subatomic particle*** and ***terra firma***.

Cosmic radiation consists basically of protons (nucleuses of Hydrogen), originated outside our solar system, which are constantly bombarding our planet at speeds close to that of light (0.995c). When a cosmic particle, while entering our atmosphere, collides with an atom of Nitrogen or Oxygen, new subatomic particles called muons (μ-mesons) are produced with a *mass* intermediate between the *mass* of the proton and that of the electron. This particle is very unstable and ***disintegrates*** into an electron and other subatomic particles. As we will see in the sequel of this book entitled *What is Reality? Einstein, Quantum*

13 D.H. Frisch and J.H. Smith, American Journal of Physics 31 (1963): 242-355.

Physics, and Folklore, the ***disintegration*** of a single atomic particle is impossible to predict, but the statistical behavior of a large population of them is predictable with great accuracy.

The phenomenon of radioactive disintegration is such that ***equal*** *fractions* of the population disintegrate in ***equal*** *times*. Mathematicians say that radioactive decay obeys an exponential law (Figure 26). This decay process is practically ***independent*** of ambient *temperature* and *pressure*, i.e. it has all the characteristics of an ***isolated*** system and, hence, could be used as a *clock*[14]. In sum, when we measure a material radioactive decay with a *clock* in relative *repose*, we can statistically define an ***average*** *lifetime* for the population, and this number is a property of any population comprising the same type of particles. For instance, the ***average*** *lifetime* for Uraninum-238 is 6,500 million years and that is why it is still found on our planet; the ***average*** *lifetime* for Radon is only 5.5 days; and the ***average*** *lifetime* for a muon is a meager 2,200 nanoseconds. This short ***average*** *lifetime* is measured producing very ***slow*** muons in particle accelerators so that its decay can be observed as if they were practically in relative *repose* with *terra firma*. Figure 26 shows that after 670 ns, 70% of the population of muons is intact (30% has decayed); after 2,200 ns, 37% remains as such; and after 6,700 ns, 95% of the muons has decayed.

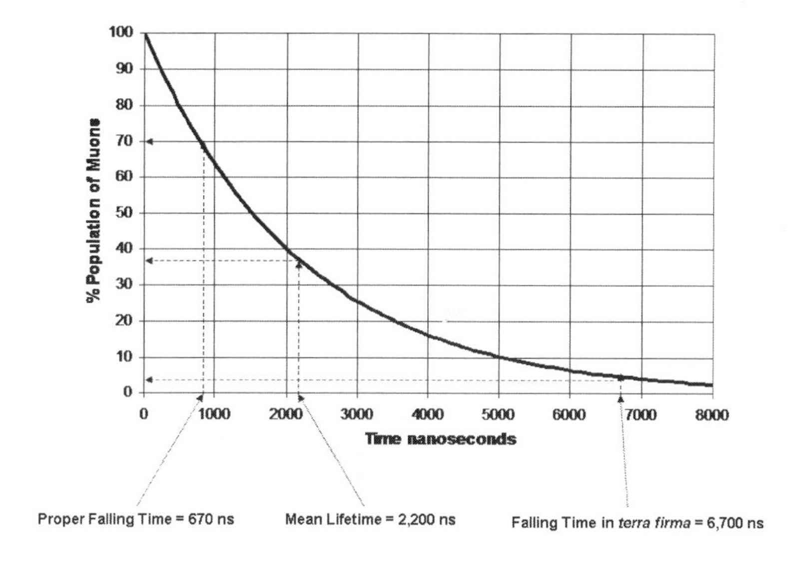

Figure 26. Disintegration of Muons Radiation

The experiment at Mount Washington (New Hampshire, USA), consisted in measuring the shower of muons, first on the apex of the mountain at 2,000 m above sea level, and then at sea level in Cambridge, Massachusetts. This cosmic shower is practically ***constant*** in *time* and does not change either with small changes in *latitude* and *longitude* -- as long as the ***altitude*** is the same. Besides, the experimenters had a detector that measured only those muons whose *speed* corresponded to 99.5% of that of *light*. The number of muons

14 In fact, Madame Curie suggested employing Radium decay to define the unit of *time*.

measured in an hour at sea level was 70% of the number measured on the mountain top, i.e. only 30% of the population of muons decayed during the time that took them to reach the floor.

Figure 27 depicts the *events* of *spatial* ***coincidence*** between the muon and the apex (left top), as well as between the muon and the floor (left bottom) from the vantage point of *terra firma*. The half on the right shows the same *events* as observed from an *Inertial Frame* in which the muon is *at rest*; the bottom part shows the *spatial* ***coincidence*** of muon and apex; the top part depicts the muon at the floor. Notice that on the left graphs, the dimension of the muon in the *direction* of *motion* is reduced as it is its ***moving*** *size* that is measured; reciprocally, the height of the mountain is reduced on the right graph as it is the mountain ***moving*** *size* that the muon 'observes'. Since the muon speed is 0.995c, the *contraction/dilation factor* is $\beta = \sqrt{1-0.995^2} = 0.0999 \cong 0.1$ (10%).

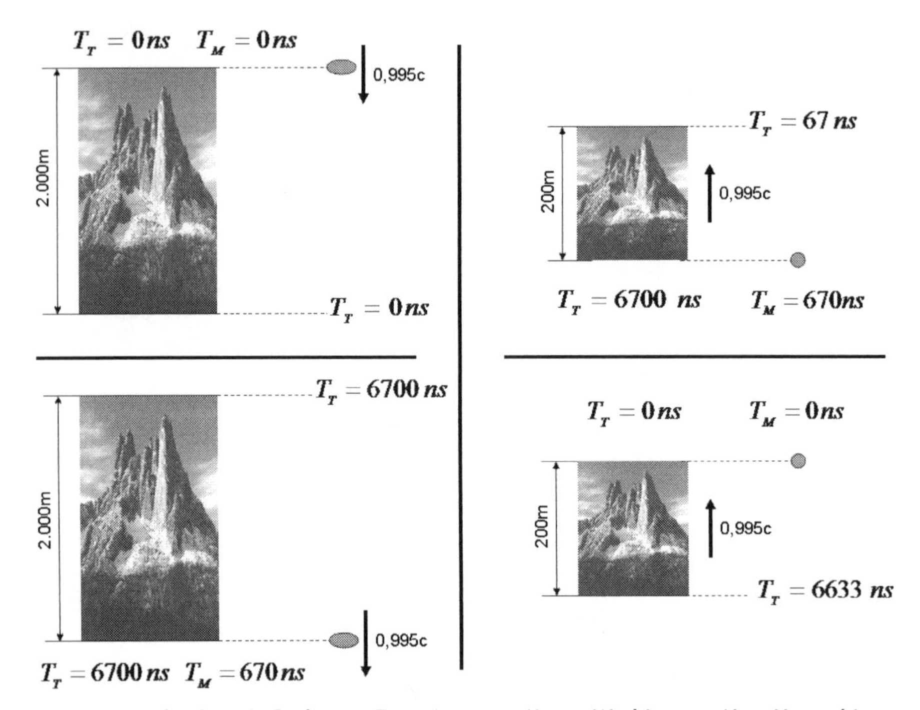

Figure 27. The Cosmic Radiation Experiment at Mount Washington, New Hampshire

Because the muon travels 2,000 m at 0.995c m/s, using *terra firma* as reference, it will take 2,000/(0.995x300,000,000[15])=***6,700 ns*** to reach the floor. Besides, according with the exponential law of decay for the muon, after 6,700 ns, only 5% of the population should have survived (Figure 26). How do we explain then that the detector measured 70% of the population still intact at sea level? The explication resides in that the decay law in Figure 26 was determined observing the population in relative *repose* with *terra firma*, while in this situation, the muons are *moving* relative to *terra firma* with a tremendous *velocity* close to that of light. Pithily: Figure 26 corresponds to the muon

15 Again, we use simply 300.000 km/s for the speed of light in vacuum.

proper average *lifetime.*

In order to properly use the experimental data in Figure 26, we need to calculate how much of the ***proper*** ***time*** of the muon elapsed when falling from apex to floor. If using the muon ***proper*** ***time*** and Figure 26, we obtain a percentage of decay which agrees with the one given by the detector, then, and finally, we should not have any further doubts as for the **reality** of the *spatial* ***contraction*** and *time* ***dilation*** -- irrespective of how strange and fantastic they may appear to us.

We already saw that the muon radioactive process provides an accurate way of measuring *time*, i.e. it is a natural *clock*. When we measure the population at the apex, we trigger the muon ***internal*** *clock* (we set it to zero) and we do the same with our *clock* on *terra firma*. Please notice that a *clock* at sea-level (properly synchronized) would also indicate ***zero***, as all clocks within an *Inertial Frame* run ***in unison***. When we measure the cosmic shower on the floor, our *clock* (at rest on Earth) will indicate (as we calculated) that T_T =***6,700 ns*** have elapsed and, for the same reason, the *clock* we left on the mountain zenith will also show ***6,700 ns*** (Figure 27, left bottom).

And... what time T_M will the muon clock (the decay process is a clock, but you can imagine another clock at rest with the muon) indicate when we measure the population on the floor? Well... we have done this already for our space travel: it will be 6,700 multiplied by the dilation factor, i.e. $T_M = \beta.T_T$ = ***670 ns*** and, according to Figure 26, after 670 ns, 70% of the population should still exist and... is it not what the detector measured? But... is 670 ns the ***proper*** ***time*** for the muon? For now, that number was calculated as the *time* indicated by the muon clock as seen from *terra firma*. Let us calculate the *time* indicated by the muon *clock* as seen from its own *reference frame* (Figure 27, right).

The muon can now be considered as immobile and 'sees' the mountain passing upwards at 0,995c m/s. When the apex coincides with the muon and we measure its population, we trigger its internal clock (begin measuring its decay), i.e. $T_M = 0$ and we set our clock to zero as well ($T_T = 0$). But, be aware that the clock on the floor does not have to display zero! And... what time is it displaying then? Let us leave that question unanswered for a little while as it requires some cogitation. The height of the mountain seen from the muon frame is the length of a ***moving segment*** so that it will be its proper length (2,000 m) multiplied by the contraction factor β, i.e. $A_M = \beta.A_T = 200m$ -- ten times ***shorter*** than seen from *terra firma*. Given that the mountain is rushing upwards at 0.995c relative to the muon, only 670 ns will pass since the muon 'saw' the mountain top until it 'sees' its bottom -- ten times ***less*** ***time*** than when seen from *terra firma*. As expected we confirm that 670 ns is effectively the muon ***proper*** ***time*** for its trip and, besides, that it agrees with the *time* that -from *terra firma*- we see it has passed for the muon -- exactly as it happened with the astronaut in the previous section.

When the floor reaches the muon, each *clock* should tell the same *time* it did when we observed the event from *terra firma*, i.e. $T_M = 670$ ns and $T_T = 6.700$ ns. And what *time* will the *clock* we left on the apex read? It has to read the *time* elapsed on *terra firma* from the muon frame, and how long is that? It will be the ***proper*** *time* of the muon times β, i.e. 670x0.1=67 ns and, ergo, the clock on the mountain zenith will read 67 ns. In summary, it is essential to understand that, due to the ***reciprocity*** of relativity, the *time* on *terra firma* measured from the muon frame is 67 ns and not 6,700 ns! From this, it is straightforward to deduce that the clock on the floor seen from the muon *frame* has to have read 6,700-67=6,633 ns at the moment when the muon coincided with the apex; only being so, that clock could have read 6,700 ns when muon and floor were contiguous.

The time elapsed on *terra firma*, according to the muon *clock* is 67 ns, but the ***proper*** *time* for us on Earth, i.e. the *time* ***lived*** and ***measured*** in a frame in relative *repose*, is 6,700 ns; the *time* passed in the muon *reference frame* according to the clock on *terra firma* and the ***proper*** *time* of the muon are, instead, the same and equal to ***670 ns***. The first two *times* ***differ***; the second two times ***coincide***. The reason behind the coincidence is that the muon is the only one of the three protagonists which is present at the two events (apex/muon and floor/muon) and, hence, it is a *frame* in which the *event interval* is purely ***temporal*** (the muon does not move relative to itself). From the vantage point of *terra firma*, the muon traverses 2000 m in 6,700 ns, and the time in the muon frame dilates to 10% of 6,700 ns, i.e. ***670 ns***; from the vantage point of the muon, it is the height of the mountain that -fully equivalently- contracts to 10% of 2000 m (200 m) so that -at the speed of 0.995c- the muon experiences 200/(0.995x300,000,000)=***670 nm*** for its trip. Obviously, the *time* elapsed in the muon *frame* has to be the same from both *frames of reference*.

The time lived by us is our ***proper*** *time*, i.e. ***6,700 ns***; the time lived by the muon is its ***proper*** *time*, i.e. ***670 ns*** and that is why 70% of the population was still 'alive' when reaching the floor (Figure 26). It is interesting the observation made by Clifford M. Will in his book *Was Einstein Right?* [10], namely that, thanks to *space* ***contraction*** and *time* ***dilation*** phenomena, cosmic radiation is considered as one of the possible causes for the genetic mutation needed for the evolution of the species on our planet.

I sincerely hope the reader is now convinced that both *space* ***contraction*** and *time* ***dilation*** predicted by Special Relativity are real and palpable phenomena, though only significant when the relative *velocities* between *frames* are a sizable fraction of the *speed of light in vacuum*. Even though the human space travel at such high speeds is still imaginary, the agreement between the theory and experimental observation for this experiment with cosmic radiation, and a multitude of others undertaken during the last 100 years, indicate that our hypothetical case of an astronaut traveling from Earth to a planet in Gliese 581 has to be correct.

We only need now to understand the fundamentals and conclusions of the General Theory of Relativity treated in the next chapter and, then, we shall be ready to imagine the twin taking off from Earth, *accelerating* up to a ***constant*** *speed* close to that of *light*, *decelerating* to land on the supposedly habitable planet of Gliese 581, rest a little, and return to Earth so he can hug his twin brother before it is too late.

8. A non-Inertial Frame of Reference is locally and momentarily Inertial

The Special Theory of Relativity is only valid for *Inertial Frames of Reference*. A *reference frame* in *accelerated motion* relative to an *Inertial Frame* is considered, in Special Relativity, as *non-Inertial*. In other words: *velocity* is ***relative***, but *acceleration* remains ***absolute*** as in classical mechanics. We also learned that in Special Relativity, *space-time* is *semi-Euclidean* so, once an *Inertial Frame* is chosen (i.e. once a particular partition of *space-time* into *space* and *time* was adopted), the tridimensional *space* so obtained is always *Euclidean*. *Light* propagates in a *straight line* and with ***identical*** *speed* in all *Inertial Frames* (its *world line* is a *straight line*); objects free of any external influence move also ***uniformly*** (their *world lines* are *straight lines*); objects under the influence of a *force* move ***non-uniformly*** (their *world lines* are ***curved***); the *distance* between two *points* in tridimensional *space* obeys the Pythagorean Theorem; the sum of the angles of a triangle is always 180°; the ratio between the length of a circumference and its diameter is always the number π (3.1416...), and so forth.

In addition, we should realize that given an *Inertial Frame* and another *frame accelerated* with respect to the former, if we consider in the latter a zone of *space* sufficiently ***reduced*** and an *interval of time* sufficiently ***short***, its *velocity* with respect to the former will be practically ***constant*** and, ergo, in that ***reduced*** portion of *space-time*, the ***accelerated*** *reference frame* can be considered as in ***uniform*** *motion* relative to the *Inertial Frame* so that it is -temporarily and locally- an *Inertial Frame*. Rephrasing: there is always a region of *space-time* sufficiently reduced to be considered *semi-Euclidean*. The idea is the same that allows us to consider a small region of our planet as planar, even though it is globally spherical (Chapter 3), or allowed the *ether hunters* (Chapter 5) to assume that our planet *moved* (locally and temporarily) with ***uniform*** *motion* relative to the Sun --when we well know that is traveling around it and, therefore, its motion is *accelerated*. Now really succinct: every *accelerated motion* can be considered as a ***succession*** of multiple ***uniform*** *motions*.

Wrapping up, every *non-Inertial Frame* is ***local*** and ***momentarily*** *inertial*, because in such a limited portion of *space-time*, it can be considered as in ***uniform*** *motion* with respect to some *Inertial Frame* and, ergo, from the *Principle of Special Relativity*, it has to be *inertial*. Here the word 'local' may mean very extended regions of *space* with respect to the human scale and the word 'momentarily' may mean an *interval of time* quite considerable for us; this is why Special Relativity, exclusively valid -in principle- for *Inertial Frames*, has enjoyed and continues to enjoy an astonishing success.

The problem that existed in classical mechanics and continues to exist in Special Relativity is that we do not know how to find an *Inertial Frame of Reference*; there is no criterion for it: the only way is to assume it is *inertial* and experimentally corroborate that the laws of Nature as described by Special Relativity are valid. This is the reason why I always started my wording of the *Principle of Galilean Relativity*, the *Principle of Newtonian Relativity*, and the *Principle of Special Relativity* with the conditional "If we find in Nature an *Inertial Frame of Reference*...". As we know, Einstein could not accept the existence of privileged *frames of reference* that -to aggravate the situation- we did not know how to find, and he challenged himself to resolve this fundamental problem without tossing away the Special Theory of Relativity.

9. What we have learned

As usual, let us summarize what I believe the reader has learned in this chapter:

• The Principle of Special Relativity (Einstein's First Postulate) asserts that all the laws of Nature are the same in all *Inertial Frames of Reference*. This is the ***absolute*** *par excellence* in the Universe, though -for now- still restricted to *Inertial Frames*;

• The relativity of *simultaneity* is independent of the relativity of *motion*. From the many possible definitions of *simultaneity* we saw in Chapter 2, Einstein chose the one providing the greatest descriptive simplicity, and the not-rigorous reader/writer soon forgot that there were many other definitions with the same epistemological legitimacy;

• With Einstein's definition and its associated ***synchronization*** of *clocks, time* can be defined for each location in an *Inertial Frame*. So defined, given two *Inertial Frames*, two events that are *simultaneous* in one frame, may not be so in another;

• The *speed of light in vacuum* is the same for all *Inertial Frames* (Einstein's Secon Postulate) and, besides, it constitutes the ***universal limit*** of *speed* for all bodies/radiation. This is the second grand ***absolute*** in Nature;

• Based on his only two postulates, Einstein deduced the *Lorentz's Transformation* that relates the *space-time coordinates* between any two *Inertial Frames* -- instead of between any *Inertial Frame* and ***absolute*** *space* in ***absolute*** *repose*, as it had been interpreted till then. The difference between the *Lorentz's Transformation* in Lorentz's theory and the *Lorentz's Transformation* in Special Relativity is not mathematical but ***ontological*** and ***epistemological***;

• With the new interpretation of Einstein, the *Fitzgerald/Lorentz's contraction* is not a phenomenon that needs a ***causal*** explanation; it is simply a ***reciprocal*** effect between any two *Inertial Frames*, due to the definition of *length* of a *moving* body we introduced in Chapter 3. The *contraction* is **not** *real* in the classical sense, because it disappears if the observer is *in repose* with the observed body; it is ***real*** and ***reciprocal*** in the relativistic sense;

• With the new interpretation of Einstein, the *time dilation* is not a phenomenon that needs a ***causal*** explanation; it is simply a ***reciprocal*** effect between any two *Inertial Frames* due to the conventionality of *simultaneity* and Einstein's definition. *Time dilation* is ***not real*** in the classical sense, because disappears if the observer is *in repose* with the *clock* being observed; it is ***real*** and ***reciprocal*** in the relativistic sense;

• For a body and radiation in *matter*, there always exists an *Inertial Frame* in which its ***proper*** *mass* is ***not zero*** and, ergo, it cannot move at *speeds* equal or higher than the *speed of light in vacuum*;

• An *electromagnetic radiation in vacuum* travels at the *speed limit* and has **no**

proper *mass*, because it does not exist an *Inertial Frame* in which the *radiation* is *at rest*;

• In this fashion, Einstein resolved his adolescence paradox, because regardless of how fast his horse was moving relative to Earth, the speed of light relative to him, would continue being the same and, hence, he would never see the *electromagnetic field* in repose;

• The *Lorentz's Transformation* and its relativistic *composition of velocities* are also valid for ***mechanical*** phenomena, once it is realized that the *mass* of a body is a relative magnitude that coincides with the classical *mass* for relative *speeds* between the body and the observer which are ***small*** with respect to that of *light in vacuum*, and that it ***increases*** indefinitely when that relative *speed* approaches the ***limit speed*** c. In this manner, *mass* and *energy* are ***relative*** and are related by the most famous equation in history $E = m.c^2$;

• *Newton's Law of Universal Gravitation* cannot be adapted so as to make it consistent with Special Relativity. Einstein had to develop a new theory for *Gravitation*: The General Theory of Relativity;

• The third grand ***absolute*** in Special Relativity is the *event interval*. Each observer sees a different partition of *space* and *time* from her/his *Inertial Frame of Reference*, but all those different perspectives have an ***absolute*** measure: the *metric* of *space-time* defined by the *event interval*;

• *Energy* and *momentum* can be combined to obtain another ***absolute*** in *space-time*. *Energy* is the *temporal* part and *momentum* the *spatial* part of this ***absolute***. ***Different*** *inertial* observers will measure ***different*** values for *energy* and for *momentum* of an object, but all of them will measure the ***same*** value for the tetra-dimensional 'momentum-energy';

• *Electric charge* and *electric current* can also be combined into a tetra-dimensional magnitude, the *temporal* part of which is the *charge* and the *spatial* part of which is the *current*. ***Different*** observers will see ***different*** values for the *electric* and *magnetic* fields, but there is an underlying **reality** common for all of them: the *electromagnetic field*;

• The ***present*** is **not** an *instant* shared by all *space*, but an *event*, i.e. an *instant* at a *place* in *space*. Likewise, the ***past*** and the ***future*** are not a collection of *instants* shared by all *space*, but a collection of *events* that correspond to a possible relation of *causal* order with the ***present event***;

• For every ***present*** *event*, its ***past*** and its ***future*** are restricted but ***absolute***. For every ***present*** *event*, there is also a kind of 'limbo', i.e. a collection of *events* (neither ***past*** nor ***future***) which *temporal* order with the ***present*** *event* will be ***relative*** to the *frame of reference*;

• An *optical clock* is a perfectly ***isolated*** system. The phenomenon of *time dilation* can be deduced straightforwardly using the Pythagorean Theorem, because the *spatial* trajectory is relative to the *reference frame* and the *speed of light in vacuum* is the same for all *reference frames* (Einstein's Second Postulate);

• Imagining an astronaut that travels at 80% of the limit velocity c towards a planet around the star Gliese 581, we confirm that the phenomena of *space contraction* and *time dilation* imply that the astronaut lived and aged 15 years, while Earthlings and inhabitants of Gliese 581 lived and aged 25 years. Under no circumstance we allowed all of them to be in relative *repose* to compare their gray hair and wrinkles, as that would violate the conditions under which Special Relativity is valid to describe the Universe;

• The real case of *cosmic radiation* measured in the famous experiment of Mount Washington proves irrefutably that, irrespective of how strange and fantastic *spatial contraction* and *time dilation* may seem to us, they constitute a physical reality that Nature zealously hid from us for millennia.

• Any *non-Inertial Frame* is ***locally*** and ***momentarily*** *inertial*, because in that ***reduced*** **space** and ***limited*** *time*, can be considered as in ***uniform*** *motion* with respect to some *Inertial Frame* and, ergo, per the *Principle of Special Relativity*, it must also be *inertial*. The word 'local' may mean large *distances* with respect with the human scale and the word 'momentarily' may mean *time* periods very considerable for us. This is why Special Relativity, even though it is only strictly valid for *Inertial Frames*, has had and continues to have an astonishing success;

• There is no definite criterion to find an *Inertial Frame of Reference*. Einstein could not accept that there would be privileged *reference systems* that -to aggravate the situation- we did not know how to find, and he challenged himself to resolve this fundamental problem without tossing away the Special Theory of Relativity.

Additional Recommended Reading

[1] d'Abro, A., *The Evolution of Scientific Thought - From Newton to Einstein*. New York: Dover Publications, 1950.

[2] Grünbaum, Adolf, *Philosophical Problems of Space and Time*. New York: Alfred A. Knopf, Inc., 1963.

[3] Reichenbach, Hans, *The Philosophy of Space and Time*. New York: Dover Publications, 1958.

[4] Wolfson, Richard, *SIMPLY EINSTEIN - Relativity Demystified*. New York: W.W. Norton & Co., 2003.

[5] Einstein, Albert, *Relativity - The Special and the General Theory*. New York: Three Rivers Press, 1961.

[6] Einstein, Albert and Infeld, Leopold, *The Evolution of Physics*. New York: Simon & Schuster, 1966.

[7] Hoffman, Banesh, *Relativity and its Roots*. New York: Dover Science Books, 1983.

[8] Poincaré, Henri, *Science and Hypothesis*. New York: Dover Publications, 1952.

[9] Reichenbach, Hans, *From Copernicus to Einstein*. New York: Dover Publications, 1970.

[10] Will, Clifford M., *Was Einstein Right? - Putting General Relativity to the Test*. New York: Basic Books, 1993.

Chapter 7

When Celestial Dynamics becomes Kinematics again

I was sitting in a chair in the patent office in Bern when all of a sudden a thought occurred to me: 'If a person falls freely he will not feel his own weight'. I was startled. This simple thought made a deep impression on me... This law [of the free-fall of bodies], that can be formulated also as the law of the equality of inertial and gravitational masses, finally impacted me with all its significance. I was astonished about its existence, and realized it contained the key to understand in depth the phenomena of inertia and gravitation... Albert Einstein referring to what he called "the happiest thought of my life".

The General Theory of Relativity

Would Einstein be pleased with the results? I think he would, since his theory has passed every one of the tests with flying colors. Clifford M. Will in his book: Was Einstein Right? - Putting General Relativity to the Test, 1993.

June 18, 2009 - Exactly 40 years ago...

The process of loading the RP-1 fuel into the Saturn V starts and will continue for six days. Unexpectedly, a Laser Ranging Retro-Reflector (LRRR) will be included in the Experiments Package (ALSEP).

1. The Law of Equality between Gravitational and Inertial Masses

In Chapters 1 and 2, we argued in detail why *Newton's second Law of Motion* and *Newton's Law of Universal Gravitation* implied that, if any two different bodies -in absence of friction- fell to ground with the ***same** acceleration*, then, *gravitational* and *inertial masses* of all bodies had to be ***equal***. And, vice versa, if those two *masses* were ***equal*** for any body, then all bodies -ignoring friction- had to fall with the ***same** acceleration*, regardless of their *mass* and chemical *composition*. This ***equality*** -considered for long as mere coincidence- made Einstein suspect that *inertia* and *gravitation* were different manifestations of the same physical phenomenon.

The universal nature of a *gravitational field* made us say in Chapter 1: "a gravitational field is equivalent to a field of accelerations, because all bodies of all materials and all masses have the same acceleration in the same place". But this affirmation is extraordinary, because a *gravitational field* is a *field of forces*, and the concept of *force* is what gave birth to the discipline of *Dynamics* as part of *Mechanics* (Chapter 4), while *acceleration* is a concept that can be fully characterized within the discipline of *Kinematics* -- which was the first part of *Mechanics* developed before the arrival of Newton (Chapter 4).

If the effect of a *gravitational field* on an object is to modify its *motion*, and the latter is the same for all objects, it seems as if -as far as *gravitational forces* are concerned- their effects could be fully described by means of exclusively-*kinematic* considerations, i.e. through relations involving only *space* and *time* -- without need of any *force* whatsoever! But... if now we think of *space-time*, and we are saying that the *space-time trajectory* of an object under the action of a *gravitational field* does not depend on the object -remembering our discussions (Chapter 3) on the *relativity of geometry*- we have to conclude that such a *trajectory* would be entirely determined by the *geometry of space-time*! Let us see how far we can go with this idea which Einstein called "the happiest thought of my life".

1.1. Experimental Confirmation of the Equality between Gravitational and Inertial Masses

In Chapter 1, we saw different experiments along history (until the arrival of Newton) which supported the ***equality*** of *inertial* and *gravitational masses*. This equality sometimes is confusedly referred to as the *Principle of Weak Equivalence* or, more appropriately, as the universality of free-fall. Friedrich W. Bessel repeated, in 1832, the measurement of the *periods* of different pendulums with equal *length* but different *masses* and *materials* and -as Newton-did not find appreciable difference between the *periods*, corroborating the ***equality*** of the *masses* with a precision of 0.1%. Loránd von Eötvös, in a series of experiments in 1889 and 1908, employing a torsion balance, measured the torsion of a wire -which sustained a bar with two bodies at its extremes- due to the combined action of *gravity* and the *centrifugal force* caused by our planet rotation. If the *inertial* (centrifugal) and *gravitational forces* acted differently for different *masses* or *materials*, it would appear a torque on the wire, causing the rotation of the bar. Using copper, water, asbestos, aluminum, etc. Eötvös confirmed the ***equality*** between *inertial* and *gravitational masses* with a precision better than 0.0000005%.

Many years after the birth of General Relativity in 1916, Roll, Krotkov, and Dicke (Princeton, USA) in 1964, and Braginsky and Panov (University of Moscow, USSR) in 1971, used ***solar*** *gravity* instead of ***terrestrial*** *gravity*, and Earth ***annual*** rotation instead of Earth ***daily*** rotation to conduct the same experiments. Princeton people attained a precision of 0.000000001% (10 parts per billion), and the Soviets achieved a precision 10 times better (0.0000000001%). We can then safely conclude -with extreme exactitude- that bodies of different *materials* and *masses* 'fall' in a *gravitational field* with the ***same*** *acceleration*. Even with subatomic particles like the neutron, in an experiment conducted at the Munich Technical University in 1975, the equivalence between *inertia* and *gravitation* was corroborated with a precision of 0.01%.

1.2. The Giant of the Road never braked: A Horizontal Gravitational Field appeared

...But, inside the vehicle, with all its windows closed, the objects were all *in repose* right before the driver pushed the brakes and, hence, their *inertia* could not explain why they started flying like a bullet given that without a *force* -according to Newton- they should have remained *in repose*. In this *reference frame* (temporarily *decelerated* with respect to ground), the only manner I could explain the apparent and momentary 'chaos' was through the spontaneous emergence of a *field of forces* that could change the state of *repose* in which bodies were before the bus braking. This field had to be ***variable*** in *time* so as to disappear once the bus fully stopped or re-attained ***uniform*** *motion*. What could be explained ***without*** the existence of a *force* from ***outside*** (although we needed the peculiar concept of humanoid reluctance (*inertia*) to *motion*), to explain it from ***inside***, we needed the existence of such a *force*! ***Inside*** the bus -while braking- in order to make sense of any phenomena, it was necessary to contemplate such a mysterious *force*. But... was this *force* **real** or fictitious? And, if it was **real**, who or, better, what exerted it? And...that outlandish humanoid reluctance... What in the world was that? Where did it come from? Why could it not be considered as a *force* as well? Or... maybe... couldn't both concepts be as artificial as unnecessary to explain all these mechanical phenomena?

The object flying forward, observed from *terra firma*, had a ***horizontal*** component of **constant** *velocity* equal to the one the bus had before braking, simply because *Galileo's Principle of Inertia* (first Newton's Law) teaches us that the object 'natural behavior' (free of any external influence) would be simply to continue with the bus original speed 'forever'. Due to *gravity*, the same object had a ***vertical*** component of ***variable*** *velocity* (with the *acceleration* of *gravity*). Observed instead from inside the bus, the object had started in *repose* to *move* with a ***horizontal*** *acceleration* equal to the *deceleration* (brake) of the vehicle with respect to ground. This is so because the flying object *moved* relative to ground at **constant** *speed* and hence (think about it a little), the *acceleration* of any third body (the bus) had to be the same with respect to any one of them.

So... using the bus as the *reference frame*, the ***horizontal*** *acceleration* of every flying object -while braking- would be ***equal*** *in magnitude* though ***opposed*** *in direction* to the *acceleration* of the bus with respect to ground and, ergo (since the latter is given by the braking process), exactly the ***same*** independently of the object *mass* and chemical *composition*! Hmm... is it not what we called a *universal field*? And... is it not what characterizes a *gravitational field*? I had to conclude that, if this mysterious ***horizontal*** *field* 'really' existed, it behaved as the ***vertical*** *gravitational field*!!! By the same token, I concluded that observing the bus braking from *terra firma* was not equivalent to observing the events from inside the bus: ***without*** external reference, ***before*** braking our motion was ***undetectable***; while braking, the 'chaos' could be interpreted as the hallmark of ***accelerated*** *motion*. Did this mean that *speed* was ***relative*** but *acceleration* was **absolute**? Are they not complementary attributes of *motion*? Why this duality? ...

As we saw in Chapter 4, it is "madman business" trying to prove that such a *field of forces* that converts chaos into order within the bus would also have appeared had it been *in repose* while the *rest of the Universe* had been moving in the ***opposite*** *direction.* Physical **reality** is that, in the only experiment we can carry out (slam the brakes), the ***relative*** *motion* between the bus and the *rest of the Universe* changes from being ***uniform*** to be momentarily ***accelerated*** and that, if we observe the flying objects from *terra firma*, we explain it with the concept of *inertia*, while if we observe them from *inside* the autobus, either we say that the laws of Mechanics are **not** valid and, ergo, the bus is a *non-Inertial Frame* where chaos reigns, or we accept the appearance of a *variable field of forces* which explains the observed *motion* of bodies (and that of the *rest of the Universe* if we open the windows). The striking feature of this peculiar *field of forces* is that the *acceleration* it imparts on ***all*** bodies (inside and outside the bus) is univocally defined by, and equal to, the relative *acceleration* between the bus and *terra firma*, i.e. it is a *field of accelerations*, ergo, it is a ***horizontal*** *field* as ***gravitational*** as the ***vertical*** *field* produced by our planet! The fact that the former is ***variable*** in *time* and the latter is not, is irrelevant.

But... if the windows were closed, while we were cruising the pampas at ***constant*** *speed*, we did not have any reason to argue we were not *in repose* and, consequently, once the driver suddenly pushed the brakes to the floor, what prevented us from interpreting the unexpected flying objects (including us) as the result of the appearance of a ***horizontal*** *gravitational field* (as those produced by material bodies), while the autobus still was *in repose* (as we thought it was while at ***constant*** *speed*)? As I mentioned in Chapter 4, Mach thought that such a ***variable*** *gravitational field* was produced by the *rest of the material universe* in *motion* relative to the bus.

If, instead, inertial and gravitational masses were ***not equal****, the motion of the flying objects could not be explained out with a gravitational field, because the latter would produce different accelerations for different bodies, while -using the braking bus as reference- all bodies (inside and outside) had the same acceleration -- equal in magnitude and opposed in direction to the acceleration of the bus with respect to terra firma.*

The ***universal*** character of the *inertial forces* -appearing while the bus is braking- leaves us impotent as to discern whether what is going on inside the bus is because we are *moving* with respect to *terra firma* (remember windows are closed), or because -for some unknown reason unrelated to *motion*- a ***horizontal*** *gravitational field* has temporarily emerged, while the bus is as 'in repose' as it was before the sudden brake. ***The latter would have certainly been the interpretation adopted by an intelligent being, had s/he been born and grown up inside the bus without any access whatsoever to the exterior world.***

But... then... it is not true that *acceleration* is ***absolute*** and *velocity* is ***relative***: once the chaos in the autobus admits a physical explanation entirely unrelated to *motion*, the latter is undetectable and, hence, it is ***relative*** -- regardless of whether that *motion* is ***uniform*** or ***accelerated*** with respect to *terra firma*. No experiment within the bus -without external reference- will enable us to distinguish between its *acceleration* relative to *terra firma* (sudden brake) and the existence of a ***horizontal*** *gravitational field* produced by causes unrelated to *motion*, while the bus keeps ***uniformly*** *moving* relative to ground (including *repose*). Everything seems to indicate that -contrary to what classical Mechanics asserted- if we knew how to describe the effect of a general *gravitational field* (variable in *space* and *time*), a *reference frame* in ***accelerated*** *motion* with respect to an *Inertial Frame*, could be as valid to describe physical **reality** as the ineffable *Inertial Frame* is.

1.3. The Carousel by the Church never rotated: A Radial Gravitational Field appeared

...However, seen from the carousel, and closing my eyes in an attempt to ignore that I knew I was rotating with respect to ground, I was *in repose* in the merry-go-round and, upon releasing my hands, I would be launched without an apparent *force* causing that motion. But, as when the bus started *braking*, according to the *Principle of Inertia*, if there was **no** *force*, I should have stayed *in repose*! In close parallel with my experience crossing the pampas, I hastily concluded that the carrousel could **not** be an *Inertial Frame* ever (the bus could, but only while at **constant** *speed*!). And which was the real difference between the two experiences? In the case of the omnibus, and seeing everything from inside, in order to explain the chaos we had to accept the action of a temporary *field* of gravitational-like *forces* that seemed fictitious to me; in the case of the carrousel, and seeing everything from inside, the analogous explaining *force* already had a name: *centrifugal force*, and I could not sensibly consider it fictitious given that I had to permanently counteract it to avoid killing myself. It was the necessity of having to counteract the *centrifugal force* what gave it the status of 'real', while it was harder to do the same with the *force* in the bus because it only appeared during the transitions between *inertial* states (before and after acceleration). But... during the braking, I also had to hold on tight so as to counteract that force and not hurt myself, so at least temporarily, it was as 'real' as the one I felt constantly in the carrousel.

In summary, if I chose the carousel as the *reference frame*, there existed a ***radial*** *field* of *centrifugal forces* whose intensity was ***zero*** at the *center*, and the greater the farther away from it. The farther away I was from the rotation axis, the tighter I had to hold on so I could keep rotating without falling. If I chose *terra firma* as the *reference frame*, when I released my hands, my body did what it had to do because of my *inertia*: fly with a *trajectory* determined by my **initial** *speed* (the carrousel *speed* before getting loose), by *gravity*, and by air friction. Once again, without further cogitation, everything seemed to indicate that ***accelerated*** *motion* was ***absolute*** while ***uniform*** *motion* was ***relative...***

As we saw in Chapter 4, it is 'madman business' trying to prove that such a *field of centrifugal forces* that converts chaos into order within the carrousel would also have appeared had it been *in repose* while the *rest of the Universe* had been revolving in the ***opposite*** *direction*. Physical **reality** is that, in the only experiment we can carry out (give the carrousel an impulse), the ***relative*** *motion* between the carrousel and the *rest of the Universe* changes from being

uniform (before the impulse) to be momentarily ***accelerated*** and that, if we observe the flying objects from *terra firma,* we explain it with the concept of *inertia,* while if we observe them from inside the carrousel, either we say that the laws of Mechanics are **not** valid and, ergo, the carrousel is a *non-Inertial Frame* where chaos reigns, or we accept the appearance of a ***variable*** *field of centrifugal forces* which explains the observed *motion* of bodies inside the carrousel.

The striking feature of this peculiar *centrifugal field* is that the *acceleration* it imparts on all bodies is univocally defined by, and equal to, the relative *acceleration* between the carrousel and *terra firma,* i.e. it is a *field of accelerations* and, ergo, it is a ***radial*** ***field*** as *gravitational* as the ***vertical field*** produced by our planet! The fact that the former is variable in *time* and *space* and disappears when the carrousel stops rotating -while the latter is *constant* in *time*- is irrelevant. It is also worth noting that the Earth *gravitational field* though locally 'vertical', is globally ***radial*** as the *centrifugal field* in the carrousel is. However, the big difference between the two is that the former ***decreases*** in intensity with *distance* from its center, while the latter ***increases*** its intensity with *distance* from the rotation center. But... (think about it a little), the ***farther*** from the center of rotation, the ***closer*** we are to the ***rest of the universe***!

But... if I closed my eyes so as not to see my adored Necochea scenery while rotating, i.e. if I had no awareness of *terra firma,* before starting to rotate, I thought I was *in repose* and, then, once I started rotating, what prevented me from interpreting the unexpected *motion* of loose bodies (including mine if I stopped holding tight) as the result of the existence of a ***radial*** *gravitational field* while the carrousel was still *in repose*? As I mentioned in Chapter 4, Mach thought that this ***radial*** *gravitational field* was produced by the *rest of the material Universe* in relative *motion* with the carrousel.

If, instead, inertial and gravitational masses were ***not equal****, the motion of the flying objects could not be explained out with a radial gravitational field, because the latter would produce, at a given distance from the center, different accelerations for different bodies, while -inside the rotating carrousel- all bodies at the same distance from the center had the same acceleration.*

The ***universal*** character of the *centrifugal forces* leaves us impotent as to discern whether what is going on inside the carrousel is because we are ***rotating*** with respect to *terra firma* (remember my eyes are closed), or because -for some unknown reason unrelated to *motion*- a ***radial*** *gravitational field* has temporarily emerged, while the carrousel is as *in repose* as it was before the given impulse. ***The latter would have certainly been the interpretation adopted by an intelligent being, had s/he been born and grown up inside the carrousel, without any access whatsoever to the exterior world.***

But... then... it is not true that *acceleration* is ***absolute*** and *velocity* is ***relative***: once the chaos inside the merry-go-round admits a physical explanation entirely unrelated to *motion,* the latter is undetectable and, hence, it is ***relative*** -- regardless of whether that *motion* is ***uniform*** or ***accelerated*** with respect to *terra firma.* No experiment within the carrousel -***without*** external reference- will enable us to distinguish between its *acceleration* relative to *terra firma* and the existence of a radial *gravitational field* produced by causes unrelated to *motion*, while the merry-go-round is *in repose* relative to ground. Everything seems to indicate that -contrary to what classical Mechanics asserted- if we knew how to describe the effect of a general *gravitational field* (variable in *space* and *time*), a *reference frame* in ***rotational*** (*accelerated*) *motion* with respect to an *Inertial Frame*, could be as valid to describe physical **reality** as the ineffable *Inertial Frame* is.

July 15, 2009: Exactly 40 years ago...

Liquid Hydrogen and Oxygen are being pumped into the three stages of the Saturn V. The Laser Retro-reflector (LRRR) has been included in the Experiments Package (ALSEP).

1.4. The Classical Gravitational Field and its Interpretation as Acceleration

In all phenomena described during my experiences with the *Giants of the Road* and the *Carrousel by the Church*, the Earth *gravitational field* was -of course- present, but it did not play any relevant role, beyond displaying its action on the flying objects during the sudden brake of the former, or when releasing my hands while rotating with the latter (*gravity* acting jointly with the emerging *inertial forces*). Because both the bus and the merry-go-round were in permanent physical contact with *terra firma*, and their *motion* was ***horizontal*** while *gravity* is ***vertical***, their mechanical resistance was exerting an equal but opposite force (Newton's third law) that constantly compensated *gravity*. All our conclusions on the *inertial forces* and their possible interpretation as ***horizontal*** and ***radial*** *gravitational forces* would have been correct, even had Earth *gravity* not been present. Let us remember that Galileo discovered the *Principle of Inertia* with inclined planes and maritime experiments and, in all of them, obviously, the Earth gravitational field was always present without detectable consequences.

Our next laboratory will be an elevator, where the Earth *gravity* (the quintessence of a Newtonian *gravitational field*) is now the central character because the elevator *motion* is also ***vertical***. We will neglect air friction (as usual), as well as the minuscule variation of *gravity acceleration* with altitude while the elevator moves (for now). In the upcoming graphs, ***accelerated*** *motion* will be indicated with an arrow accompanied with the value of the *acceleration*. The existence of a *gravitational field* will be indicated with an arrow accompanied with the inscription 'F(a)' where 'F' stands for 'Field' and 'a' is the *acceleration* characteristic of the *field*. Now, instead of interpreting the *acceleration* as a *gravitational field* (as we did so far), we shall interpret the terrestrial *gravitational field* as an *acceleration*.

We are inside the elevator completely incommunicado from the exterior; we are *in repose* on first floor; the body, spring hanging from the ceiling, stays stable (Figure 28, top left). Suddenly, we observe the suspended body going ***down*** with the spring extending to a new stable *length*. Evidently, the body is pulling the spring with a stronger *force* than before; how can we explain what happened? There are three ***fully equivalent*** ways: a) the elevator is ***ascending*** with an *acceleration* 'a' and the body *inertia* produces the additional spring *length* (Figure 28, top center); b) the elevator is ***at rest*** but, the vertical *gravitational field* has **increased** (Figure 28, top right) from 'g' to 'g+a'; the ***longer*** spring *length* is now produced by a ***greater*** *gravitational* intensity due, perhaps, to the appearance of nearby additional *matter*; and c) any combination of a) and b). It is impossible, **without** external reference, to know which of these phenomena has occurred (though the first one is the best bet by far). ***Accelerated*** *motion* and *gravitation* are equivalent and, ergo, we prove once again that -thanks to the *Law of Equality of Inertial and Gravitational Masses-* ***accelerated*** *motion* is as ***relative*** as ***uniform*** *motion*.

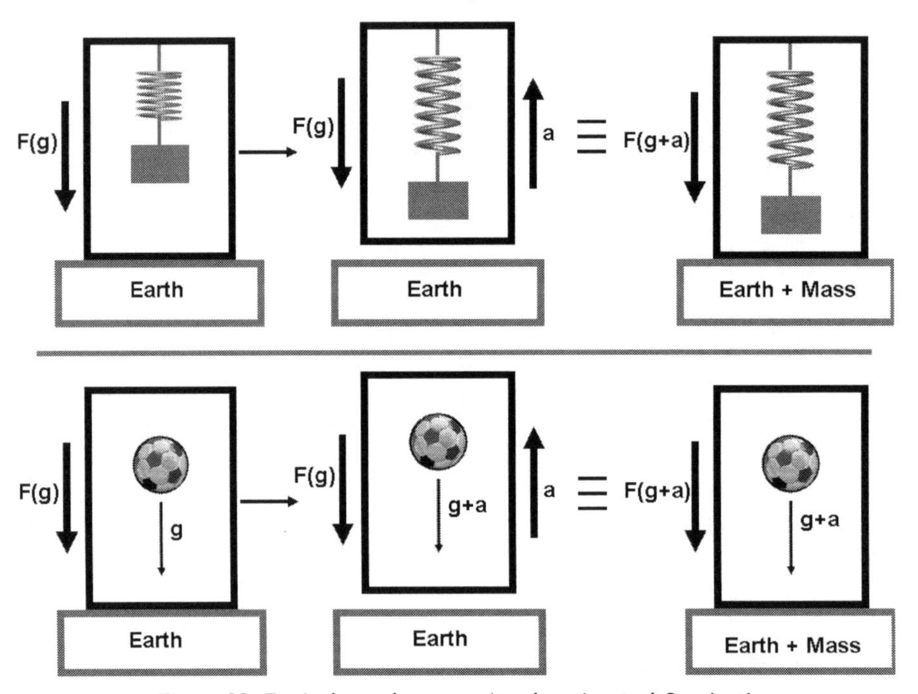

Figure 28. Equivalence between Acceleration and Gravitation

If, instead, inertial and gravitational masses were ***not*** *equal, the new spring length due to the elevator upward acceleration could not be explained out with a gravitational field, because the latter would produce different spring expansions for different bodies, while experience shows that the additional length when the elevator accelerates is always the same, regardless of the mass and chemical composition of the hanged body.*

Now imagine that, whichever the real case was (a, b, or c), we heat up (give *energy* to) the body. What do you think will happen? The most famous equation in history ($E = m.c^2$) tells us that the body *inertial mass* shall *increase* and, hence, the spring *length* shall *increase* (an imperceptible amount) but... this ***increase*** of spring *length* may again be interpreted -at leisure- as the result of a), b), or c)! Furthermore, given that this alternative explanation is independent of how I gave the extra *energy* to the body, if we relax a little our language precision, we may say that all types of energy 'fall' with the same *acceleration*. This is why scientists can get so easily in trouble, enunciating statements that -if not accompanied with the appropriate rationale- sound to the layperson as mystical as incomprehensible. It is the ***object*** (with different amounts of *energy*) that falls, **not** the *energy*.

The bottom half of Figure 28 demonstrates the same concept by letting a ball fall while we are at rest on first floor. Our experimentation confirms that the ball falls to the floor with the well-known acceleration of g=9.81 m/s^2 (bottom left). During one of our repetitions of the experiment, we found with surprise that the *acceleration* has increased to 'g+a' arriving the floor ***earlier***. How can we explain what happened? There are three fully equivalent ways: a) the elevator is ***ascending*** with *acceleration* 'a' and, for exclusively-*kinematic* reasons (and so fully ***independent*** of the ball *inertia*), the *distance* between the ball and the floor ***decreases*** faster, i.e. with *acceleration* ***greater*** than that of *gravity* (Figure 28, bottom center); b) the elevator is ***at rest*** but, the ***vertical*** *gravitational field* has increased (Figure 28, bottom right) from 'g' to 'g+a' due, perhaps, to the appearance of nearby additional *matter*; and c) any combination of a) and b). It is impossible, **without** external reference, to know which of these phenomena has occurred (though the first one is the best bet by far). ***Accelerated*** *motion* and *gravitation* are ***equivalent*** and, ergo, we prove once again that -thanks to the *Law of Equality of Inertial and Gravitational Masses*- ***accelerated*** *motion* is as ***relative*** as ***uniform*** *motion*.

*If, instead, inertial and gravitational masses were **not equal**, the greater acceleration of the ball could not be explained out with a gravitational field, because the latter would produce different accelerations for different balls, while experience shows that all balls fall with the same additional acceleration when the elevator accelerates upwards, regardless of their mass or chemical composition.*

1.5. The Free-Fall - The Fundamental Concept of Local Inertial System

But... then, if the existence of an additional *gravitational mass* is totally equivalent to an additional *acceleration* in the opposite direction, all Earth *gravitational mass* is -at least locally- fully equivalent to an *acceleration* and, if our planet (hypothetically) stopped exerting its *gravitational field* and our elevator kept ***ascending*** with an *acceleration* of 9.81 m/s^2, our muscular sensation would be indistinguishable from our daily experience on solid ground, so that it would be impossible for us (at least for some time) to know whether we are *moving* (and the planet disappeared) or our dear planet still is with us and the elevator is simply *at rest*. In fact, it is this *equivalence* between *gravitation* and *acceleration* that was employed in the space station depicted in the legendary movie "2001: A Space Odyssey", in which the *centrifugal field* due to the station *rotation* produced on human beings the same muscular sensation, and the impression of 'being upward', that we experience here daily on Earth.

And... if the Earth *gravitational field* exists and the elevator, instead of ascending, ***descends*** with exactly the *acceleration* of *gravity* (think of a macabre joker cutting the elevator cable), what do you think will happen? Due to the ***equality*** between *inertial* and *gravitational mass*, the elevator and all bodies inside it (including us) will fall with exactly the ***same*** *acceleration* with respect to ground (Figure 29, left) and, ergo, all relative *distances* between them will remain *constant*, so that inside the elevator -**without** external reference- all bodies will be *in repose*! Moreover, the muscular sensation due to our contact with the elevator floor will have disappeared and all bodies (inanimate and animate) will be ***floating***! Everything occurs as if the elevator were a spaceship in the middle of intersidereal *space* far away from any *matter*, and with its motors off (Figure 29, right) -- i.e. entirely ***isolated*** from the *material* Universe. As soon as we turned the motors on, the *acceleration* would be equivalent to a *gravitational field*, we would be in contact with one of the ship walls, the muscular sensation would appear, and we would think that we are 'upwards'.

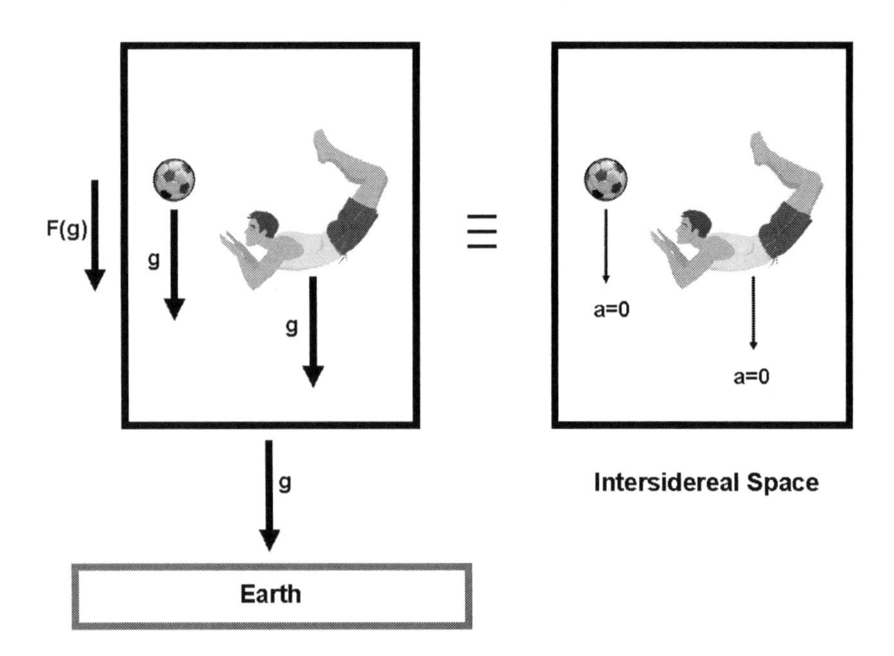

Figure 29. The Free-Fall: The Concept of Local Inertial Frame

July 16, 2009: Exactly 40 years ago...

The 5 F1 motors are consuming 15 tons of fuel per second. At 10:32 of Cape Canaveral, the Saturn V majestically leaves the launch platform. Houston takes control. About 10 min later, the Apollo 11 is in free-fall orbiting our planet at 215 km of altitude. The tanks still have 60 tons of fuel. A load not less valuable for the credibility of the General Theory of Relativity is a Laser Ranging Retro-Reflector (LRRR).

Returning to our macabre experience, using *terra firma* as reference, we and the elevator are in ***free-fall*** but, inside the elevator -**without** external reference- we are ***in repose*** and there is **no** *gravity* as we can easily confirm by letting 'fall' a ball which never falls but stays where we leave it and, if we impart an ***initial*** *velocity* to the ball, it will trace a ***rectilinear*** *trajectory* with that ***initial*** *velocity* as long as it does not interact with other object. Inside the elevator in ***free-fall***, besides the terrestrial *gravitational field* having disappeared, the *Principle of Galilean Relativity* (Newton's first law) seems to be valid. The elevator, as a *reference frame* to describe **reality**, is then an *Inertial Frame* where there is **no** *gravitational field* due to the trick of being in *motion* with exactly the ***same*** *acceleration* as the one produced by that *field*. Special Relativity is consequently valid. Recalling what we learned in Chapter 6, any other *reference frame* ***uniformly*** *moving* with respect to our elevator in ***free-fall*** shall also be an *Inertial Frame*, Special Relativity shall be valid, and the transition from one frame to the other is achieved through the *Lorentz's Transformation.*

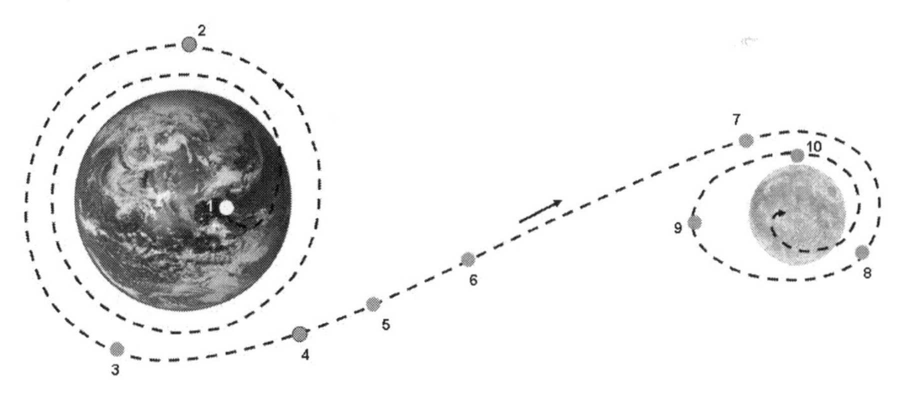

Figure 30. 'Free-Fall' Stages for the Trajectory of Apollo 11

Realizing that ***free-fall*** simply means that we are ***accelerated*** under the sole non-neutralized action of the terrestrial *gravitational field*, it is not difficult to see that the International Space Station is permanently in free-fall (what it 'falls' is compensated with the Earth roundness, so that it stays in orbit) and, hence, the astronauts float inside and have an ideal *Inertial Frame* at their disposal. Likewise, the Apolo 11, while orbiting before turning its motors on in stage 2 (Figure 30), was in free-fall towards the Earth.

So much for the 'reality' of the Earth *gravitational field*! It turns out that, if we use *terra firma* as reference, *gravity* ***exists***; if we use the International Space Station (or the Apollo 11 orbiting), *gravity* ***does not exist***! And it would not exist either inside any other *reference frame* subject to a generic *gravitational field* which action is not counteracted (free-fall). For example, in the Apollo 11, when its motors are off (Figure 30, between small circles), all objects are subject to the ***same*** *acceleration* produced by the external celestial *gravitational field*, so that their relative *distances* are ***constant*** and, ergo, they ***float***. From outside, the *motion* of the spaceship (and of all the objects inside) is determined by the *gravitational field*; from inside, there is **no** *gravitational field*!

1.5.1 Tidal Effects

As we know from classical mechanics, the *gravity acceleration* varies in *space* producing, in the case of lunar *gravity* interacting with Earth *gravity*, the well-known phenomenon of ***tides***, and that is why the generic *spatial* variation of a *gravitational field* is referred to as ***tidal*** effects.

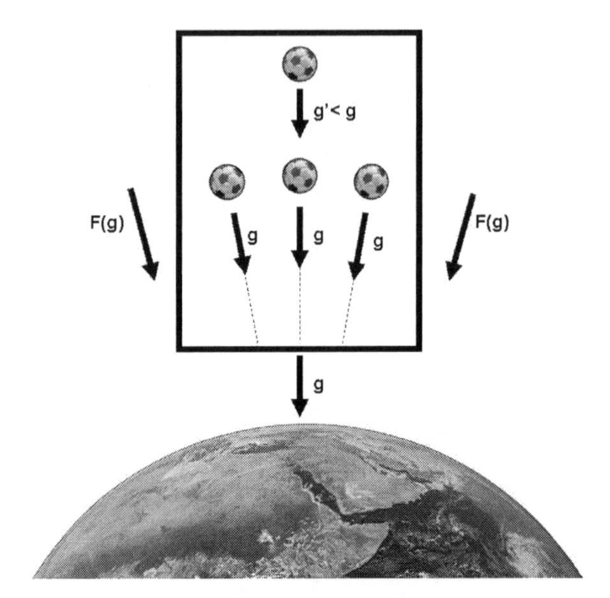

Figure 31. Tidal Effects

The Earth *gravitational field* is ***radial*** and of ***variable*** intensity so that, if the elevator in ***free-fall*** were large enough, watching the three balls (supposedly in relative *repose)* for an extended period of *time* (Figure 31), the 'horizontal' *distances* between them would ***decrease***![1]. Likewise, the 'vertical' *distance* between the upper ball and one of the lower balls would gradually ***increase*** because the former would move with a slightly ***lower*** *acceleration* than the latter.

In summary: inside the gigantic elevator, ***without*** external reference, objects ***in repose*** and free of any influence would eventually ***move***, the *Principle of Inertia* would not be valid, and the elevator in *free-fall* would not be an *Inertial Frame*. From all this, we infer that a *reference frame* in ***free-fall*** is only ***locally*** and ***momentarily*** *inertial*, i.e. it constitutes an *Inertial frame* only in a limited region of *space-time*. Tersely: a *gravitational field* is ***relative*** -- but only ***locally***.

In the previous section I said that, in the free-fall elevator (Figure 29, left):

... Everything occurs as if the elevator were a spaceship in the middle of intersidereal space far away from any matter, and with its motors off (Figure 29, right) -- i.e. entirely isolated from the material Universe. As soon as we turned the motors on, the acceleration would be equivalent to a gravitational field, we would be in contact with one of the ship walls,

1 The elevator would have to be about 5,000 km wide for the tidal effect to be noticeable.

the muscular sensation would appear, and we would think that we are 'upwards'.

This assertion is a little confusing if, given that we are in deep space far away from the gravitational influence of other celestial bodies, we do not specify with respect to which *reference frame* the *acceleration* produced by the motors is given. Clearly, inside the ship with the motors on, there is a *gravitational field* that keeps us in contact with a wall but -***without*** external reference- we do not have any reason to infer that we are *moving*. From outside the ship, there is no Earth (as in Figure 29, left) to refer our *motion* to, but there is the *Inertial Frame* that the very spaceship constituted when the motors were off and, it is respect to this frame that the *acceleration* produced by the motors acquires ***local*** meaning.

From the above considerations, it is inferable that we could use a series of local *inertial frames* to describe global phenomena where transition from one *Inertial Frame* to the next one would be determined by the variations of the gravitational field (tidal effect) in *space-time*. In each one of those *Local Inertial Frames*, Special Relativity would be valid because such a limited region of *space-time* can be approximately considered as semi-Euclidean, i.e. with null intrinsic curvature (Chapter 3). In each *Local Inertial Frame*, the local internal *gravitational field* would be internally ***cancelled*** out, but still represented by its equivalent local *external acceleration*. This local *external acceleration*, being exclusively ***kinematic*** (only *space* and *time*; no *forces*), can be mathematically represented, as we shall see, through a *transformation of coordinates* in *space-time*.

Wrapping up, Einstein saw in the *reference frames* in ***free-fall*** what he had been seeking for a long time: in any region of *space-time*, it was possible to establish a *Local Inertial Frame*, if we imparted to the latter the *acceleration* needed to cancel out the *local gravitational field*. He had finally found a well-defined criterion to establish an *Inertial Frame*, though only ***locally***. Being Special Relativity valid in that *Local Inertial Frame*, physical **reality** could be described from such a vantage point, while the same **reality** from any other frame in ***accelerated*** *motion* with respect to the former, now could be inferred applying a transformation of *space-time* coordinates between frames.

The notion of *Local Inertial Frame* is crucial to understand Nature and, in particular, General Relativity. Notwithstanding, very few popular science books (not even textbooks) emphasize enough its fundamental character. It is in these *Local Inertial Frames*, so conceived by Einstein, where Special Relativity holds, and not in the astronomical inertial frames postulated by Newton for his laws of *motion*.

July 18, 2009: Exactly 40 years ago...

The maximum speed of 45,000 km/h the spaceship had when accelerated to get out of the Earth orbit, had been reduced (by Earth gravity) to about 3,700 km/h, and now is lunar gravity which causes it to accelerate to about 9,000 km/h, following the so-called free-return trajectory (reaching stage 7 in Figure 30): if they decided to abort landing on the

Moon, this trajectory would take them behind our satellite and back to Earth without the need for fuel.

2. Einstein's General Principle of Equivalence

I am confident that, after the detailed presentation of the multiple physical situations in which it is impossible to discriminate between *accelerated motion* and the existence of a *gravitational field* created by a *material* body, the reader will understand why Einstein concluded that the *Law of Equality between Inertial and Gravitational Masses* implied the ***equivalence*** between *accelerated motion* and *gravitation*. There was **no** *mechanical* experiment within a generic *reference frame* that could allow determining whether the local *motion* of bodies was due to the *accelerated motion* of the *reference frame* itself, or to the existence of an external *gravitational field*, or to a combination of both. By the same token, this ***equivalence*** endowed *accelerated motion with* the ***relative*** character so hard to achieve for so long. At last, *accelerated motion* and *uniform motion* were both ***relative*** to the *reference frame*.

But, as we are already accustomed to, Einstein's mind did not stop at proclaiming this ***equivalence*** experimentally demonstrated only for *mechanical* phenomena: with his usual intellectual courage, he made another giant epistemological step, comparable to the one he had made with his *Principle of Special Relativity*. When enunciating the latter, he took what he knew with great experimental certainty about *Electromagnetism* and extrapolated it to *mechanical* phenomena. When enunciating his *Principle of General Equivalence*, he took what he knew about *mechanical* phenomena, and extrapolated it to *electromagnetic* phenomena. I would articulate it as:

> *Acceleration and gravity are equivalent not only with respect to mechanical phenomena, but with respect to all physical phenomena. In particular, there is no electromagnetic experiment realizable within a given reference frame that would allow to determine whether such a frame is in accelerated motion with respect to the Local Inertial Frame or, instead, is in uniform motion with respect to it (including repose), but subjected to the action of a gravitational field.*

Armed with this new general principle and much before he had his General Theory of Relativity finished, Einstein concentrated on investigating what new conclusions he could attain, and what new physical phenomena -susceptible of being experimentally confirmed- he could predict. I shall present the two most conspicuous predictions he made while he was giving mathematical flesh to his General Relativity Theory in the period 1907-1915.

2.1. The Deflection of Light by a Gravitational Field

Figure 32 (left) shows again the situation of Figure 29 (right). This magic elevator is in deep *space* with its 'motors' turned off and, thus, it constitutes an ideal *Inertial Frame* where Special Relativity holds. On the left wall (top), there is a *light* source whose beam, according to Special Relativity, traverses a *straight line* in *Euclidean space* (a *geodesic* in the *semi-Euclidean space-time* of Minkowski), hitting the opposite wall -after a tiny (but not nil) period of *time*- at a place whose *distance* from the ceiling is the same as it was at its departure. The same would have happened, had we given a horizontal impulse to the ball: it would have described a ***rectilinear** trajectory* in the *Euclidean space*, i.e. a *geodesic* in the *semi-Euclidean space-time* of Minkowski.

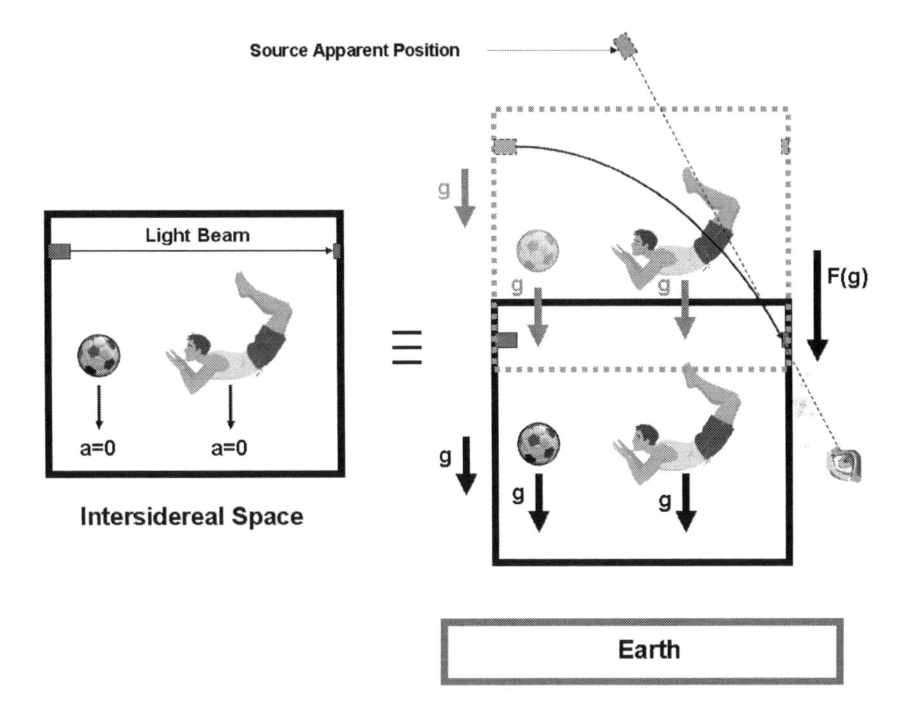

Figure 32. The Trajectory of Light is affected by a Gravitational Field - I

Now, if the *Principle of Equivalence* of Einstein is correct, then the situation of Figure 29 (left) is equivalent to that of the Figure 32 (left), not only from the *mechanical* point of view but also from the *optical* vantage and, ergo, there is no measurement whatsoever we could do inside the elevator in ***free-fall*** in Figure 32 (right) that would -**without** external reference- give a different result from the obtained when in the situation of Figure 32 (left). The sheer existence of a ***different*** result could be used to discriminate between ***free-fall*** and the ***absence*** of *gravity*. We conclude then that using the ***free-fall*** elevator as the *reference frame*, the *light beam* has to travel on a ***rectilinear** trajectory* reaching the opposite wall at exactly the same *distance* from the ceiling than for the case on the left.

As we expected, being both *Inertial Frames*, they are completely equivalent. But... then... for strictly ***kinematic*** reasons, in the *time* the *light beam* took traversing the elevator, the latter was *descending* with **non-zero** *acceleration* and, ergo, the *trajectory* for the *light beam* -if it is to hit the opposite wall at exactly the same spot- cannot be seen from *terra firma* as ***rectilinear***! The same would happen to the ball. Remember that *electromagnetic waves* and *material* bodies were grouped in Chapter 1 under the category of 'objects'. *Radiation* and *matter* seemed to behave in the same way.

In sum, if we choose the elevator as the *reference frame*, there is **no** *gravi-*

tational field and *light* travels along a *geodesic* in the *Euclidean space*; if we select *terra firma* as the *reference frame*, there exists a *gravitational field* and *light trajectory* is **not** a *geodesic* of the *Euclidean space*. Given that Special Relativity assumes that *light* travels along *straight lines* (geodesics) in *Euclidean space*, it is clear that this theory is **not** valid when there are *non-negligible gravitational fields*. This is why it is said that General Relativity is a theory of *gravitation*.

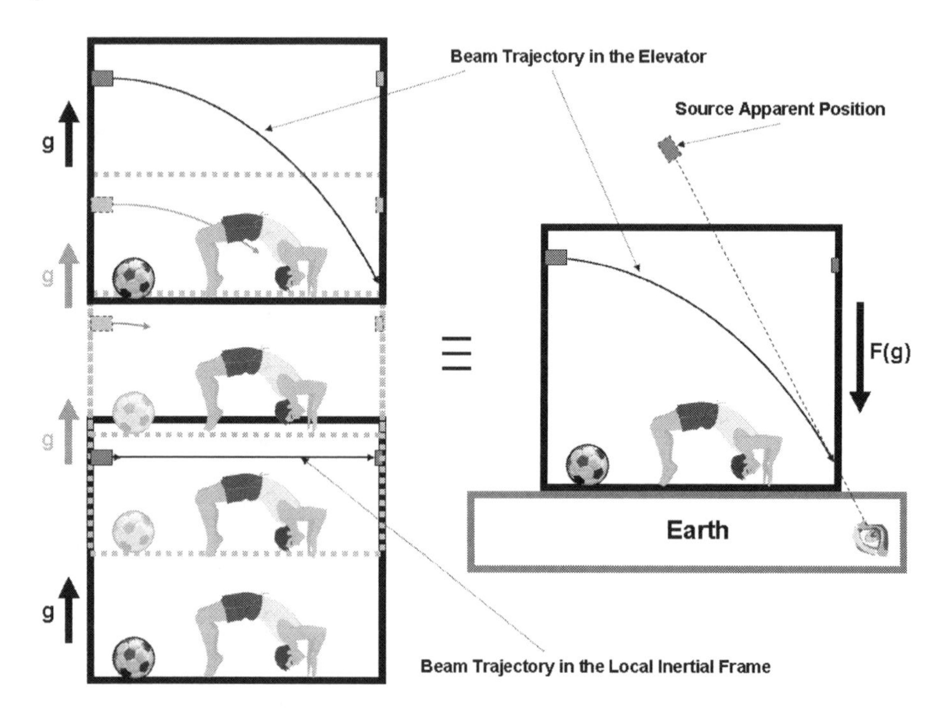

Figure 33. The Trajectory of Light is affected by a Gravitational Field - II

And... what happens if the elevator, instead of being in ***free-fall***, is ***in repose*** relative to *terra firma* (Figure 33, right)? Now the elevator does **not** constitute an *Inertial Frame* as the *gravitational field* is **not** cancelled out any longer and, thus, the ***equivalence*** displayed in Figure 29 does **not** hold. It is to be noticed that now ball and human being are in contact with the floor instead of *floating*. However, in accordance to the *Principle of Equivalence* of Einstein, the *gravitational field* existing in the elevator is ***equivalent*** to ***not existing*** at all if, instead, the elevator is ***ascending*** with the *acceleration* of local *gravity*. But, just a moment! If the Earth disappears... what is the elevator *acceleration* relative to? Einstein responds: with respect to the *Local Inertial Frame*. It is only in the latter where light *propagates* on a *straight line* in *Euclidean space* (*light world line* in Minkowski *space-time* is ***rectilinear***).

If we imagine that, when the *light beam* departs form the source (bottom black), both elevator and *Local Inertial Frame* coincide and are in relative *repose*, then the elevator ***climbs*** and Figure 33 (left) depicts two intermediate positions (in gray) with the final position again in black. Once again, for purely

kinematic reasons, the *light trajectory* (***rectilinear*** in the *Local Inertial System*), when seen relative to the elevator, has to be ***curved*** -- arriving at the same result that when the elevator was in ***free-fall***: *light* follows a curved trajectory relative to *terra firma* (where now the elevator is ***at rest***). But, if *light* is affected by a *gravitational field*, and given that we know that (by virtue of the most famous equation in history) all *radiation* has *inertial mass*, *light* has to have *gravitational mass* as well... but if that is the case, then *light* has to generate a *gravitational field*! Little by little we are discovering that what we thought as different and independent concepts (and, because of that, we gave them different names), in reality, were different manifestations of the same essential phenomenon.

The eye-brain of an observer on *terra firma* receiving the *light beam* (Figures 32 and 33-right) would project the beam à la Euclid[2] and would see a *luminous source* coming from the wrong place -- as we know it occurs when inserting a stick in water. This latter resemblance with *light* ***refraction*** at the interface between air and water was of great concern to Einstein. And, why is that? Because the change of travel *direction* of *light* when passing from air to water is due to its ***different velocities*** of propagation in ***different*** *media*, so that the ***curved*** *trajectory* of *light* had to imply that the *velocity of light in vacuum* was **not** constant! The sacrosanct ***universal constancy of light velocity in vacuum*** -revered in his Special theory- had to go! Did this indicate that Special Relativity was wrong? Not at all! It simply pointed out that Special Relativity has a limited scope and, in particular, that it is only valid when the influence of a *gravitational field* can be ignored.

Needless to say that, due to the tremendous *speed of light* and the *smallness* of a real elevator, the ***curvature*** of the *light beam* due to Earth *gravity* is imperceptible and immeasurable. We shall see later how this phenomenon was astronomically measured.

July 19, 2009: Exactly 40 years ago...

From the far side of the Moon, during 4 minutes, hypergolic[3] fuel motors get the spacecraft away from the free-return trajectory (stage 8 in Figure 30). After half an hour of being incommunicado with Houston, the Apollo 11 reappears confirming that it is orbiting our natural satellite at about 200 km of altitude. In a few hours, the Laser Ranging Retro-Reflector (LRRR) shall be installed on the lunar surface.

2.2. Gravitational Time Dilation or Gravitational Red-shift

Red-shift is a confusing term that refers to the difference between the *frequency* of a ***received*** wave and the *frequency* of the ***transmitted*** wave. As we saw in Chapter 5 ('Galloping with Sound'), this phenomenon is called *Doppler Effect* and is present with any type of wave. In the case of *electromagnetic* waves, the color ***red*** corresponds to the ***lowest*** *frequency* (***longest*** *wavelength*) in the visible spectrum and, thus, *red-shift* means that the ***received*** *frequency*

2 This is what our brain subconsciously does, after our evolutionary adaptation to a space-time ridiculously limited with respect to the whole Universe.
3 Release of chemical energy is attained through a catalytic reaction without the need for ignition.

is ***lower*** than the ***transmitted*** *frequency*. If the latter is the ***lower*** one, then we say there is a *blue-shift*. To complicate things further, this terminology is also used for transmitted *frequencies* outside the visible spectrum, and which are ***lower*** than that of the color ***red*** -- so that a *red-shift* corresponds in reality to a shift away from ***red***. To avoid all confusions, the term *red-shift* simply means that the ***received*** *frequency* is ***lower*** than the ***transmitted*** *frequency*, and the opposite for *blue-shift*.

Let us recall the *prima donna* role that the *propagating medium* played with ***sonic*** waves and, besides, that the specious notion of *ether* was abolished by Special Relativity, simplifying the interaction for *light* to only two protagonists: the ***transmitter*** and the ***receiver***. The classical *Doppler Effect* with *electromagnetic* waves occurs when there is *motion* between ***transmitter*** and ***receiver*** even at speeds much ***lower*** than that of *light in vacuum*. If ***transmitter*** and ***receiver recede*** from each other, the latter will detect a ***lower*** *frequency* that the one ***transmitted*** with a *red-shift* taking place; if they ***approach*** each other, the ***received*** *frequency* will be ***higher*** than the ***transmitted*** one and a *blue-shift* will take place. It is to be understood that this classical *Doppler Effect* does not occur when the *direction* of relative *motion* is ***perpendicular*** to the line joining *transmitter* and *receiver*, as we saw when I was by the tracks and the train passed by me -- at which moment the *frequency* ***emitted*** by the whistle was equal to the *frequency* ***perceived*** by me (Chapter 5).

We should recall as well that the process of ***atomic*** *electromagnetic* emission constitutes the ***standard*** process *par excellence* for measuring *time* and, because of this, a *red-shift* at the ***receiver*** is equivalent to ***dilation*** of the observed *time* because, if the ***receiver*** measures a ***lower*** *frequency*, it 'sees' the ***transmitter*** *clock* running *slowly* with respect to an *identical clock in repose* with the ***receiver***. We shall now analyze the behavior of a *light beam* traveling from ***emitter*** to ***detector*** in comparison with the behavior of two conventional *clocks*; but it has to be understood from the aforesaid that the difference between an *electromagnetic* vibration and a conventional *clock* is exclusively semantic, and that we include the *light beam* so as to be able to appeal to our everyday intuition and, at the same time, to get rid of our prejudices regarding the behavior of a conventional *clock*.

If the relative *velocity* between ***transmitter*** and ***receiver*** is a significant fraction of that of *light in vacuum*, there is a relativistic *red-shift* associated with the *dilation of time* in accord with Special Relativity (Chapter 6), which is governed -as we saw- by the factor $\beta = \sqrt{1 - V_{A/B}^2 / c^2}$. Contrariwise to the classical *Doppler Effect*, the relativistic *Doppler Effect* exists even when the relative *motion* between ***emitter*** and ***detector*** is *transversal* to the *direction* defined by them.

Finally, Chapter 5 ('Galloping with Sound') also showed us that, even when *source* and *receiver* are in relative *repose*, if the pair is in ***accelerated*** *motion* relative to the *propagating medium*, there will be also a *Doppler effect*. For the case of *electromagnetic radiation*, this effect exists as well, but we have to be very careful because we eradicated the concept of a *propagating medium*. It is again the notion of *Local Inertial Frame* that we shall employ to understand the so-called *gravitational dilation of time*. *Dilation of time* in presence of a *gravitational field* was the ***first*** prediction that Einstein made as a direct consequence of his *Principle of Equivalence* in 1907, though it was the ***last*** in being experimentally confirmed.

Figure 34 (right) schematizes a tower with a *luminous* ***source*** on the ***floor*** and a ***detector*** up in the ***attic***. Besides, there are two identical conventional *clocks*, one at each place. The tower is, of course, subjected to the Earth *gravitational field*, so that the *Local Inertial Frame* is defined by the tower hypothetically in ***free-fall***. In this frame, Special Relativity holds and *light* travels in *straight lines* at **constant** *speed* 'c'. Given that the *distance* between ***transmitter*** and ***receiver*** is L, *light* would take a *time* equal to L/c to reach the ***detector***. Equally, if we wanted to compare the ***clock*** on the ***floor*** with the *clock* in the ***attic***, an observer at the latter would compare the ticktack of the *clock* by his side with the ticktack of the *clock* on the floor, and this comparison would entail the *delay* between the *clock* being ***observed*** and the ***observer***.

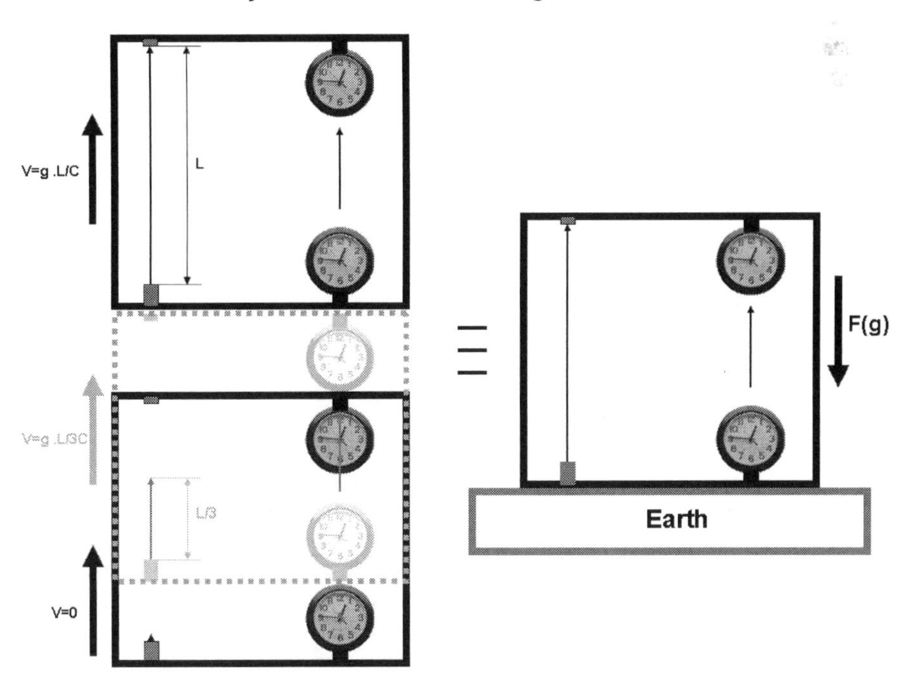

Figure 34. The Gravitational Dilation of Time

Applying the *Principle of Equivalence*, the tower ***in repose*** with Earth, and subjected to the Earth *gravitational field*, is ***equivalent*** to the tower ***moving upwards*** with respect to the *Local Inertial Frame* in which there is **no** *gravity*. If when the *light beam* departs from the ***source***, the tower coincides with the *Local Inertial Frame* and is *in repose* with it (Figure 34, left, in black); then, as the *beam* progresses towards the ***detector***, the tower *moves* upwards with ***increasing*** *velocity* (*acceleration* 'g' of gravity). As an example, the box in gray broken-line corresponds to the tower when the *beam* traversed one third of the *distance* necessary to reach the ***detector*** and, ergo, the latter is moving away at a *velocity* equal to the *acceleration of gravity* multiplied by the *time* that light takes to cover that *distance* according with Special Relativity (g.(1/3).L/c). When the *beam* reaches the ***receiver***, the latter is receding with respect to the *Local Inertial Frame* at a *speed* equal to g.L/c and -by virtue of the exclusively ***kinematic*** *Doppler Effect*- the ***detector*** will indicate a *frequency* which is ***lower*** than the ***transmitted*** *frequency*, i.e. there will be a *red-shift*.

Similarly, if the observer in the ***attic*** wants to compare the *clock* on the ***floor*** with the *clock* by his/her side, the process involves necessarily the interchange of signals: imagine that with each ticktack of the *clock* on the ***floor***, it automatically sends a luminous pulse towards the observer in the ***attic***. With the same rationale we used for the *light beam*, we conclude that the observer in the ***attic*** will assign to the *clock* on the ***floor*** a *frequency* which is ***lower*** than that of the *clock* by her/his side and, hence, the latter will run ***ahead*** the former.

But then, from the *Principle of Equivalence* between *acceleration* and *gravity*, when the tower is rigidly attached to *terra firma* and subject to *gravity*, the *clock* which is ***closer*** to Earth will run ***behind*** the one that is farther away! It is crucial to understand that this *gravitational dilation of time* between two *clocks* is ***not reciprocal*** as the *dilation of time* due to the relative ***uniform*** *motion* in Special Relativity. In this case, both *clocks* are in relative *repose* in a *gravitational field* which is ***more*** intense for the *clock* on the ***floor*** than for the *clock* up in the ***attic***. If, in Figure 34, the ***transmitter*** were in the ***attic*** and the ***receiver*** on the ***floor***, the ***receiver*** would ***approach*** the *light beam* instead of ***moving away*** and, due to the *Doppler Effect*, the ***received*** *frequency* would be now ***higher*** (blue-shift) than the ***transmitted*** and an observer on the ***floor*** would likewise conclude that it is the *clock* ***closer*** to Earth (***higher*** *gravitational* intensity) the one that runs ***behind*** with respect to the one ***farther*** away. Evidently, it is an effect due to the *gravitational field* completely different to the *dilation of time* in Special Relativity (due to the relative ***uniform*** *motion* between *Inertial Frames*).

The inescapable question is: is the rhythm of the *clocks* **really *different*** or simply the result of being ***observed*** from ***different*** places in the *gravitational field*? Rephrased: is the ***difference*** in the clocks paces intrinsic to them or just associated with the ***interchange*** of signals needed to ***observe*** them? If we bring the *clock* on the ***floor*** (which, when observed from the ***attic***, is running ***behind***) to the ***attic*** with us, both *clocks* now run ***in unison***; if we bring the *clock* in the ***attic*** (which, when observed from the ***floor***, is running ***ahead***) to the ***floor*** with us, both *clocks* now run ***in unison***. Hmm... It seems as if it is impossible to know! But, wait a minute: what happened with the ***accumulated*** *time* for the *clock* that was running ***ahead***, and the ***lost*** *time* for the *clock* that was running ***behind*** before they were taken together?

Imagine two ***identical*** *clocks* side by side on the tower ***floor***, and that we *synchronize* them, i.e. they not only run with the **same** *rhythm* but they indicate the **same** *time*. Let us take one of them very slowly to the attic and leave it there for some *time*, at the end of which we take it back carefully to the ***floor***. While the *clock* was ***farther*** away from Earth, its *rhythm* was ***higher*** (ran ***ahead***) than the one for the *clock* on the ***floor*** so that, even though when brought back to the ***floor***, both *clocks* run again in sync, the ***accumulated*** *time* cannot disappear, ergo, the traveler *clock* will continue being ***ahead***, though at the **same** *rate* as the other *clock*. *Mutatis Mutandis*, if we start with two identical synchronized *clocks* in the ***attic***, slowly move one of them to the ***floor***, leave it there for a while and take back to the ***attic***, the traveler *clock* will continue running ***behind*** with respect to the *clock* which stayed in the ***attic*** -- in spite of both now running with the *same pace*. Evidently, if it can be experimentally proved that what I just described is true, we have to conclude that the intensity of a *gravitational field* really modifies the *rhythm* of a *clock* -- be it electromagnetic, atomic, mechanic, or biological!

It goes without saying that Einstein, because this effect -even though in principle had to exist in any building on Earth- was so minuscule, he would have never dreamt of measuring it with the technological state-of-the-art at the time, and suggested to measure it astronomically. In 1960, 40 years later, researchers at Harvard University would astonish the world demonstrating that we do age ***more slowly*** when living on the first floor than when living at the top of a skyscraper.[4]

3. Einstein's Principles of General Relativity and Covariance

As shown in Chapter 6, the *Principle of Special Relativity* implies that all the laws in Nature must be ***covariant*** under the *Lorentz's Transformation*. Reiterating, this ***covariance*** in Special Relativity means that, for a given law of Nature, its two mathematical expressions in two *Inertial Frames* have to be ***structurally identical*** (the difference is to be only nomenclatural). Pithily: the law transforms into itself when passing from one *Inertial Frame* to another. This requirement is certainly quite restrictive on the set of possible laws of Nature and, had we found a physical law not transforming into itself between *Inertial Frames*, Special Relativity would have to be abandoned.

We know already that Einstein could not accept that the laws of the Universe could be only valid in ***privileged*** *frames of reference* (the inertial class). What about all those *reference frames* in ***accelerated*** *motion* with respect to the privileged class? Should humans resign from observing and describing Nature from 'forbidden' vantage points? Something like that was repugnant to Einstein; it had to be our fault and not the fault of the 'Old Man', as he used to refer to his God of Spinoza.

If we accepted the existence of the *inertial forces* (Chapter 4), then both the autobus suddenly ***braking*** and the carrousel ***rotating*** seemed to be valid *reference frames*, i.e. frames where the laws of Mechanics would hold. And, why we only say "both seemed"? Because all we could ascertain was that the emergence of these *inertial forces* would explain the chaos inside the bus and the carrousel, but we did not know if that 'explanation' was comprehensive and consistent with the laws of Mechanics and Electromagnetism in the framework of Special Relativity. And, if it was... would the *Lorentz's Transformation* still leave all physical laws ***covariant*** so that there would not be any privileged *reference frames*? It would certainly not! Because the *Lorentz's Transformation* assumes that the relative *velocity* between the two *reference frames* is

4 If the reader is considering moving to a lower floor to live longer, I remind her/him that we are talking about a fraction of a second for our whole life!

constant.

The *Principle of Equivalence* clearly indicated that *acceleration* was as ***relative*** as *velocity* and, besides, that the action of a *gravitational field* in a given *reference frame*, could be analyzed by ignoring it as such, while imparting to the *reference frame* a suitable ***accelerated*** *motion* with respect to the *Local Inertial Frame*. This ***accelerated*** *motion* is mathematically attained by applying to the *space-time* of the *Local Inertial Frame* -where Special Relativity is valid- a *transformation* much more general than the *Lorentz's Transformation*. But that was not enough: the laws of Nature had to be expressed in such a manner that they could hold in **any** *reference frame* or, equivalently, that those laws were ***covariant*** under a transformation general enough so as to include all physically possible *gravitational fields* or, equivalently, all arbitrary *motions* between the *Local Inertial Frame* and the desired *reference frame*. *Newton's Law of Universal Gravitation* was ***not covariant*** with respect to general *coordinate transformations* in *space-time*.

To achieve such a feat, Einstein -instinctively convinced of its validity- pronounced his *Principle of General Relativity*, which would guide him throughout his intellectual adventure, and that I would express thus:

The Laws of Nature must hold in any reference frame, regardless of its state of motion.

And its mathematical version which he called the *Principle of General Covariance*:

The Laws of Nature must be covariant under arbitrary continuous transformations of the space-time coordinates.

It is not difficult to imagine that a principle so general as this one -which only demands the mathematical continuity of a transformation without restricting its structure- must drastically restrict the set of possible laws of the Universe -- much more drastically than the *Principle of Special Relativity*, which only demanded the ***covariance*** of the laws of Nature under the *Lorentz's Transformation*. But, as a recompense, those laws -if found- would be ***universally*** valid; the *reference frame* called 'Earth' would deliver the same laws of planetary motion as the reference frame called 'Sun'. We would not talk any longer about the simplicity of the Copernican description (Sun as *reference*) -for which Newton's laws are ***valid***- as opposed to the complexity of the Ptolemaic description (Earth as *reference*), for which Newton's laws are ***not valid***.

The new general laws of motion -that Einstein was looking for- had to be ***structurally invariant*** for every physically possible *frame of reference* and, besides, given their great success, Newton's laws had to be a very good approximation to those general laws, in those situations where they had reigned and continued reigning. Such is how Science makes progress: not destroying the past, but learning from it, and building on it.

3.1. The Geometry of Space-Time within the Carrousel

We have learned that in a *reference frame* where a significant *gravitational field* exists, neither *light* nor ***free*** bodies travel uniformly in a *straight line* in Euclidean *space* and, therefore, Special Relativity is ***not*** valid. In other words:

in both the braking autobus and the carrousel, even assuming the emergence of a horizontal *gravitational field* in the former and the *centrifugal* and *Coriolis forces* in the latter (so as to convert chaos into order), those frames continue being *non-Inertial*. This is why General Relativity Theory is needed. Let's now further strengthen our understanding of the aforesaid by imagining that my infancy carrousel is much bigger than it was.

In Figure 35 we see two *Cartesian Coordinate Systems* sharing their origins: SC-A (thin line) attached to *terra firma* so that we can consider it approximately inertial, and SC-B (thick line) which is *in repose* with the merry-go-round. Given that SC-B rotates with respect to SC-A, the *space-time* coordinates for the *center of rotation* are the same in both frames, while the respective SC-A and SC-B *space* coordinates of an arbitrary off-center point differ from each other during a single full turn, repeating themselves over and over. Please realize that an off-center object *at rest* in SC-B is *at rest* with respect to an object located at the *rotation center* when using SC-B as the *reference frame*, but it is *in motion* with respect to the same centered object when the *reference frame* is SC-A -- and *mutatis mutandis* for an off-center object *at rest* in SC-A. Yeah, now you know it: two objects in *space* can be in *relative repose* or in *relative motion* depending upon the *frame of reference*! That's the weirdness of a *center of rotation*.

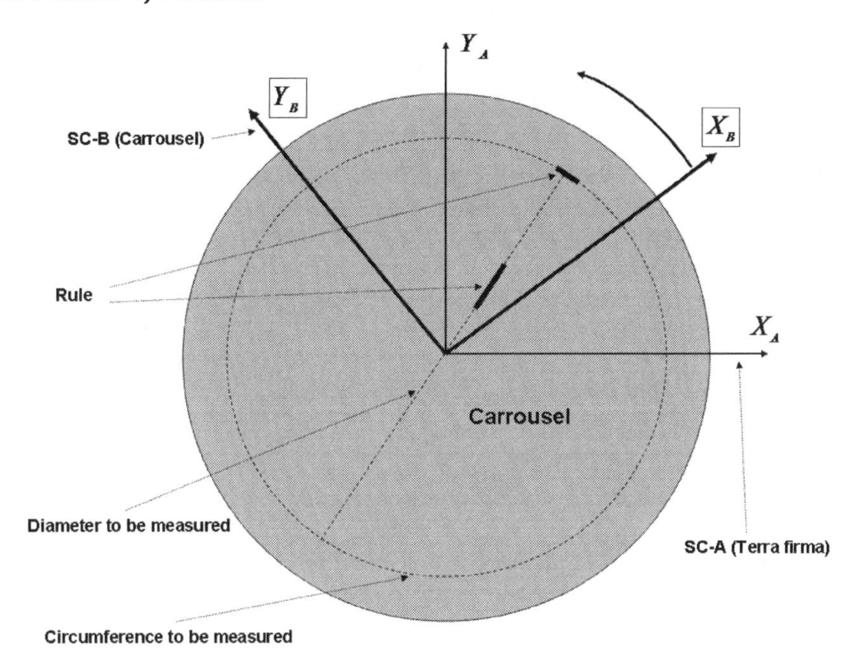

Figure 35. The Space-Time within the Merry-go-round

I was born inside the huge carrousel completely isolated from the exterior, and am aiming to prove that the Euclidean geometry I learned at school holds globally true in this world which -locally- looks to me perfectly 'normal'. Since I was a baby, I am used to think we are ***at rest***, and to subconsciously counteract what adults taught me to interpret as a ***radial*** *gravitational field* whose intensity increases gradually as we get away from a peculiar place of our world where that *field* fully disappears. Because we started our civilization close to that peculiar place, the *gravitational field* is very tenuous and Einstein's Special Relativity holds pretty accurately in our well-known -though limited-world. However, we have now developed traveling technology which allows us to explore the validity of our science much farther away from home.

As we saw in Chapter 6, relative to *terra firma*, a ***small*** portion of the carrousel during a ***short*** period of *time* can be viewed as moving at a ***constant*** *velocity*, so that Special Relativity is ***locally*** and ***temporarily*** valid -- even far away from the center (Figure 35). Let us start by analyzing *time* in the carrousel.

Given that *terra firma* can be considered as globally *inertial*, we know that *time* in SC-A is defined by identical *clocks* at each place in *space* and synchronized with Einstein's technique. Let us now use identical *clocks* to attempt defining *time* in SC-B. One of them, at the center of the carousel, is also *at rest* in SC-A and, ergo, its pace is identical to that of all *clocks* in SC-A. However, as I was going farther away from home carrying another identical *clock*, my subconscious counteracting of the *gravitational field* became little by little conscious and, in addition, something was not right with my *clock*. Let's see why.

Looking from SC-A, an off-center *clock* in the carrousel is in *motion* -- with greater velocity the farther away it is from the center. As a result, locally applying Special Relativity, we conclude that an off-center *clock* in SC-B -due to the ***kinematic*** *time dilation* effect- could never be synchronized with the *clocks* in SC-A. The ***farther*** from the center a *clock* is in SC-B, the ***slower*** it runs with respect to the *clock* at the center.

Looking now from SC-B, this strange behavior of the *clocks* in the merry-go-round is nothing else but the manifestation of the ***gravitational*** *dilation of time*: the greater the *field* intensity (the farther from the center), the further behind the *clock* runs -- as it happened in the tower (floor and attic) subject to the Earth *gravitational field*. This is what was going wrong with my *clock* as I traveled farther away from home. Consequently, I found that it was impossible to globally define *time* in the same way our Science had done it near home (where the gravitational field was negligible) for centuries.

As for *space* exploration, I am anxious to experimentally confirm that the ratio between the *length* of any circumference and its *diameter* is always the number π (3.1416...), as I know it is true close to home. I decide to carry my rule and measure circumferences of increasing radii, all centered at that mysterious spot where the *gravitational* intensity is zero. Figure 35 displays one of those circumferences in broken-line. As we know, had those circumferences been *at rest* in SC-A, that ratio would have been equal to π, regardless of their *diameters*. We will prove that, being those circumferences on the carrousel, from the vantage point of *terra firma*, such a ratio cannot be π.

From the *rotation center* vantage point (*at rest* in SC-A), every ***piecewise*** segment of any circumference will be ***shorter*** than its proper value by the *contraction factor* $\beta = \sqrt{1 - V_{A/B}^2 / c^2}$, where $V_{A/B}$ is the temporarily ***constant*** linear *velocity* of the ***piecewise*** segment of carrousel. Summing up all those

shortened segments will give a circumference *length* ***shorter*** than its *proper length*. However, all ***piecewise*** segments that make up the circumference *diameter* are ***perpendicular*** to their direction of *motion* and, hence, there is ***no*** *kinematic contraction*. As a result, being the *moving* and *proper* circumference *lengths* ***different*** while the *moving* and *proper diameters* ***equal***, the ratio between the *moving* circumference *length* and the *moving diameter* ***cannot*** be equal to π. Pithily: from the SC-A vantage point, SC-B is ***not*** Euclidean.

In sum, clocks in SC-B cannot be synchronized and Euclidean geometry does not hold true. Once again we find that Special Relativity is ***not*** valid in the carrousel as a whole. Even though the *gravitational field* has the potential to explain physical phenomena, and the *Principle of Equivalence* allows us to locally replace the *gravitational field* with an accelerated motion of SC-B with respect to SC-A, we ***cannot globally*** define *time* and *space* in SC-B in the same fashion as we do in SC-A. A new class of *coordinate system* that would entitle us to transform a Euclidean geometry into a non-Euclidean geometry is necessary for us to find laws of Nature which are valid in ***any*** *reference frame*.

3.2. The Absolute or Tensorial Calculus - Scalars, Vectors and Tensors

In 1912, Einstein returned to his *alma mater*, the Zurich Federal Polytechnic Institute (ETH) -- this time as a Professor. He soon realized that the problem he wanted to solve required mathematical tools and skills he did not have but that, fortunately, had been already developed by the Italians Gregorio Ricci-Curbastro and his pupil Tullio Levi-Civita, who recommended Einstein the use of the Absolute Differential Calculus (also known as Tensorial Calculus) to develop his theory of gravitation. Marcel Grossmann (1878-1936), a friend of Einstein's since they were students at ETH, and who was by then a Professor of Mathematics and an expert on non-Euclidean geometries, introduced him to the subtleties of Tensorial Calculus and began to work together on the subject.

The term 'absolute' in Absolute Calculus refers to the mathematical ideas of *invariance* of a magnitude and *covariance* of a formula with respect to different *coordinate systems*. For example, the *distance* in *Euclidean space* is an ***absolute***, i.e. invariant under *rotations* and *translations* of the coordinate system: nobody expects to come up with a different *length* for a metal bar, after *rotating* or *translating* our coordinate system. As we saw along this book, classical physics postulated that the *distance* at a given *time* was also ***absolute*** when we changed from a *reference frame* to another *frame* in ***uniform*** relative *motion* with the former. However, Special Relativity and its abundant experimental evidence demonstrated that our intuition was wrong, and the *length* of a metal bar is ***relative*** to the *state of motion* of the *reference frame*. This search by Einstein for ***absoluteness*** was successful when Minkowski proved that the ***absolute*** was neither in *space* nor in *time* but in *space-time*. In this fashion, the *event interval* replaced the Euclidean *distance*, the former being an ***invariant*** within the class of *Inertial Frames of Reference*. The class of *reference frames* for which the term 'absolute' had a meaning had been extended and, as a result, *space* and *time* were not ***absolute*** any longer.

Both the Euclidean *distance* and the *space-time (event) interval* are what mathematicians call 'real'[5] numbers and, if ***invariant***, they are also called 'scalars'. The concept of 'vector' is an extension of the notion of *scalar*. As we saw in Chapter 1, a *vector* in the three-dimensional Euclidean *space* has ***magnitude*** and ***direction*** and is geometrically represented with an arrow (*force*, *velocity*, etc.). The idea can be extended to higher-dimension *spaces*, e.g. the *space-time*. For a given coordinate system, we need as many real numbers as the dimension of the *space* to define a given *vector*, and these numbers are called the *components* of the *vector* in the coordinate system. For instance, a *vector* in *space-time* has ***four*** components.

The numerical values for the *components* of a *vector* depend upon the *coordinate system* within which they are represented, i.e. those values ***change*** from one to the other in accord with a well-defined *transformation*. In experimental physics, we always ***measure*** a magnitude with respect to a chosen *frame of reference* (which in turn is mathematically represented with a coordinate system); however, a *vector* is unique in the sense that it has a set of intrinsic features which would remain the same despite its ***changing*** *components*, had we selected another coordinate system. It is those innate features -independent of the *reference system*- that constitute the physical **reality** the *vector* represents, and not just a mere perspective as its *components* are. The *coordinate system* is arbitrary and, ergo, selected to maximize descriptive simplicity; physical **reality** is beyond *perspectives*, and has to be always the same.

A *tensor* is an extension of the notion of *vector*. A *tensor* has a number of *components* different from the dimension of the *space* within which is defined. A *scalar* is a *tensor* of ***zero*** order; a *vector* is a *tensor* of ***first*** order; a *tensor* of ***second*** order in *space-time* has $4^2 = 16$ components, etc. A *tensor components* ***change*** with the *coordinate system*, but the *tensor* as such (the set of all its intrinsic properties) does ***not change*** and, hence, the *tensor* is an ***absolute*** magnitude which represents a physical **reality** beyond the *perspective* provided by the chosen *coordinate system*.

As a basic example of the ***absolute*** character of a *tensor*, the ***equality*** between two *tensors* transcends the *reference frame*: their *components* ***change*** from one frame to the other, but its ***equality*** holds in both. As another example, if a *tensor* is ***nil*** in a given *reference frame*, it will be ***nix*** in ***all*** *frames*. Because of these considerations, the laws of Nature, if they are to objectively represent **reality**, they are to be expressed through *tensorial equations* which are ***covariant***: the *equation* will continue to hold for any *reference frame* because, even though the numerical *value* produced by each side of the *equation* will ***change*** from frame to frame, those *changes* will be such that the *equation* remains ***valid***. This is why we say magnitudes are *invariant* and laws are *covariant*. The concept of *tensor* was, doubtlessly, what Einstein needed to develop his General Relativity Theory.

3.2.1. The Metric Tensor: From Necochea to Madrid and from Madrid to Space-Time

... It was joyfully easy for a child to walk around Necochea! The coordinate system designed by its founders is Cartesian: two axes X and Y (sea and river)

5 Not much to do with the concept of **reality**.

mutually perpendicular; the difference between the house numbers (coordinates) on both axes, not only indicates their degree of contiguity (topology) but precisely their *distance* in meters; the *distance* D between any two *points* with coordinate differences ΔX and ΔY is given by the Pythagorean Theorem:

$$D^2 = 1.\Delta X^2 + 0.\Delta X \Delta Y + 0.\Delta Y \Delta X + 1.\Delta Y^2 \qquad [I]$$

As a reminder (Chapter 3), the coefficients of the products of coordinate increments (explicitly shown even if zero) constitute the *components* of the so-called *metric tensor* of Necochea and, in this case, they could not be any simpler: 1, 0, 0, and 1. Besides, and very importantly, all of them are ***constant*** for the whole *space* (city). This simple *metric* allowed me to know the *distance* between any two places, even on the few diagonal streets. However, had the designers been lovers of complexity, they could have used different ***units*** of *length* for each axis (e.g. meters for X and centimeters for Y) or, even worse, they could have chosen as axes the ocean and an arbitrary direction to it, the river and another arbitrary direction, or even two arbitrary directions forming an arbitrary angle. In the first case, the coefficient of ΔY would be 0.01 instead of 1, and in the other cases -the axes not being perpendicular- the formula for the *distance* between two *points* would have required a generalization of the Pythagorean Theorem in which the cross-products $\Delta X \Delta Y$ and $\Delta Y \Delta X$ would have been necessary, with their coefficients reflecting the non-right angle between the axes:

$$D^2 = g_{1.1}\Delta X^2 + g_{1.2}\Delta X \Delta Y + g_{2.1}\Delta Y dX + g_{2.2}\Delta Y^2 \qquad [II]$$

In such a hypothetical case, the internal *curvature* of Necochea *space*, calculated from its *metric tensor* $g = \{g_{1.1}, g_{1.2}, g_{2.1}, g_{2.2}\}$ would have been still zero because all the *tensor components* were still constant throughout the city. It is this zero value for the internal *curvature* that makes Necochea *space Euclidean*. The symbol for the *metric tensor* is simply $g_{i.k}$ where it is implicit that sub-indexes 'i' and 'k' take all possible values between 1 and the space dimension (2 in this case). It is easy to see that, because the product is a commutative operation, we can make $g_{1.2} = g_{2.1}$ so that we say the *metric tensor* is symmetric, and only three components are to be determined. The value $g = \{1,0,0,1\}$ corresponds to the real case in Necochea of a Cartesian coordinate system and can be obtained form the general case (Equation II) by a *transformation of coordinates*.

Let us remember as well that the concept of intrinsic (internal) curvature of a *space* does not correspond exactly to the intuitive idea of *curvature* that our Euclidean sub-conscience imposes upon us. It does exist however a relation: what a zero intrinsic *curvature* is telling us is that a *coordinate system* with *Euclidean metric* -as the one in Necochea- could not be suitable to represent e.g. a *sphere*, because given that the Earth is round -eventually- as the city grew up, the efforts to maintain a ***constant*** *metric tensor* throughout the city would eventually collapse.

The *geometry* in Necochea is *Euclidean* and its *metric tensor* and *curva-*

ture are *invariants* under a *coordinate transformation*. The *components* of the *metric tensor* $g_{i.k}$ change under rotation, translation, or change of the angle between the axes, but the *tensor* as such, together with the *curvature*, reflect immanent attributes of Necochea *space*. Pithily: the *metric tensor* of Necochea (and of any other space) is much more than its *components* in a chosen *coordinate system*.

...I walked and walked without seeing any number and I became very doubtful of getting to the *place* on *time*. After more than what in Necochea would have been two blocks, I finally saw the number '63' which confirmed the *topological* rule of my childhood in the New World. But once again I became frustrated by my miserable failure in estimating the considerable *distance* I had had to walk so as to reduce the portal addresses by just three numbers in this strange *coordinate system* of the Spanish capital.

By then, it was clear to me that the portal numbers (coordinates) were nothing but mere indicators of the *topological* properties of Madrid's space, and that they had very little to do, directly at least, with its *metric* properties. Clearly, places ***close*** to one another in *space* had assigned numbers ***close*** to one another; the farther away the portals, the more different the numbers but... to use the portal numbers to estimate their mutual *distance*, required of a formula which, if existed, would not be simple and, even worse, it would have -apparently- to change from *place* to *place*. Naturally, thus, the Pythagorean Theorem was utterly useless to estimate *distances*...

The general expression (Equation II) for the *distance* between two *points* is as valid in Madrid as is in Necochea. The difference resides in the *metric tensor* which is ***constant*** for the latter and ***varies*** from *place* to *place* for the former. Because of that, Necochea *space* is Euclidean and Madrid space is *non-Euclidean*. If we limited ourselves to a small region of Madrid, we could define a *local coordinate system* in which the coefficients of formula II were approximately ***constant***, so we could consider that region as Euclidean and, with a suitable *coordinate transformation*, we would attain the *metric tensor* $g = \{1,0,0,1\}$ of the Cartesian system in Necochea. As soon as we left that ***small*** region in Madrid, we would need to recalculate a new *metric tensor* to estimate *distances*.

If we now imagine all those *piecewise coordinate systems* for different zones of Madrid and try to get their axes spliced, we would see a global coordinate system in which its axes, instead of ***rectilinear***, would be ***curvilinear***! - in the same way that when we look at a sphere with our Euclidean preconception we see that the *coordinate system* based on ***meridians*** and ***parallels*** has its axes ***curved***, but that they can be regarded as *spliced rectilinear axes* defining coordinate systems which are locally valid. General curvilinear coordinate systems as these are also called *Gaussian coordinate systems*.

The generalization of formula II to the third dimension is straightforward:

$$D^2 = g_{1.1}\Delta X^2 + g_{1.2}\Delta X\Delta Y + g_{1.3}\Delta X\Delta Z + g_{2.1}\Delta Y\Delta X + g_{2.2}\Delta Y^2 + g_{2.3}\Delta Y\Delta Z + g_{3.1}\Delta Z\Delta X + g_{3.2}\Delta Z\Delta Y + g_{3.3}\Delta Z^2 \quad \text{[III]}$$

Once more, if the *metric tensor* $g_{i.k} = \{g_{1.1}, g_{1.2}, g_{1.3}, g_{2.1}, g_{2.2}, g_{2.3}, g_{3.1}, g_{3.2}, g_{3.3}\}$ is ***constant*** for all *space*, the latter is Euclidean and, by means of a *transfor-*

mation of coordinates, we can eliminate the cross-products (annul their coefficients), reemerging the Pythagorean Theorem. Due again to the symmetry of the *metric tensor*, of those 9 components, only 6 are independent.

As for the Minkowski *space-time*, the ***invariant*** defining the *metric* is not the spatial distance but the event (space-time) interval that we defined as (for the sake of simplicity, terms with nil coefficients are not explicit):

$$s^2 = c^2.\Delta T^2 - (\Delta X^2 + \Delta Y^2 + \Delta Z^2) = c^2.\Delta T^2 - D^2 \qquad \text{[IV]}$$

The *metric tensor* for *space-time* is $\{-1,0,0,0,0,-1,0,0,0,0,-1,0,0,0,0,c^2\}$. Again, by symmetry of cross-products, of those 16 *components*, only 10 are independent. All *components* are also constants as for a Euclidean space but, because the *component* associated with *time* has the opposite sign to the ones associated with *space*, this *metric* is not the generalization of the Pythagorean Theorem and, because of that, we say that *space-time* is *semi-Euclidean*. As we know (Chapter 6), to change from one *Inertial Frame* to another is equivalent to *change* the *angle* between the *temporal* axis and the *spatial* axes so that cross-products appear in the expression for the *event interval* without changing the *semi-Euclidean* character of *space-time*. This means that -in Special Relativity- for any *Inertial Frame*, the *space* part of *space-time* is always *Euclidean*.

On the other hand, we also discovered that, when the relative *motion* between two *reference systems* is ***accelerated*** (merry-go-round or braking autobus), while one of them was Euclidean (*terra firma*), the other was *non-Euclidean* and, ergo, if we want to find laws of the Universe that are valid in both *reference systems*, we ought to accept that, as for Madrid, the *metric tensor* of *space-time* could change from one region to another, i.e. that *space-time* -though locally Euclidean- may not be Euclidean as a whole. From this, it is now easy to understand why the General Theory of Relativity does not invalidate the Special Theory but, on the contrary, the latter is locally employed to achieve the global description of the Universe.

3.2.2. Eureka: The Metric Tensor includes the Gravitational Field!

Remembering Chapter 3, and only as a partial and intuitive analogy, a great circle of a sphere is **not** a *geodesic* of the Euclidean tri-dimensional space in which the sphere is immersed but, viewed the latter as a self-contained bi-dimensional space, it is non-Euclidean and its great circles are its geodesics. Correspondingly, if -as we learned- the *trajectory of light* and of bodies within a gravitational field is not a geodesic of the Euclidean space, then... if we assumed that *space-time* is globally *non-Euclidean* (we assume that even its *space* part by itself is globally *non-Euclidean*), that same *trajectory* could now be a ***geodesic***...

Rephrasing: if we supposed that *space-time* is *non-Euclidean* in such a way that the *trajectory of light* was always a ***geodesic***, then, we could stop considering *gravitation* as a mysterious and instantaneous *force* at a distance and, instead, say that the *local geometry* in the neighborhood of a material body changes with respect to that in regions without *matter* and, in this fashion, generalize the *Principle of Inertia* (Newton's First Law) by saying that "a body free of any disturbance shall move along the ***geodesics*** of *space-time*". Here the term 'geodesic' replaces the term 'straight line' which semantics is inseparable from Euclidean geometry. With this new interpretation, there is **no** *force* exerted on the objects so that they change their *motion*; it is simply that the *local geometry* is different and the trajectories adopted by free objects change (in the Euclidean sense) with the distribution of *matter* in their surroundings. In the general *non-Euclidean* sense, the *trajectory* is always a *geodesic* whether *matter* is ***present*** or ***not***. Far away from *matter*, the trajectory will be a *straight line* in the Euclidean sense; close to *matter*, the trajectory will be *curved* in the Euclidean sense. In the general *non-Euclidean* sense, all trajectories are ***geodesics***; period.

And finally: the *Principle of Equivalence* clearly indicates that *acceleration* is as ***relative*** as *velocity* and, besides, that the action of a *gravitational field* in a given *reference frame*, can be studied ignoring it as such, while imparting to the *frame* an ***accelerated*** *motion* with respect to the *Local Inertial Frame*. But... this ***accelerated*** *motion* is mathematically achieved by applying to the *space-time* in the *Local Inertial Frame* (where Special Relativity holds) a *transformation of coordinates*. But... the sole effect of that *transformation* is to change the components of the *metric tensor*! Ergo... the *metric tensor* represents the *gravitational field* and, given that it obviously represents the ***geometry*** of *space-time*, the *gravitational field* defines the ***geometry*** of *space-time* and, vice versa, the latter represents the former.

A region of *space-time* entirely free of *matter* will have *semi-Euclidean* ***geometry*** (*space* will be *Euclidean*), and Special Relativity will be valid; a region of *space-time* where there exists *matter*, will have a *non-Euclidean* ***geometry***, and it is General Relativity -as an extension of Special Relativity- that applies to objectively describe physical **reality**. A planet moving around the Sun is as free of disturbances as it would be if it moved in deep *intersidereal space* -- but its *trajectory* would be different because *space-time* nearby the Sun is 'curved' (non-Euclidean) while *space-time* in deep *space* far away from *matter* is 'flat' (Euclidean). In both cases, the planet *trajectory* is a ***geodesic***.

3.3. The Field Equation: From Newton to Poisson and from Poisson to Einstein

Einstein needed to find the new law of gravitation, a *tensorial equation* (i.e. ***covariant***) which could predict the *gravitational field* associated with a given distribution of *matter* (e.g. the *space-time* surrounding the Sun). Once the *metric tensor* (*gravitational field*) is known, the ***geodesics*** in *space-time* are known and, hence, we can determine all the planetary orbits in our solar system. To carry out such a feat, and knowing that the new law had to become *Newton's Universal Law* for those cases in which the latter had been so successful, Einstein grabbed inspiration from *Poisson's Equation* in classical physics. In classical physics, the *vectorial field of forces* called *gravitational* can be univocally determined by means of a *scalar field* called the *gravitational potential* which has to obey Poisson's Equation:

$$\Delta\ \Phi = 4\pi.G.\rho \qquad [V]$$

This equation is the *continuous* version of Newton's Universal Law[6]. With our non-mathematical approach, this equation simply states that the density of *gravitational mass* (ρ) produces the gravitational potential (Φ) and it is the variation of this *scalar field* in *space* that produces the *vectorial* field of *gravitational forces* at each point in *space*. Different distributions of *gravitational mass* in *space* will produce a different *potential* and, ergo, a different *field of forces*. If the *potential* is ***constant*** in all *space*, then the *gravitational force* is ***nil*** in all *space*. The symbol $\Delta.\Phi$ stands for a mathematical operation[7] on the *potential* that relates its value at a point in *space* with those in its neighborhood. G is the so-called *Universal Constant of Gravitation*. With this equation, we can determine the *gravitational field* inside and outside a material body. Outside the object, ρ is zero and the equation becomes *Laplace's Equation*. We see that in classical *gravitation* theory, a *scalar field* (a single real number for each point in space) is enough to univocally determine the *gravitational field* (a *vector* for each *point* in *space*). Both *scalar* and *vectorial fields* are assumed to exist in the Euclidean *space*.

But, according to Einstein, the *gravitational field* is **not** a *field of forces* actuating in the Euclidean *space*, but the *metric tensor* that gives *space-time* its ***geometry*** which, in turn, is determined by the distribution of *matter* in *space-time*. From this interpretation, it seems that the 10 *components* of the *metric tensor* play a role similar to the one played by the single *potential* of Newton. Let us see if this analogy makes sense: a) far away from gravitational *masses*, *space-time* is Euclidean, Equation IV of Special Relativity holds, and the *metric tensor* is ***constant*** in *space-time*; a body *in repose* stays *in repose*; a body ***uniformly*** *moving* continues doing so. This corresponds to the single Newtonian *potential*: when the latter is ***constant*** in *space*, bodies do ***not*** experiment any *force*; and b) there are gravitational *masses* in the neighborhood and, then, *space-time* is ***non-Euclidean***, the *metric tensor* is variable in *space-time*, and a free body travels on a ***geodesic*** which, from the Euclidean perspective, will be ***curved***. This corresponds to the single Newtonian *potential*: when the latter varies in *space*, bodies experiment a *force* that deviates their *trajectory* from a *straight line*.

It looks, thus, as if the 10 *components* of the *metric tensor* replace the single Newtonian *potential*. Does this mean that Einstein complicated things so much that now we need 10 *potentials* instead of one? NO! Let us recall that the *metric tensor* also defines the ***geometry*** of *space-time* so, in reality, Einstein simply eliminated the Newtonian *potential* because, in General Relativity, the *gravitational field* is not a *field* actuating in *space* but the *space* itself!

Extending the conceptual framework behind Poisson's Equation, we see that the vectorial operator Δ on the single Newtonian *potential* would have to be replaced with a tensorial operator on the *metric tensor*, while the right-hand-side term that describes the distribution of matter in space with only one real number (the density), would have to be replaced with a *tensor* called $T_{i.k}$ which represents the distribution of *matter* in *space-time*. Given that neither *mass* nor *energy*, nor *velocity* are invariant in relativistic physics, this *tensor* combines *mass*, *energy*, *internal stresses*, and *momentum* of *matter* so as to

6 It deals with *continuous* distributions of matter, instead of with discrete particles.

7 It is called the 'Laplacian'.

achieve the ***covariant*** character that every *tensor* has to enjoy. It is clear, then, that in General Relativity, a *gravitational field* is not defined simply by the distribution of *mass*, as it is the case in Newtonian physics.

Einstein and Grossmann worked hard to obtain such a *tensorial* operator which would allow them to predict the *space-time* variations of the *metric tensor* for a given distribution of *matter*. They were on a dead-end line of work for some time because they had not realized that one of the mathematical entities they were using was not a *tensor*. They published a sort of progress report in 1913[8], and another in 1914, but their collaboration ended when Einstein moved to Berlin.

Once in Berlin, Einstein resumed his quest for *tensorial* entities. In *tensorial* calculus, the *variation of a tensor* is not necessarily another *tensor*; that was the problem he was confronting. Such a *variation* had a spurious component that changed from one coordinate system to another. This spurious variation had to be filtered out to make the entity a *tensor*, but, unfortunately, this remaining ***covariant*** part was identically ***nix***. Because of that, a *tensor* that included the *metric tensor* and its *variations* had necessarily to include the *curvature tensor*[9] (Chapter 3) as well.

But something was still missing: the *curvature tensor* had 20 independent *components* while both $T_{i.k}$ and $g_{i.k}$ had only 10 *components*, so that there were more unknowns than equations relating them. Einstein only needed to add the conditions that both *energy* and *momentum* had to be ***conserved*** and... Presto! The equation was entirely *tensorial* (ergo, with objective meaning), and with as many ***unknowns*** as ***equations***. And so did Einstein arrive to the glorious equation for the *gravitational field*:

$$R_{i.k} - (1/2)g_{i.k}.R - \lambda.g_{i.k} = -(8\pi.G/c^4)T_{i.k} \qquad \text{[VI]}$$

In this equation, $R_{i.k}$ is the *curvature tensor*, $g_{i.k}$ is the *metric tensor*, R is the scalar *curvature* of Gauss (Chapter 3), G is the *Universal Constant of Gravitation*, and $T_{i.k}$ is the *tensor of matter* distribution. The term λ is an arbitrary constant, the *cosmological constant* that Einstein introduced to attain a static Universe and later on, when it was discovered that the Universe was expanding, he referred to it as the greatest blunder of his life. Lately, it has been resurrected. These cosmological topics are beyond the intended scope for this book.

To wrap up, it is worth noting that, even though Einstein *field equation* appears as a single compact equality (due to the *tensorial* notation), the latter represents what mathematicians call 10 'partial differential equations' that allow us to determine the 10 *components* of the *metric tensor*, if we know the *components* of the *matter* distribution *tensor*. Knowing the *metric tensor*, mathematicians may use the so-called variational calculus to determine the ***geodesics*** of *space-time*.

8 *Outline of a Generalized Theory of Relativity and of a Theory of Gravitation*, CPAE, vol. 4, pp. 302-343, 1913.
9 The curvature tensor had been independently developed by Riemann and Edwin Christoffel.

4. Experimental Confirmation of General Relativity Theory

As Clifford M. Will (1946-) says in his detailed compendium of the multiple experimental tests to which General Relativity has been put to in the 20th century, it is ironic that while Einstein's legend grew beyond limit, relativity science was becoming stagnant and sterile. Starting with the second half of the 1920-1930 decade, Einstein had embarked on a project he would never finish: integrating *electromagnetism* into his *gravitation theory* as part of the *space-time metric field*. It is equally or more ironic, I think, that having *electromagnetism* been the *raison d'être* for his developing Special Relativity Theory, it was precisely *electromagnetism* that vehemently ***resisted*** its integration into his General Theory -- to the point of taking him to the tomb without significant progress.

The state of General Relativity by the beginning of the 60s is summarized by Will thus: a) it was, without doubt, a fundamental and very important theory of Nature; b) its contacts with **reality** were severely limited; c) it was not easy to understand; and d) it was practically impossible to make useful calculations. In 1960, a new theory of gravitation had emerged -the Brans-Dicke theory- which was taken very seriously for a while, after the appearance of new experimental data which seemed to invalidate General Relativity. It was also in that decade when young scientists as Kip Thorne (1940-), Roger Penrose (1931-), Stephen Hawking (1942-), James Hartle, Igor Novikov (1935-) and others not-so-young as John A. Weeler (1911-2008)[10], entered the field with new mathematical tools, and the desire to reconnect General Relativity to the **real** world, in particular to the macrocosm -- giving the theory the impulse and attractiveness necessary to stimulate other young physicists as well as the general public. This reconnection with **reality** could be achieved in physics in only one manner: with carefully conceived ***experiments*** purposely designed to: a) discard the theory as ***incorrect***, or b) allow it to continue its reign **until** the next ***experiment***.

Einstein knew very well the importance of ***experimental*** corroboration, and proposed in 1916 three basic tests: a) the *precession of Mercury perihelion*[11] for which experimental data were available and in agreement with his theory; b) the *gravitational deflection of light*, with also favorable agreement with theory (though not with a high accuracy); and c) the *red-shift* or *gravitational dilation of time*, whose experimental results remained inconclusive for 40 years. The Brans-Dicke also predicted those effects, differing very little from General Relativity. For this reason, and because the magnitude of all them was minuscule, excellent accuracy was mandatory to proclaim the success of one, or the other... or none. This was General Relativity situation in 1960, waiting for the technological revolution that quantum physics, superconductivity, semiconductors, computers, lasers, etc. were about to instigate.

4.1. Precession of Mercury Perihelion

The first success of General Relativity was the experimental confirmation of its predicting a well-known anomaly of Mercury orbit -- a phenomenon that Newtonian celestial mechanics had not been able to fully explain since the

10 John A. Weeler was one of Einstein's collaborators, and coined the terms 'black hole' and 'worm hole'.

11 The perihelion is the point where the planet is closest to the Sun.

middle of the 19th century. We saw that the *Principle of Equivalence* by itself allowed Einstein to predict (well before the completion of his theory) that the trajectory of a *light beam* would be affected by a *gravitational field,* as well as the phenomenon of *gravitational dilation of time* but, under no circumstances, the *Principle of Equivalence* could by itself predict that the *perihelion* of a planet trajectory had to slowly rotate -- as it had been observed for the closest planet to the Sun. It would finally be General Relativity -through its *field equation*- that would predict with great accuracy the *precession* of Mercury *perihelion.*

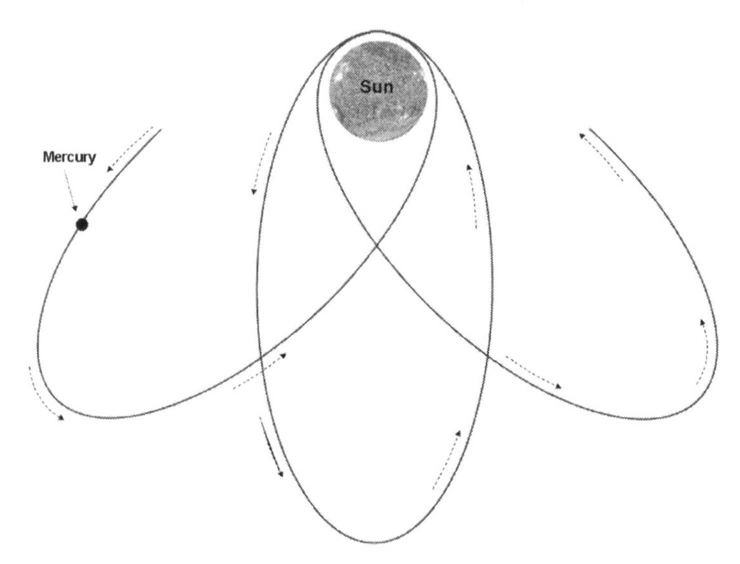

Figure 36. Precession (exaggerated) of Mercury Perihelion

By the middle of the 19th century, the *motion* of the great majority of the known planets had been explicated with great accuracy by virtue of Newton mechanics. In 1846, due to some anomaly in Uranus trajectory, Urbain Le Verrier (1811-1877) surprised the world providing the *coordinates* in our planetary system for a ***new*** planet that had to exist to explain out the anomalous behavior of Uranus orbit. Johann G. Galle at the Berlin observatory oriented his telescope to the indicated place and... Neptune was discovered. Newton's theory became a legend and the paradigm of truth and exactitude.

Le Verrier found that the precession of Mercury *perihelion* (Figure 36) could not be entirely accounted for with Newtonian mechanics. After all classical interactions were considered, there was a deviation from the ***measured** precession* of 38 seconds of arc per century. He tried to increase the *mass* of Venus by about 10% but created other anomalies on Earth orbit and, after other failed attempts, and encouraged by his success with Neptune, Le Verrier proposed the existence of another planet between Mercury and the Sun, which he baptized as 'Vulcano'. This theory was rejected in 1882 and replaced with that of the existence of an asteroid ring. Simon Newcomb later demonstrated that the excess *precession* (with respect to Newtonian mechanics) was 43 seconds of arc per century (instead of 38"/century) -- with several theories coming into and going out of fashion to explain this inadequacy of Newton's theory.

Einstein knew that, in order to convince himself he was on the right track, as well as to convince the world of the soundness of his incipient theory, he had to put it to the test as soon as possible and, in 1913, he chose the already available experimental data on Mercury anomalous orbit. According to this theory, the ***farther*** the planets from the Sun, the ***smaller*** the *curvature* of *space-time* would be, and their trajectories would be closer to Kepler's closed ellipses. Because Mercury was so close to the Sun, it was the one whose *precession* had to be the most pronounced and, in fact, the one that astronomers had had the most difficulty to explain with Newtonian mechanics. Einstein's calculations (he had not still realized his mistaken employing of a *non-tensorial* magnitude) predicted an excess precession of only 18"/century, instead of the measured 43"/century. This, naturally, constituted a big disappointment for him. Finally, in November of 1915, a week before his completing of General Relativity, Einstein recalculated Mercury *precession*, and this time he obtained a value of 43"/century. He told a friend that, when comparing his predictions with the existing data, he had had palpitations. Current measurements give a more accurate value of 42.98±0.04 seconds of arc, and the full theoretical prediction is 42.98". For a nascent theory, this excellent concordance with astronomical measurements gave the theory a considerable boost and stimulated the scientific world to conduct -even in war times- a second crucial experimental test in 1919: the *deflection of light* in a *gravitational field*.

4.2. Gravitational Deflection of Light

Mercury *precession* had been measured much before Einstein accurately predicted it with his theory. Contrariwise, the *deflection of light* passing close by a celestial body was a phenomenon that had been predicted and calculated (with Newtonian mechanics) in 1801 by the Austrian Johann von Soldner (1776-1833), but had never been measured. Henry Cavendish (the discoverer of Hydrogen) had made the same calculations 15 years before, but he had not published them. The English John Michell (1724-1793) also thought that, if *gravitation* was *universal* for *material bodies*, it had to influence the 'corpuscles' of *light* as well. Remember that, in those days, Newton's *corpuscular* theory of *light* was the only accepted.

The fact that *light* seemed not to have *mass* (after all, 'light' also means 'without weight') was not a major impediment for it to be affected by a great *gravitational mass* because, precisely due to the ***equality*** between *gravitational* and *inertial masses*, the *acceleration* produced by a *gravitational field* is always the same for ***any*** *inertial mass*![12]. If Newton's theory was correct, Soldner and Michell calculations were impeccable: a *light beam* passing tangentially to the Sun surface would change its *direction* by an angle of 0.875 seconds of arc and, if it passed close by Earth, its deflection would be a meager 0.001". Even the deflection caused by the Sun was too small to be accurately measured.

When the *undulatory* theory of *light* acquired preeminence, the idea that

12 No matter how small is the *inertial mass*, as long as it is not exactly zero, it cancels out with the *gravitational mass* (Chapter 1), so that the *acceleration* produced by the *gravitational field* will be ***independent*** of the body *inertial mass* (as we extensively discussed throughout the whole book).

light could be affected by a celestial body was discarded without major difficulty. In a few words: Newton's theory was sufficiently flexible to accept -without falling- one or the other of the two hypotheses with respect to the interaction between *light* and *matter*. The undeniable power of ***experimental*** evidence was not available to separate **reality** from *theory*, and the phenomenon was forgotten until Einstein resurrected it in 1907 when, without knowing about Soldner and Michell calculations, he reproduced them as an inevitable consequence of his *Principle of Equivalence*. Besides obtaining the same deflection of 0.875", he suggested measuring the effect when the Moon eclipsed the Sun, so that the absence of the Sun glare -that impedes seeing the stars during the day- would allow the accurate comparison of a star apparent *position* when passing near the Sun on its way to Earth, with its apparent *position* when not doing so. Figure 37 depicts the situation during an eclipse.

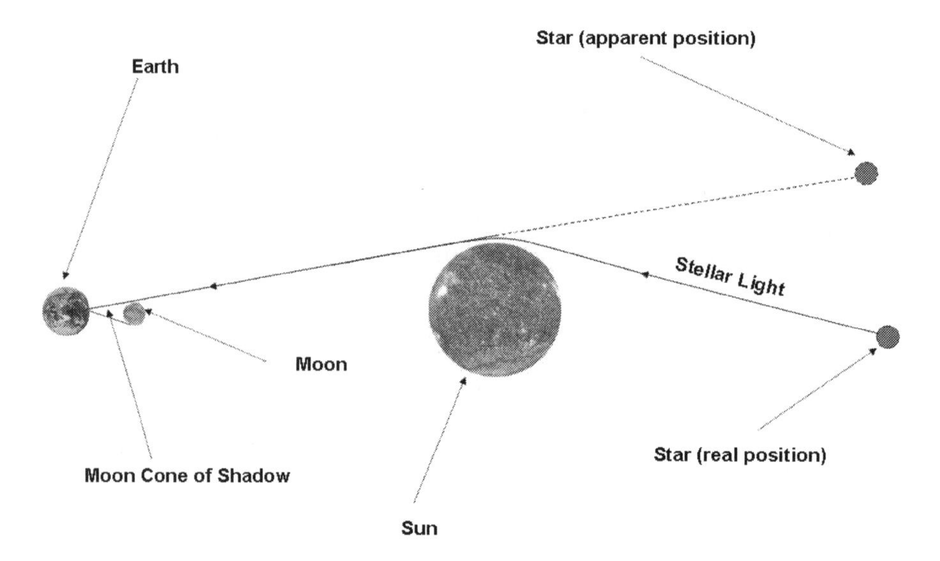

Figure 37. Deflection of Light by the Gravitational Field of the Sun

Edwin Freundlich at the Prussian Royal Observatory in Berlin, together with Charles D. Perrine, Director of the Astronomical Observatory in Cordoba (Argentina), began working together in the problem: there was going to be a solar eclipse in South America on October 10, 1912 and delegations from Brazil, Argentina, France, and England congregated in Mina Gerais (Brazil); weather was rainy and frustrated all efforts, particularly those of Perrine who wanted to measure light deflection. A new eclipse, uniquely suited to ***measure** light deflection*, would occur in 1919. Freundlich, in turn, was trying to get funds to observe the eclipse to occur in Russia in 1914, and met in August of that year with Perrine in Crimea for a new attempt at ***measuring** light deflection*. Once again misfortune came about: Germany declared war to Russia, and both were taken prisoners. The Director of California Lick Observatory, William W. Campbell, and his assistant Heber D. Curtis were not arrested because of their nationality.

By the end of 1915, Einstein -in possession of his full General Theory- recalculated *light deflection* by the Sun, obtaining a value which was ***double*** the one originally calculated. General Relativity predicted a deflection of 1.75" instead of the 0.875" predicted by classical physics. This additional *deflection* was the result of the *curvature* of *space-time* around the Sun. This fresh and so different *prediction* was very important because: a) being ***double***, it could be ***measured*** with ***higher*** accuracy; and b) the big ***discrepancy*** between the predictions of General Relativity and Newton's theory established a clear ***experimental*** criterion to discriminate between both theories -- even in the case that the ***accuracy*** of the experimental technique would not be sufficiently high to proclaim the indisputable validity of one or the other. As pointed out by Arthur S. Eddington, the results of *light deflection* measurements had to -at least- enable the scientific community to enunciate one, and only one, of the following possible conclusions: a) there is **no** *deflection* (Newton would win); b) there is a *deflection* of the order of 0.875" (Newton would win); or c) there is a deflection of the order of ***double*** 0.875" (Einstein would win).

Curtis prepared for the upcoming eclipse in Goldengale (Washington State, USA) in June, 1918. Due to the war, the instrumentation (still in Crimea) did not arrive on time and he had to use inferior quality lenses. Weather was not optimal either, but some pictures were taken that allowed ***measuring*** the *deflection*. The results were analyzed by Campbell and Curtis with the interpretation that the *deflection* was closer to 0.875", i.e. Newton was triumphant; Einstein was wrong. Many astronomers -who considered General Relativity incomprehensible- delightfully received the news.

The favorable conditions offered by the eclipse to occur in May 29, 1919 were optimal. An expedition headed by Eddington established their observation camp in Principe Island on the west coast of Africa; another expedition went to Sobral, a little villa in the northeast of Brazil. Needless to say that the objective interpretation of the photographic plates obtained in both places, and their comparison with those obtained before or after the eclipse, constituted an intellectual enterprise of cyclopean proportions. The conclusions were that, using the best photographs, there was a deviation of 1.98±0.12 seconds of arc and, including other plates no so good, deflection was 1.61±0.30 seconds of arc. Other photographs showed deviations of 0.92" but their inferior quality did not enable an estimation of their measurement precision.

The picture of Newton on the main wall gave the ceremony an aura of historical and dramatic transcendence. Hegemony of British science for more than two centuries was -painfully but proudly- about to be transferred as a result of respect for scientific objectivity and the quest for truth -- even when -in war time- a theory developed by a scientist on the enemy camp was about to be ratified, while dethroning a British giant. Under the presidency of Joseph J. Thomson (the discoverer of the electron), at the joint meeting of the Royal Society and the Royal Astronomical Society, on November 6, 1919, Frank W. Dyson (the Royal Astronomer) speaking on behalf of the Sobral team, said: "After careful study of the photographic plates, I can say that there are no doubts they confirm Einstein's predictions". Eddington, on behalf of the Principe Island team said that their results were consistent with those of Sobral. Einstein became an instant celebrity.

Light deflection added to Mercury *perihelion precession*, and the number of scientists that began believing General Relativity was increasing. Nevertheless, the notorious propensity that Dyson and Eddington had in favor of General Relativity, triggered a long scientific debate which clearly indicated that more ***experimental*** *evidence* was needed. Such is the way Science makes progress.

In 1922, a new expedition by scientists from the Lick Observatory confirmed the results of 1919, but neither Curtis nor Perrine was willing to believe in a "purely philosophical", "strange", and "sagacious" theory, but that "it was not a truthful representation of the physical universe". The original article from Soldner was republished by Philipp Lenard, a scientist favorable to the Nazi movement, who promoted the ridiculous idea that Einstein had plagiarized Soldner, and that he was a complete fraud.

After several failed attempts due to bad weather, Freundlich led one expedition to Sumatra in 1929 during which he obtained a deflection hard to accept (2.24") because exceeded considerably Einstein's value of 1.75" and, hence, differed even more from the Newtonian value of 0.875". By the beginning of the 1950s, Freundlich's results continued covering General Relativity with a veil of uncertainty but, even more pernicious was that -after almost 40 years- the only contact with **reality** still was the original Mercury *perihelion precession* and *light deflection*. As the reader may have surmised after almost reaching the end of this book, the validation process for a theory in Physics is an extremely complex process which, in reality, never ends.

Beginning the 1970s, the measurement of eclipses lost its relevancy due to the much greater exactitude provided by measuring the *deflection of radio waves* emitted by faraway galaxies known as *quasars*, using radio-interferometers. In 1969, General Relativity was confirmed with an accuracy of 15%, comparable to the exactitude achieved in the past with the measurement of eclipses. Every year between 1969 and 1975, multiple experiments were conducted with accuracies gradually increasing up to an agreement with General Relativity of 0.1%.

4.3. Gravitational Time Dilation

Einstein had suggested that the *red-shift* of the luminous spectrum due to a *gravitational field* could be measured comparing the *wavelength* of the *radiation* emitted by some chemical elements in the Sun with that of the same element here on Earth. The Sun was taking the place of the ***floor*** of my tower in Figure 34 (***stronger*** *gravitational field*); the Earth was taking the place of the ***attic*** of my tower (***weaker*** *gravitational field*). Let us recall that when the ***transmitter*** was on the ***floor*** (Sun), and the ***receiver*** in the ***attic*** (Earth), the latter received a *frequency* that was ***lower***, i.e. there was a *red-shift*. This means that, e.g. the ***yellow*** *light* emitted by an atom of Sodium on the surface of the Sun (***floor***), whose *wavelength* here on Earth (***attic***) is 589.3 nm, should be measured by us a little ***redder*** (*frequency* a little ***lower***, i.e. the *wavelength* a little ***longer***). This 'little redder' amounts to only 0.00125 nm ***longer***.

Measurements carried out between 1915 and 1919 were negative, including the famous expeditions to Sobral and Principe Island. From 1919 on, French and German researchers attained results that tended to corroborate Einstein's prediction, but not unequivocally. Einstein had said to Eddington: "If this effect is proven non-existent in Nature, then, my dear theory will have to be abandoned". The effect for the Sun was too small; a celestial body with a much *stronger gravitational field* was needed. The discovery of the so-called 'white dwarfs' with a density thousand times greater than the Sun's (the effect would be considerably amplified) made it possible to measure the *red-shift* with higher accuracies. In 1925, astronomers at the Mount Wilson observatory in California detected a *red-shift* of the same order of magnitude than Einstein's prediction but -after 40 years- the effect still had not been measured with the required accuracy to eliminate all doubts. Other phenomena occurring on the surface of the Sun produced other *Doppler Effects* which could not be separated so as to identify and quantify the *gravitational red-shift.*

4.3.1. Pound-Rebka Experiment

Unexpectedly, in 1960, the required technology to measure the *red-shift* improved dramatically and, thus, my ridiculous example of a tower on our own planet became a reality, precisely embodied in the Jefferson Tower of Harvard University with a height of only 22.5 meters! The experiment was carried out by Pound and Rebka who -using gamma radiation- confirmed the *gravitational red-shift* predicted by General Relativity with an accuracy of 10%. Repetitions of the experiment in 1965 delivered an accuracy of 1%. This achievement is even more impressive if we realize that the ***change*** of *frequency* between the ***attic*** and the ***floor*** is incredibly small: 0.0000000000002%.

The American Journal of Physics announced with great enthusiasm: "This is an exciting era: Einstein's Gravitation Theory of 1915 is moving from the realm of pure mathematics to that of physics. After 40 years of astronomical deficient results, new terrestrial experiments are possible and are being planned". A new era of interest on, and accurate corroboration of, General Relativity had been born.

4.3.2. Hafele-Keating Experiment

In October 1971, an exceptional experiment was carried out which included the ***measurement*** of both the ***kinematic*** (Special Theory) and ***gravitational*** (General Theory) *dilations of time*. Even though the achieved accuracy was lower than that of Pound and Rebka experiment, this experiment showed irrefutably the cumulative effect of minuscule delays experienced by the *clocks* when in different *states of motions* and *gravitational* intensities. Given that the effect is independent of the type of *clock* (inanimate or biological), understanding this experiment and its results, will allow us to finally comprehend why the so-called twin paradox is not such. Likewise, it will enable us to conceptually understand how Global Positioning System (GPS) technology works.

To grasp the concepts which explain the results, we shall simplify things maximally. Imagine a *clock* on the Equator and an airplane up there at 10,000 m of altitude (Figure 38). Because the Earth *gravitational field* at that altitude (***attic***) is ***less*** intense than on *terra firma* (***floor***), the ***flying*** *clock* experiences a *blue-shift*, i.e. it will ran ***faster*** than the one on the ground. This is the *gravitational dilation of time* and, given that it only depends upon the *clock position* in the *gravitational field*, this *dilation* exists regardless of whether the airplane travels East or West.

The application of Special Relativity is a little more subtle. We need an *Inertial Frame* the *paces* of the ***flying*** *clocks* and of the *clock* on *terra firma* can be referred to. Only so, their relative behind/ahead operations can be estimated. During the experiment, our planet ***translation*** *velocity* can be regarded as ***constant*** and, thus, if we imagine a ***master*** *clock* ***at rest*** in a *reference frame moving* ***in sync*** with our planet around the Sun, the only *motion* producing a ***kinematic*** *time dilation* will be our planet ***daily*** rotation. On the other hand, as we know, given that *rotation* is an ***accelerated*** *motion*, the application of Special Relativity may only be achieved *space-time* ***locally***, and these ***local*** effects will have to be integrated to obtain the ***global*** effect.

Let us look at the aircraft flying towards East (Figure 38, bottom). Because plane and Earth move in the ***same*** *direction* and the plane moves relative to Earth, the ***local*** *speed* of the ***flying*** *clock*, relative to the *Inertial Frame* in which the ***master*** *clock* is *at rest*, will be ***greater*** than the ***local*** *speed* of the *clock* on *terra firma*. This means that the ***flying*** *clock* runs ***more*** *behind* the *master clock* than the ***ground*** *clock* and, ergo, the former runs ***behind*** the latter. Because this ***kinematic*** effect is the ***opposite*** of the already discussed *gravitational* effect, the net result will depend upon the flight ***altitude*** (*gravitational* intensity) and upon the airplane *ground speed* (Chapter 4).

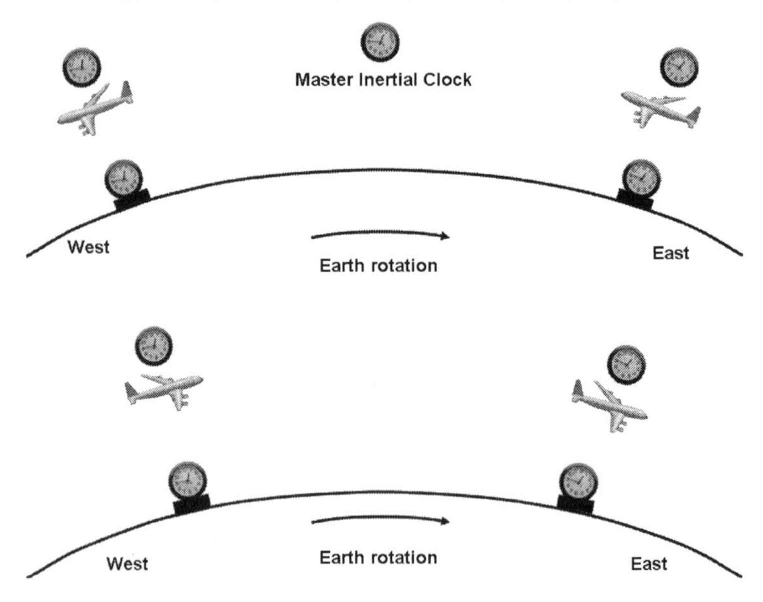

Figure 38. The Experiment of the Flying Clocks

As for the plane flying West (Figure 38, top), with the same rationale, it has a ***local*** *velocity* relative to the ***master*** *clock* which is ***lower*** than that of the ***ground*** *clock* and, hence, the former runs ***less behind*** relative to the ***master*** *clock* than the latter, i.e. the ***flying*** *clock* runs ***ahead*** of the ***ground*** *clock* -- as with the *gravitational* effect. When flying towards West, then, the ***flying*** *clock* always runs ***ahead*** of the ***ground*** *clock* regardless of its *altitude* and *ground speed.*

Integrating these ***local*** effects, if we conceive an experiment in which we start with three identical ***synchronized*** *clocks*, leave one 'steady' on Earth and send a second around the world towards ***West***, while the third does the same towards East, when they get back together, the one that circumvolved the planet toward West shall be ***ahead*** ('aged' more) the one steady on Earth, and the one that flew ***East*** shall be ***behind*** or ***ahead*** depending upon its ***altitude*** and ground ***velocity***. This is the experiment that J.C. Hafele and Richard Keating conducted with Cesium atomic clocks in 1971.

Carefully accounting for flight conditions (they employed several commercial flights), the researchers calculated the contribution of each one of the local effects to arrive at the global prediction. For the ***West*** flight, the prediction was that the ***flying*** *clock* had to be ***ahead*** by 275±21 nanoseconds, of which 179±18 nanoseconds corresponded to the *gravitational* effect. In reality, the ***flying*** *clock* ended up being 273±7 nanoseconds ***ahead***. As for the flight towards ***East***, the prediction was that the ***flying*** *clock* had to be ***behind*** the ***ground*** *clock* by 40±23 nanoseconds; the *clock* actually was 59±10 nanoseconds ***behind***. If we consider that the total experimental uncertainty (due to the uncertainty in the flight conditions combined with the atomic clock accuracy) was ±20 nanoseconds, we can affirm that the cumulative effect of ***kinematic*** and ***gravitational*** *dilation of time* was successfully confirmed. The experiment was repeated on its 25th anniversary with even better agreement between *theory* and **reality**.

In 1976, a similar experiment was conducted, though this time an atomic *clock* was sent in a rocket permanently communicated with another *clock* on *terra firma*. The much greater *velocity* and *gravitational* differential provided much ***higher*** *accuracy* to the experimental data. Two hours of rocket flight generated information for two years of analysis. Once the latter was finished, the ***measured*** *net dilation* agreed with theory within 0.007%.

4.3.3. The Global Positioning System (GPS)

During the Gulf War in 1991, the US Defense Department distributed more than 10,000 GPS receivers to Allied forces, so that soldiers and pilots could know their *positions* with a precision of 30 meters. Today, it is a popular accessory in all automobiles, with a precision of a few meters.

The system utilizes a set of between 24 and 32 satellites circumvolving Earth at an ***altitude*** of about 20,000 km, with a ***velocity*** of about 14,000 km/h. These satellites continuously transmit radio signals detected by the receivers in our cars, which are always within the range of at least four of those satellites. Knowing the *transit time* of the signals, the receiver calculates its *distance* to each satellite and, using trilateration[13], determines the *position* and *velocity* of our car. Due to the high speed of the radio signal, a small error in the *clocks* produces a big error in the calculated *distances*. In some cases, navigation is acceptable even when the receiver has the signals from only three satellites.

13 Trilateration is a method to determine the intersection of three spheres given their centers and radii.

As a result of the ***gravitational*** and ***kinematic*** *dilations of time* described in the previous section, the *clocks* in the satellites run about 38 microseconds ***ahead*** per day, with respect to the *clock* on our cars. Please note that, at 20,000 km of altitude, gravity is about four[14] times weaker than on ground. Had these relativistic effects not been included in the mathematical processing of the signals received by our cars, in less than an hour, our *position* would be estimated with an error high enough to make our navigation impossible.

4.3.4. The Twin Paradox is not a Paradox

Finally, we are in a position to explain and understand why the so-called twin paradox is **not** a paradox. First, I need to clarify that a paradox is, in essence, an apparent logical contradiction, but that the word is also used to indicate the apparent repudiation of our *common sense* which, subconsciously and often incorrectly, we consider to be 'logical'. In fact, this latter popular acceptation of the word is listed first in most dictionaries and, according to it, the twin paradox will always be a paradox -- unless we come to terms with the fact that *time* is relative to the *reference frame*, and that it is our ***proper*** *time* that determines our ***aging***.

During my imaginary trip to Gliese 581, I was very careful not to include the ***accelerated*** *motion* that would inevitably occur ***taking off*** and ***landing***; otherwise, I could not have estimated the *times* by using only Special Relativity. It is clear though that, I could have ideally supposed that the *periods of acceleration* were so ***short*** as compared to those of ***uniform*** *motion* that my conclusions attained exclusively with Special Relativity had to be approximately correct. Let us recall that, given that the traveler twin took 15 years to reach the star (while 25 years had elapsed on Earth), and would have taken another 15 years to return to Earth (while another 25 years would have passed on Earth), the 'nomad' twin would have found the 'sedentary' twin 20 years ***older*** than s/he was. If, after having put up with me for seven long chapters, you are still not convinced that *time* is **relative**, and the ***proper*** *time* for each twin is ***different***, then, the twin paradox -in the sense of ***negating*** your *intuition* and *common sense*- will continue being a paradox -- until your *intuition* is readjusted.

But the detractors of Relativity Theory usually brandish the existence of another paradox, this time in the sense of a ***logical*** *contradiction* (which I mentioned while discusing the trip to Gliese 581 in Chapter 6): given that *motion* is clearly ***relative***, in the same manner that we conclude the traveler aged ***less*** because s/he developed *speeds* comparable to that of *light in vacuum*, while the other less fortunate twin stayed home; what impedes us from viewing the trip as one in which the former twin stays *in repose* while it is the latter (together with the Earth-star system) that *moves* in the ***opposite*** *direction*? By now, we know that such a vantage point is legitimate but... then... the relation 'younger than' would be reverted! Let us recall that, from the vantage point of the traveler, the *time* elapsed on Earth was (for the round-trip) only 18 years, against the 30 years elapsed on his/her spaceship... and this seems to be a real paradox because, once the twins are embraced face to face, only one of them can have more gray hair and wrinkles than the other. We know that *space-time*

14 Earth radius is about 6,370 km so the satellites are at a distance from the center of the Earth which is more than double the distance the automobiles are and, by Newton's Law of Gravitation, gravity is less than a quarter.

coincidences are ***absolute*** (they cannot depend on the vantage point). Either the 'nomad' twin or the 'sedentary' twin is older. There must be an explanation to this logical contradiction.

The paradox can be resolved in two stages: first, as we saw during the one-way trip to Gliese 581, what really matters is the ***proper** time* for each twin -- **not** how each one measures the other's time from his/her respective *reference frame*. The *time **lived*** by the earthling twin until the nomad twin arrived at the star, is his ***proper** time*, i.e. 25 years; the *time **lived*** by the astronaut twin is his ***proper** time*, i.e. 15 years; *ceteris paribus* -and allowing for individual differences- it is the sedentary twin that has more wrinkles! Because the same happens during the return trip, the total ***proper** time* for the astronaut is 30 years, while the total ***proper** time* on Earth is 50 years. But now we are in big trouble: when both twins finally get together, only one can be older or younger than the other, regardless of what *reference frame* we use to calculate it. Using the frame Earth-star we calculated that the traveler is 20 years younger (50-30), while using the spaceship as the *reference frame*, we calculated that the twin who stayed home (now traveling in the ***opposite** direction*) is 12 years younger (30-18). Given that, from the comparison of ***proper** times*, we know that the traveler has to be 20 years ***younger***, the only way out is that, when we see things from the vantage point of the astronaut, the twin on Earth -for some still unknown reason- ***aged*** an additional 32 years (-12+32=20).

This is the apparent paradox wielded for over 100 years to declare Relativity Theory as absurd and false. And why is the paradox only apparent but not real? Because many people still do not understand that, if we only apply Special Relativity, the two *reference frames* are **not** equivalent: for those two vantage points to be equivalent, we need to apply General Relativity as well, and the ***gravitational** dilation of time* cannot be ignored. Even though some authors emphasize this need, they are short of explaining how both Special and General Relativity theories blend to rationally explain the paradox out. Some say that the 'traveler' twin does experience *acceleration* while the other does not, and that makes the whole difference. That is not true: from the traveler's vantage point, it is the *rest of the Universe* (including the 'sedentary' twin) that experiences *acceleration*! Not surprisingly, the best explanation was made by Einstein himself in 1918 [12], and soon forgotten (or simply not fully understood).

We know that *motion* is **relative** only when the whole Universe is considered so as to achieve symmetry between two opposite vantage points: if we want to consider the spaceship as a *reference frame*, then the *rest of the Universe* has to *move* in the ***opposite** direction*. We also know that *gravity* and *acceleration* are equivalent. When the *rest of the Universe* is used as *reference* (our usual way of thinking), the spacecraft moves with a given *acceleration* while the *rest of the Universe* is assumed *in repose*; when it is the spaceship that we assumed to be *in repose*, the *rest of the Universe* is in *motion* with a ***common** acceleration* (***free-falling*** in the equivalent *gravitational field*). This ***free-falling*** disappears when the relative *velocity* is ***constant***. Recalling my experience with the *Giants of the Road* crossing the pampas, when things were seen from inside the braking bus, there was a *gravitational field* that explained all phenomena inside and outside; when things were seen from *terra firma*, such *gravitational* *field* did ***not exist***. This is why General Relativity is necessary when we choose the spacecraft as the *reference frame*, but is not needed when choosing the Earth-Gliese 581 frame.

Using, thus, the spaceship as the *reference frame*, there exists a *gravitational field* variable in *time* so that: a) when the trip starts, the *rest of the Universe* is in *free-fall* until ***cruise*** *speed* is reached (0.8c) so that the *distance* between ship and star ***decreases*** (Earth goes farther away); b) while at ***cruise*** *speed*, the *gravitational field* ***disappears***; c) when ship and star are close, the *rest of the Universe* starts braking and again a *gravitational field* of ***opposite*** *direction* appears until its *velocity* is ***zero***, *motion* is ***reversed***, and its *velocity* now *increases* to reach ***cruise*** *speed* (-0.8c), with the *distance* between ship and Earth now ***decreasing*** (Gliese 581 going farther away); d) while at ***cruise*** *speed*, the *gravitational field* ***disappears***; e) getting close to Earth, a *gravitational field* appears again to ***decelerate*** the *rest of the Universe* ***free-falling*** until ship and Earth are in relative ***repose*** and the twins embrace.

Let us now identify the situations where there is ***gravitational*** *dilation of time* due to the fact that the *clock* on the spaceship and the *clock* on Earth find themselves in ***different*** *positions* in a *gravitational field* (***attic*** and ***floor*** in Figure 34). Stages b) and d) are discarded because while the relative *speed* between the two frames is ***constant***, there is **no** *gravitational field* and, hence, there is **no** ***gravitational*** *time dilation* (only ***kinematic*** *dilation*). If we assume -for the sake of simplicity- that upon start the ***cruise*** *speed* is attained in ***negligible*** *time* (stage a) and that this ***cruise*** *speed* is reduced to ***repose*** on Earth also in a ***negligible*** *time* (stage e), then the *clocks* on the ship and on Earth are at the ***same*** *position* on the *gravitational field* in both cases and, ergo, they run *in unison* so that stages a) and e) can also be ignored. It only remains stage c) during which *motion* is ***reversed*** and, even though ***forward*** *deceleration* and ***backward*** *acceleration* can be also assumed to occur in ***negligible*** *time*, Earth and ship are in very ***different*** *locations* of the *gravitational field* so that the respective *clocks* run with ***different*** *paces*.

Leaving aside the specific numerical calculations (not corresponding in a book like this), let us qualitatively see which one of the *clocks* (ship or Earth) is the one that runs ***behind*** at the end of the trip. We know that the *clock* on the **floor** of the tower runs ***behind*** the *clock* up in the ***attic*** (Figure 34) because the *gravitational field* is ***stronger*** on the ***floor*** than in the ***attic*** (the *field acceleration* is directed from the ***attic*** towards the ***floor***). During stage c), the *acceleration* is directed from Earth *towards* the ship -- both while the *rest of the Universe* brakes to ***reverse*** its *motion*, as well as when it finally ***heads*** back to the ship. This means that the ship is in a ***stronger*** *field* (***floor***) than the Earth (***attic***) is and, hence, the ***ship*** *clock* runs ***behind*** with respect to the ***clock*** on Earth. The calculations show that the traveler's ***clock*** runs 32 years ***behind*** so that, when combined with its 12 years of running ***ahead*** (*kinematics*), ends up running a net 20 years ***behind*** the *clock* on Earth -- which is the same conclusion attained as when **reality** is seen from the Earth-star *reference frame*.

It is only when we include the ***gravitational*** *dilation of time* that occurs when we choose to describe the trip from the astronaut vantage point, that the relation 'younger than' between the two twins admits an unequivocal interpretation, regardless of the *reference frame*, i.e. the 'nomad' twin is ***20 years younger*** than the 'sedentary' twin no matter how you look at it. Neat, Eh?

4.4. The Fourth Experimental Proof: Light Propagation Delay

In 1959, the echo of a radar signal was received from Venus and started the application of radar technology to astronomy. Signal distortions provided useful information on the rotation and topography of the planets. Transmitters had powers in the order of 400 KW, while the returning echoes had powers in the order of a quadrillionth of a watt! And, even so the information was invaluable. Besides, the round-trip time for the signal made possible a more accurate determination of the planetary orbits.

As we saw, if the *trajectory* of an *electromagnetic wave* experiences *deflection* nearby the Sun, such a *deflection* is associated with a ***change*** of *speed* that produces a ***delay*** in the ***transit*** *time*. As with *deflection*, this ***delay*** is a consequence of the *Principle of Equivalence* on one side, and of the *curvature* of *space-time* on the other. Einstein had not thought of measuring this phenomenon to strengthen the experimental evidence for General Relativity.

Irwin Shapiro, in 1961, applied General Relativity to predict how much the round-trip ***delay*** of the radar signal would be, when its path approached the Sun. For a signal sent to Mars while in its superior conjunction (when Earth and Mars are diametrically opposed relative to the Sun), the round-trip is about 45 minutes, and the predicted ***delay*** (relative to Newtonian mechanics) was about 250 microseconds. In order to measure such a ***delay*** -called the Shapiro delay- a precision in the measurement of a few kilometers was needed, and the interplanetary distances at superior conjunction were too large for the returning echo to be detected.

By the end of 1966, technology had improved enough to measure Venus superior conjunction about to occur on November 9th, and the three Mercury conjunctions to occur in January, May, and August of 1967. Results agreed with General Relativity with an accuracy of only 20%. To increase accuracy, it was decided to carry out the experiments with spaceships instead of planets (they are not good reflectors of radio waves), namely the 'Mariner 6' and 'Mariner 7' during their missions to Mars. Measuring *distances* with a precision of 15 meters, agreement between experiment and General Relativity increased to 3%. To further increase accuracy, both techniques were combined by means of a spaceship orbiting Mars or, even better, depositing a reflector on its surface. The Mariner 9 was the first aircraft used for the purpose while orbiting Mars, attaining an agreement with theory of 2%.

In the period 1976-1977, spacecrafts for the Viking space program landed on Mars and were utilized to reflect radar signals and acquire enough information to determine the Shapiro ***delay***. With such an optimal configuration, agreement between measurement and theory was excellent (0.1%). The fourth test -never imagined by Einstein- corroborated beyond doubt the soundness of its masterpiece. Finally, in 1978, the observation of a binary pulsar[15] made possible to confirm another conspicuous prediction of General Relativity: the

15 A binary pulsar is a pair of stars in reciprocal rotation.

existence of *gravitational waves*.

Returning now a little to the past, since 1965, a fifth test -not imagined by Einstein either- had been in the works waiting (by chance) to crystallize during the greatest technological achievement in human history: the descend on the Moon of the Apollo 11.

4.5. Houston... Tranquility Base here... the Eagle has landed

Nothing visible could separate that Sunday of the Argentinean winter in Bahía Blanca as different from any other -- but our generation will never forget it. As the *Little Prince* in Antoine de Saint-Exupéry's famous novella said: "...Anything essential is invisible to the eyes". While walking down the 'Avenida Alem' towards the student restaurant, so as to have the necessary-but-never-fancied gastronomic daily episode, I was experiencing a sense of pride for our human species; a magnificently strange feeling of passive participation in an act of cosmic transcendence. In a few hours, humans would step on another celestial body for the first time since they erected walking on this Earth -- millions of years ago. The radio transmission in Spanish of the 'Voice of America' was our only communication with the events about to occur that glorious 20th of July of 1969 -at about 384,000 km of *distance*- exactly 40 years before the moment I am writing this paragraph[16]:

Neil Armstrong (1930-) and Buzz Aldrin (1930-) enter the Eagle. Once again on the far side of the Moon, Mike Collins (1930-) disengages the Eagle (stage 10 in Figure 30) and backs off so it can leave the lunar orbit initiating its descend. The flight computer seems to be taken the Eagle to a dangerous crater; Armstrong takes manual control of its lateral motion, searching for a suitable place to land. With barely 30 seconds of fuel remaining, the Eagle lands at only 38 meters from a 24 meter diameter and several meter deep crater.

It was virtually impossible to concentrate on something else, let alone on study matters; the window blinds of our apartment on the second floor of 'Rodríguez 532' -very visible from 'Avenida Alem'- would be open well into the night, as a previously accorded signal to indicate the absence of any extra-curricular private activity, and that all our friends were welcome. Our apartment would be the place of reunion to carry out that thrilling passive participation in the human landing on the Moon.

On my third college year, still trying to understand Newton, and far from comprehending Einstein, what was about to take place was beyond my human power of assimilation and, while listening to the radio -likewise in the prior Christmas while experimenting with the *Carrousel by the Church*- I silently promised that I would dedicate my life to Science, so as to try to understand this mysterious Universe we live in. I had soon realized that it was neither possible nor necessary to understand everything, but that I could certainly develop the skills to comprehend anything that I would consider necessary.

At 03:17 pm of Houston (04:17 of Bahía Blanca), everyone in the Mission Con-

16 This chapter for the Spanish Edition of this book was written in June-July 2009, precisely during the 40th Anniversary of the Apollo 11 Mission to, and landing on, the Moon.

trol Room heard with a momentary momentous disconcertment: ***"Houston... Tranquility Base here...the Eagle has landed"***. Armstrong had unexpectedly changed the call sign. A new human base outside our planet -unknown to Houston- had been born. After a few seconds of indescribable stupor, they realized that the Eagle was on *Luna firma*, specifically on the *Mare Tranquilitatis*.

During the many hours that passed until Armstrong got out of the Eagle, we interchanged insights, opinions, venturing hypotheses on the technological and human aspects of such an incredible voyage, and wondering about what we were soon to witness -- all of it with the unfounded self-confidence and flagrant naïveté characteristic of our intellectual virginity. Some of us were cognizant of the uncertainties and risks associated with such a space journey; risks as extraordinarily high as the intellectual capacity, courage, discipline, physical and mental training of the astronauts, as well as of the 300,000 individuals that had participated in the project. Others, who had been born with less capacity for self-astonishment and less reverence for the contingent, thought that everything that could happen had been anticipated in detail, and the voyage was simply a mere confirmation of what had been rehearsed during training on Earth.[17]

Finally, I also heard with disconcertment somebody saying (I do not know whether because he really believed it, or because he had heard about it) that everything we thought that was taking place on the Moon, was actually being filmed on -and broadcasted from- Earth, i.e. made up by the United States government so as to win the space race to the Soviet Union. Already at that early age, I had developed a severe intellectual allergy to those individuals who issue grandiose statements without the slightest foundation to back them up. The perfect vulgar adjective for them is irreproducible in a book like this, but everyone would understand me -and laugh with me- if I give its academic version: *intelligent rectum*.

Despite the magnificence of the feat, it was instinctively ridiculous to me that somebody could think that setting up such a theatrical stage in secret (to be kept by 300,000 people!) could be easier than really landing on the Moon. At that moment, I could not have foreseen how prominent three conspicuous manifestations (though not so magnificent) of human nature would become during the upcoming 40 years: a) the phenomenon of humble pseudoknowledge (innocuous *folklore*); b) the phenomenon of pseudointellectualism (*folklore* as irritating as pernicious); and c) the phenomenon of the conspiracy theories (really harmful *folklore*). Thirty three years later, in 2002, a filmmaker would call Buzz Aldrin "a coward, liar, and a thief" because he had said he "walked on the moon, when he didn't". Aldrin punched him in the face.

Just before midnight, Armstrong finally opened the hatch stepping down the nine-rung ladder, activating the TV camera that transmitted images around the world, and pronouncing his famous comparison between the insignificance of his first step on lunar soil and its gigantic meaning for mankind. As equally or more impressive was Aldrin spontaneous reaction when meeting a "magnificent desolation". The cord which kept them attached to the module was quickly cut off starting their excursion into the new world.

17 Obviously they did not know that Armstrong almost had lost his life during one of the Moon landing rehearses.

Besides the USA flag, a commemorative plaque and a disk with messages from all nations, they installed a screen to receive particles from the 'solar wind', a seismometer, and -our protagonist in this chapter- a *Laser Ranging Retro-Reflector* (LRRR) (Figure 39) to conduct an experimental study of the ***lunar orbit*** that lasted for over 20 years, and that confirmed "with flying colors" the General Theory of Relativity. As Clifford M. Will [1] points out: the 'small step' given by Armstrong constituted a huge step forward for the science of experimental gravity.

Figure 39. The Laser Ranging Retro-Reflector left by the Apollo 11 on the Moon.[18]

4.5.1. The Brans-Dicke Theory and the Nordtvedt Effect

Since Einstein published his General Relativity Theory in 1916, several alternative theories of *gravity* phenomena were developed whereby *space-time* was not given a *non-Euclidean* ***geometry*** whose *metric* replaced the classical *gravitational field*, but retained the *semi-Euclidean* ***geometry*** of Special Relativity throughout *space-time*, while *gravity* being a classical external *field of forces* -- exactly as the *electromagnetic field*. The idea of a *space-time* -unfortunately called 'curved'- was very difficult to assimilate. Nonetheless, none of those theories was successful.

In 1962, the Brans-Dicke theory was published and, for quite some time, was considered as a serious alternative to General Relativity. This new theory accepted the *curvature* of *space-time* as representing *gravity* à la Einstein but, in contrast with the latter, allowed the *Constant of Universal Gravitation* G to vary in *space* and *time*. Kenneth Nordtvedt -at the Montana State University, USA- wondered if the *Law of the Equality of Inertial and Gravitational Masses* would also hold for bodies of astronomical size, i.e. -as it is risibly suggested in Figure 40- if a feather, Commander David Scott's hammer (Apollo 15), and the planet Saturn were inserted into a *gravitational field* (intense enough so as to not being modified by the bodies, and the latter sufficiently separated so they

18 Public domain picture in http://www.hq.nasa.gov/office/pao/History/alsj/a11/AS11-40-5952.jpg.

did not mutually interact), would all three (human scale and celestial) objects reach the 'floor' at the same time? Or, equivalently, would they all 'fall' with the same *acceleration*? Einstein probably never asked such a question because, for him, the *Law of Equality of Masses* had to be independent of the subatomic or astronomical size of *matter*. This law sometimes is confusedly identified with the *Principle of Equivalence* which, as we saw, is a principle based on the law. Because of this, when it is assumed that the law is also valid for astronomical masses, people refer to the 'Strong Principle of Equivalence'. Semantic precision is not some scientists' forte.

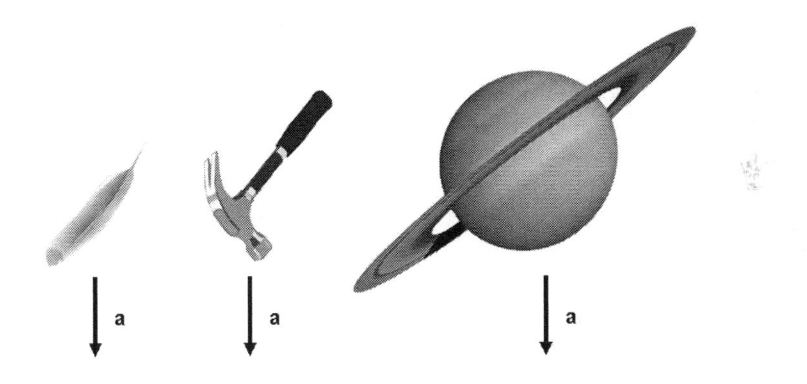

Figure 40. The 'Strong' Principle of Equivalence

The human-scale bodies used on Earth to validate this law (with the very high accuracy reported at the beginning of this chapter) maintain their physical integrity by virtue of *electromagnetic* and *nuclear* forces. *Gravity* is incredibly ***weaker*** than *electromagnetism* and *nuclear* forces but, being always ***attractive*** (in contrast with *electromagnetism*) and ***long*** range (in contrast with *nuclear* forces), when the *masses* acquire astronomical values, *gravity* becomes the dominant effect and celestial bodies retain their physical integrity by virtue of their very ***inner*** *gravity*. Contrariwise, in a small sphere like the ones used in Eötvös' experiments, their internal own *gravity* is a thousand trillion trillion times ***smaller*** that their total internal *energy*. Given this disparity, would it be not possible that the *gravitational energy* contributed only to the *gravitational mass* and not to the *inertial mass*? If that were true, this contribution (negligible for laboratory objects) could be significant for celestial bodies (for which *gravitational energy* is so important). For this reason, it seemed legitimate to wonder whether the *Law of Equality of Inertial and Gravitational Masses* were still valid in the macrocosm or not.

The legitimacy of this question acquired even more sense when Nordtvedt, comparing General Relativity with Brans-Dicke's theory (and others), discovered that while the former was consistent with the *Strong Principle of Equivalence* (the *Law of Equality of Masses* was valid in the macrocosm), the latter violated it! Evidently, here there was a solid criterion to determine which of the two theories was correct: if the so-called *Nordtvedt Effect* was observed, i.e. if the *Law of Equality of Masses* did not hold in the macrocosm, Brans-

Dicke's theory would be triumphant; if the Nordtvedt Effect was absent, Einstein's General Relativity would, once again, be the winner. But... how could they have both a celestial body and a small laboratory object (or also celestial but with much less internal *gravitational energy*) subjected to the same *gravitational field* so as to compare their *accelerations*?

In the winter of 1967-1968, Nordtvedt purposely took the same plane in which he knew Robert Dicke would be, and bluntly told him that his theory breached the *Principle of Equivalence* for astronomical size objects. Dicke was initially skeptical but, as a good scientist should do, he kept an open mind and, finally, became convinced that Nordtvedt was right, so that the eponymous effect had to be either confirmed or refuted -- even when the latter conclusion would certainly invalidate his own theory of gravitation.

Nordtvedt thought of the Moon and the Surveyor space program which was sending robotic spacecrafts to the Moon in preparation to the Apollo program. One of the objectives in the Surveyor program was to deposit a Laser Ranging Retro-Reflector (LRRR) on the lunar surface. A laser retro-reflector is a mirror specifically designed to reflect a laser beam in the direction opposite to that of its arrival. ***Measuring*** the round-trip *time* from Earth, they expected to know the ***lunar orbit*** with an accuracy higher than the available at the moment. Nordtvedt asked himself how the ***lunar orbit*** would be, if the Earth and its natural satellite did not 'fall' with the ***same*** *acceleration* in the solar *gravitational field* (as the Brans-Dicke's theory predicted). The ***lunar*** *gravitational* internal *energy* per unit mass is about 25 times ***lower*** than that of the Earth, so that according to the Brans-Dicke's theory, the effect had to be ***measurable***.

Nordtvedt's calculations indicated that the ***lunar orbit*** would have to show an *elongation* of about 1.3 meters, while General Relativity predicted none. Was the *distance* between Moon and Earth ***measurable*** so accurately as to discriminate 1.3 m from 384,000,000 m (i.e. 0.00000034%)? Laser technology[19] was expected to measure the *distance* with a precision of 15 cm! If the Nordtvedt Effect was **real**, it could be detected and ***measured***. As soon as Nordtvedt published his calculations, the quest for his ***effect*** to corroborate or refute General Relativity became the principal target of the Lunar Laser Ranging program.

Unfortunately, by the end of 1968 the Surveyor program, before it could place the LRRR on the lunar surface, was cancelled because NASA decided to focus on the Apollo program. The Apollo Lunar Surface Experiments Package (ALSEP) was overloaded with instrumentation and did no have room for an LRRR. But, right before launching the Apollo 11, NASA determined that the experimental work planned for the astronauts on the Moon could not possibly be completed during their brief stay, so that the ALSEP was simplified. This was a stroke of luck for the LRRR because its installation on the Moon surface was going to be a child play as compared to the diversity and complexity of the original plans. In this fortuitous way, did a thorough study -still in place- of reflected laser signals from the Moon start. Subsequently, two more reflectors were placed by Apollo 14 and Apollo 15 missions, and two more, of French manufacture, were deposited during the Soviet robotic missions Luna 17 and Luna 21.

19 A laser beam is a *light* signal so its wavelength is much smaller than that of radar signals. Because of that, much shorter (in time) pulses can be sent, and the distance can be measured with much higher precision.

4.5.2. The Fifth Experimental Test: The Nordtvedt Effect - Einstein was right once again

Barely 10 days after the LRRR was active on the lunar surface, were astronomers at the Lick Observatory measuring the *distance* to the Moon with a precision of a ***few meters***. In October 1969, the project was transferred to the McDonald Observatory in Texas, and other groups starting taking measurements in France, Soviet Union, Japan, Australia, and Hawaii. In December of 1971, the expected precision of 15 cm was attained, and the assessment on the existence or not of the *Nordtvedt Effect* started. By May 1975, more than 1,500 successful ***measurements*** had been obtained and, after considering all known *gravitational* interactions, the remaining effect would be attributed to the *Nordtvedt Effect*.

Two independent analyses of the experimental data were carried out, one of them by a group of 17 scientists in 9 institutions; the other by a group at the Massachusetts Institute of Technology (MIT). Both studies concluded that, if the *Nordtvedt elongation* existed on the *lunar orbit*, it had to be ***smaller*** than 30 cm. In the 1990s further analyses determined that, if the *elongation* was real, it had to be ***smaller*** than 2 cm and, ergo, Earth and Moon 'fell' towards the Sun with *accelerations* which were equal to each other within 0.00000000007%. This accuracy is even greater than the accuracy with which Braginsky and Panov proved the ***equality*** of ***inertial*** and ***gravitational*** *masses* for laboratory-size objects (described at the beginning of this chapter).

Realizing that the Brans-Dicke prediction for the *elongation* of the *lunar orbit* was 130 cm, it is crystal clear that the experimental data refuted such theory and, given that General Relativity predicts a ***nil effect***, once again Einstein was right and his theory was ratified with "flying colors".

Figure 41. The Eagle taking off the Moon to rendezvous with the Columbia.[20]

20 This picture is public domain in http://nssdc.gsfc.nasa.gov/imgcat/html/object_page/a11_h_44_6642.html. The photo was taken by Mike Collins from the Columbia.

July 21, 2009: Exactly 40 years ago...

After more than 21 hours working and resting on the Moon, leaving the Eagle lower stage on the surface, its ascent stage elevates to enter into orbit and rendezvous with the Columbia (Figure 41). The astronauts complete the transfer of lunar soil samples, abandon the Eagle, and the Columbia fires its motors to initiate a free-fall trajectory (geodesic) which will take them back to Earth after 60 hours.

July 24, 2009: Exactly 40 years ago...

The capsule enters Earth atmosphere whose friction increments external temperature to 3,000°C and reduces the capsule speed from 40,000 km/h to a few hundred km/h, at which the parachutes are open. The grandest exploration voyage ever attempted by human beings since they walked on the face of this Earth had successfully ended.

Additional Recommended Reading

[1] Will, Clifford M., *Was Einstein Right? - Putting General Relativity to the Test*. New York: Basic Books, 1993.

[2] d'Abro, A., *The Evolution of Scientific Thought - from Newton to Einstein*. New York: Dover Publications, 1950.

[3] Grünbaum, Adolf, *Philosophical Problems of Space and Time*. New York: Alfred A. Knopf, Inc., 1963.

[4] Reichenbach, Hans, *The Philosophy of Space and Time*. New York: Dover Publications, 1958.

[5] Wolfson, Richard, *SIMPLY EINSTEIN - Relativity Demystified*. New York: W.W. Norton & Co., New York, 2003.

[6] Einstein, Albert, Relativity - *The Special and the General Theory*. New York: Three Rivers Press, 1961.

[7] Einstein, Albert and Infeld, Leopoldo, *The Evolution of Physics*. New York: Simon & Schuster, 1966.

[8] Hoffman, Banesh, *Relativity and its Roots*. New York: Dover Science Books, 1983.

[9] Reichenbach, Hans, *From Copernicus to Einstein*. New York: Dover Publications, 1970.

[10] Eisenstaedt, Jean, *The Curious History of Relativity*. Princeton: University Press, 2006

[11] Goldsmith, Donald, *The Ultimate EINSTEIN*. New York: Byron Press Multimedia Books, 1997.

[12] Einstein, Albert, *Dialog About Objections Against the Theory of Relativity*, Lexington, KY, World Library Classics, 2009.

EPILOGUE

The Seduction of Pseudoknowledge

If you don't read the newspaper, you're uninformed; if you do read the newspaper, you're misinformed. Attributed to Mark Twain (1835-1910).

I think it is possible. I know that the websites that speak of this problem are websites that have the highest number of visits... And I tell myself that this expression of the masses and of the people cannot be without any truth. Christine Boutin (1944-), French Housing Minister, 2006.[1]

When Folklore becomes Pseudoscience

Is $E = m.c^2$ *a sexed equation? Perhaps it is. Let us make the hypothesis that it is insofar as it privileges the speed of light over other speeds that are vitally necessary to us. What seems to me to indicate the possibly sexed nature of the equation is not directly its uses by nuclear weapons, rather it is having privileged what goes the fastest.* Luce Irigaray (1932-), philosopher, linguist, psychoanalyst, 1987.[2]

The sleep of reason produces monsters. Etching n° 43 in 'Los Caprichos' of Francisco de Goya, 1799.

The consequences of scientific illiteracy are much more dangerous today than have ever been in the past. Carl Sagan (1934-1996).

How is it possible that Christine Boutin, a major Christian democratic figure in France, a center-right party in French politics -and many others with diametrically opposed political and social views- would rationally accept that the number of visitors to a website in the Internet could imply the verisimilitude of a factual statement?

How is it possible that Luce Irigaray, and many others with doctoral degrees in philosophy, sociology, linguistics, economy, psychoanalysis, etc. may confuse the possible sexism of a scientist with the sexism (?) of a law of Nature? How is it possible that they do not understand (or don't want to) that, regardless of the subjective motivations that the scientist who discovered such law could have had, the supposed law of Nature will be tossed away (not for being sexist, though!) as soon as experimental facts disqualify it?

1 Response given by the minister when asked if she believed that the President of the United States had been involved in the terrorist attack to the twin towers on September 9, 2001. As cited in Thomson, Damian, *Counterknowledge*, New York: W.W. Norton & Co., 2008.
2 As cited in Sokal, Alan & Bricmont, Jean, *Intellectual Impostures*, India: Replika Press Pvt. Ltd., 2004.

These are fascinating human phenomena beyond the realm of Physics *per se*, but that -as we have seen in this book- generate profuse amounts of *folklore* about Science in general and Physics in particular, distorting the nature of the *scientific method*, and promoting a total despisement for the notion of ***truth*** and the irrefutable power of ***experimental*** *evidence*. The technological revolution that took place during the last 50 years, and its effects on the news media -with the Internet being its prominent example- have facilitated the proliferation and strengthening of this pernicious *folklore*.

Needless to say, the positive effects of the Internet surpass by far its negative facets, and it is our responsibility to contain and, if possible, eliminate the latter. As I warned the reader in my Introduction, if s/he is reading this book in its electronic version, s/he has access -through hyperlinks to Wikipedia and the whole Internet- to a virtually limitless world of ***objective*** *knowledge* and... of *pseudoknowledge* (*folklore*). The capacity to discriminate between them is, today more than ever, paramount for the spiritual and material stability of our society. I cannot finish this book without attempting to offer the reader -responsible denizen of the Universe- a *frame of reference* from which to distinguish ***Science*** from *folklore*.

1. Science, Epistemology, and Linguistics

The meaning of words and the respect for the concept of ***truth*** (independent of the subject) are crucial to understand a statement of any nature, be it scientific, philosophical, social, political, economical, etc. ***Science*** without *semantics*, *epistemology*, and respect for the *factual* is *pseudoscience*, and that person who systematically practices it is, at best, simply an erudite intoxicated with his/her own verbosity -- a *naked emperor who thinks he is fancily dressed*.[3]

1.1. Science and the Concepts of Reality and Truth

As Jacob Bronowski says in [1] (emphases are mine), "Science is ***fact*** and ***thought*** giving strength to one another" and, in the same book, he also says: "Science is a great many things, and I have called them a great many names; but in the end they all return to this: Science is the ***acceptance*** of what ***works*** and the ***rejection*** of what ***does not***. That needs more ***courage*** than we might think."

The idea of ***objective reality***, the philosophical position that there is an ***external*** world (a ***factual*** world), is essential to the scientific attitude and, in order to capture such an ***objective*** world, Science has developed reproducible techniques for ***measurement*** and ***discernment*** with the purpose to attain *conclusions* which are ***independent*** of the ***subject*** who executes them.

Another critical concept, intimately related to the notion of **reality**, is that of **truth** existent regardless of our *volition* and *subjectivity*: the possibility of pronouncing ***rational*** assertions about that ***factual*** world which can be **universally** accepted. Let us halt for a minute and meditate about what William K. Clifford said: "Remember, then, that scientific thought is the guide of action; <u>that the truth at</u> which it arrives is not that which we can ideally contemplate

3 Hans Christian Andersen's fable in which the emperor's vanity prevented him from realizing that his majestic and expensive dress was not invisible but that, instead, he was naked.

without error, but that which we ***may act upon without fear...***"[4]. Combining the concepts of **reality** and **truth**, the objective of Science is clearly stated by Bertrand Russell in his book *ABC of Relativity* as:

> *Science does not aim at establishing immutable truths and eternal dogmas; its aim is to approach the truth by successive approximations, without claiming that at any stage final and complete accuracy has been achieved.*

A widely-spread prejudice is that scientific activity -due to its ***objective*** nature- is an impersonal chore completely unrelated to the individual who practices it. This is patently false; a *scientist* is as an *artist* as a sculptor is, but overwhelmingly more ***limited*** in his/her possibilities of personal expression: it is the **reality** ascribed to the ***external*** world, not his/her imagination that imposes those limitations. Even with such constraints, as Einstein wrote (and we confirmed along the historical facts related in this book): "Physical concepts are free creations of the human mind, and are not, however it may seem, uniquely determined by the external world".

Experimental observation, *induction*[5], and formulation of *theories* which are to be subject to experimental *falsification* or *validation*, constitute the quintessence of the ***scientific method***. Beyond what I just have said, it is impossible to be specific as to what the ***scientific method*** is all about. Peter B. Medawar (Induction and Intuition in Scientific Thought) said:

> *Ask a scientist what he conceives the scientific method to be, and he will adopt an expression that is at once solemn and shifty-eyed: solemn, because he feels he ought to declare an opinion; shifty-eyed, because he is wondering how to conceal the fact that he has no opinion to declare.*

The nexus between a *theory* and **reality** is attained in Science in only one way: with carefully designed ***experiments*** for a) discard the *theory* as ***incorrect***, or b) allow the *theory* to continue its reign until the ***next*** experiment. According to Karl Popper (1902-1994), a prominent philosopher of science in the 20th century, our confidence on the *verisimilitude* (truth value) of a *theory* cannot have a basis more solid than our failed attempts to *refute* it. One single *experimental **fact*** could be (though rarely is) sufficient to ***reject*** a *theory* or an assumed natural law. In contrast, absolute certainty about the soundness of a *theory* is unreachable. The result of each new experiment (contrast between *theory* and **reality**) simply increases or decreases the probability for the *theory* to be correct. Science is ***self-corrective*** and ***progressive***, and carries out its business within the ethnic, religious, political, and cultural diversity of the scientific community.

The assessment of a *theory* ***verisimilitude*** is a very complex process because, as we saw in this book, every experimental result is assessed in the context of its accuracy and its epistemic relevance, when compared to the existence of prior experimental results and their respective accuracies. For instance, Mercury *perihelion precession* (Chapter 7) was considered for a long time as an anomaly within Newton's theory of gravitation, but it never was considered with sufficient weight as to dethrone the theory -- until the advent of Relativity Theory. Even so, as we know, today, Newton's theory is not

4 As cited in Bronowski, J. *The Common Sense of Science*, Harvard University Press, 1994.
5 Induction is the mental process of inferring a generalized conclusion from particular instances.

regarded as *false* but, instead, *approximately valid* and extremely useful in a great variety of practical situations. In short, the ***range of validity*** and accuracy with which the theory represents **reality** was ***readjusted***. As an another example, the futile ***ether hunt*** (Chapter 5) did not cause the consternation and epistemic crossroads it did simply because of the negative results of Michelson-Morley experiment; it was the *convergence* of copious ***experimental** evidence* -equally negative- that forced the scientific community to reconsider its most precious philosophical beliefs so as to create the Special Theory of Relativity.

1.2. Science and Language

We also learned along the book about the intricate relation between *knowledge* and *language* and, as a pathetic example, the frustrating search for a physical *medium* that had to *undulate* so we could 'understand' and justify the *undulatory* theory of *light*. Associated with the term 'ether' -fictitious archetype of absolute immobility- new antinomic adjectives like 'real' and 'apparent' for the *length* of bodies and for the *time* measured in different *frames of reference* were created. Special Relativity tossed away all that unnecessary and confusing ***linguistic** folklore*.

Similarly, the use of a word like *curvature* -so imbued with our Euclidean prejudices- to represent the characteristic property of a *non-Euclidean space*, generated a conceptual thicket so entangled with Euclidean and non-Euclidean language that even scientists get logically trapped when trying to explain what it means for a *space* to be *curved*. I sincerely hope that my efforts to show that a 'planar' city as Madrid (Spain) has a *non-Euclidean geometry* (non-zero *curvature*) illuminated the reader to the point of realizing that it is useless to attempt visually imagining the *curvature* of our tri-dimensional *space*, let alone of our tetra-dimensional *space-time*. I also trust the reader will not infer that this impossibility reduces the validity of General Relativity in any way.

Incorrect or ambiguous meanings of words, as well as the confusion between *cognition* and *arbitrariness* were pointed out over and over as great creators of *folklore*, and a big impediment to in-depth understanding of Relativity Theory. It is not the reverence for words, but for their ***meaning*** that determines our deepness of comprehension of a given assertion about Nature. For instance, who of you did not feel thunderstruck when learned that in order to know the *velocity* of an object we have to establish the *simultaneity* of *distant* events and that, in order to establish the latter we need to know the *velocity*? How can the ***meaning*** of a word like ***speed*** -so engraved in our daily lives- be logically ***vitiated*** without most of us knowing it? Only a *definition* (convention) with a certain degree of *arbitrariness* can break out such a vicious cycle. This fundamental fact is not mentioned in any of the hundreds of popular science books I have read; but their authors pretend that you will submissively accept that *simultaneity* is ***relative***!

Likewise, was it not much easier for you to accept that the *mass* of an object is ***relative*** to the *reference frame* and that it ***increases*** with its *velocity* with respect to that *frame*, once you learnt that *mass* is **not** *quantity of matter*? Was it not much easier for you to assimilate the body ***contraction*** that accompanies its ***increase*** of *mass* when you were explained that the meaning of the word *length* has to be revised when ***object*** and measurement ***rule*** are in relative *motion*, and that such a revision requires, in turn, the notion of *simultaneity*? Many of the apparent paradoxes of Relativity Theory are only such until we realize that our ***semantics*** is not clear. We define the concepts appropriately (provide them with unambiguous ***meaning***), we ***reeducate*** our *intuition* in accord with the correct ***meaning***, and all paradoxes go away.

Another palmary example of the epistemologically fictitious difficulties into which Science falls for *linguistic* causes is the use of verbs which are inadequate to describe physical **reality**. Newton believed *light* **was** *corpuscular* until the scientific community became convinced that *light* exhibited the *behavior* of a ***wave*** (Chapter 5) and... what did that community do? Abusing the language (as I did while relating the historical events), *light* was declared to **be** a *wave*. Ironically, at the same time -and in the same scientific magazine- that Einstein presented his Special Relativity Theory in 1905 (in which *light* is treated as a *wave*), he also published another article[6] on the *photoelectric effect* whereby he introduced the ***quantum*** nature of *light*, creating the concept of what would later be called a *photon*. In essence, Einstein postulated that *light* did not behave as a *wave* but as a stream of *photons* (discrete packets of energy). Did he go back to Newton? Were his two simultaneous papers incompatible? **Is** *light*, thus, a ***wave*** or a shower of ***particles***? Is the verb 'to be' correctly applied?

Here is another curious example: Joseph J. Thomson won the Nobel Prize in 1906 for discovering a 'particle' called *electron*; his son George P. Thomson won the Nobel Prize in 1937 for proving the electron was a 'wave' What is it? Who was right, father or son? Is the verb 'to be' correctly used? According to the precise meaning of the verb 'to be' neither the father nor the son was right. If, instead, we use the verb 'to behave' instead of 'to be', then both were right. In **reality**, the father demonstrated -observing the scintillation on a fluorescent screen- that the *electron* ***manifested*** to us like what -in our macroscopic world- we call a 'particle', and the son showed -observing the typical diffraction rings of a wave- that the *electron* also manifests to us like what -in our macroscopic world- we call a 'wave'. There is no paradox in this duality; the electron is not schizophrenic: the apparent incongruence resides in attempting to describe a subatomic entity like the *electron*, using concepts which -in our macroscopic world- are mutually exclusive.

And what about the verb 'to exist'? Does the *electron* ***exist***? With regard to the *electron*, all we can say, paraphrasing Rudolf Carnap in [5], is that there are ***reproducible*** and ***observable*** phenomena that, in scientific language, we call *electrons*. This latter assertion may not be quite gratifying for many people but... can we really say something different about the existence of a *pebble*? The *existence* of both (*electron* and *pebble*) depends upon the context created by our *thoughts*, our *language*, our *theories*, and our interaction (***experimentation***) with our ***external*** world.

1.3. The Validity of the Theory of Relativity

Clifford M. Will in [6] points out that it is hard to imagine our modern world

6 A. Einstein, *On a Heuristic Viewpoint Concerning the Production and Transformation of Light*, Annalen der Physik 17: 132-148, 1905.

without the Special Relativity Theory. Physics in general, chemistry, quantum physics, evolution of species, GPS, particle accelerators, etc. are directly or indirectly related with Special Relativity to a point that, having been so extensively tested, we could venture it to be more undisputable reality than theoretical speculation. Likewise, after the detailed historical review I did in Chapter 7 of how General Relativity has been put to the test in the last 90 years, and despite the plethora of additional testing and cosmological applications that the theory is awaiting for in the future, we can conclude (using Will's rhetoric) that General Relativity:

→ *Bent* and *delayed light* by the ***measured*** amount;

→ *Advanced* Mercury's *perihelion* at the ***observed*** rate;

→ Gravitationally *dilated time* by just the right amount to prove the Twin Paradox ***was not*** a paradox;

→ *Made* the Earth and the Moon *fall* with the ***same*** acceleration towards the Sun;

→ *Caused* a binary star system to *lose energy* to *gravitational waves* at the right rate.

Despite the aforementioned, there is a marginal group of scientists and pseudoscientists who persist in denying the validity of Einstein's theory. Some of them, like the Argentinean Marcelo A. Crotti in his book La Relatividad Conceptual [8], accept the correctness of all Special Relativity equations, but not its epistemic and ontological interpretations, proposing a return to ***absolute*** *space*. Regardless of his philosophical position, Crotti's book is serious, well-written, and interesting to read. Other authors, like Kamen G. Kamenov in [7], assert that Relativity Theory "is one of the greatest fallacies of the twentieth century". Faithfull to my scientific spirit, I went through the trouble of carefully reading Kamenov's book entitled *Space, Time, and Matter, and the Falsity of Einstein's Theory of Relativity* and, without exaggeration, I can assert that in each attempt the author makes to -as he says- "demolish piece by piece" Einstein's theory, I found at least one flagrant instance of the logical *fallacy* known as *petitio principii*. This Latin-for-begging-the-question *fallacy* consists in having hidden in the premises -or using during a logical proof- the very proposition to be proved. This type of crass logical deficiency is inacceptable in the scientific community; it was a common place with all detractors of the theory since its appearance in 1905 and, as we can see, people continue falling into it. A *theory* may ***oppose*** to our *intuition* and *common sense* and still -as an axiomatic logic system- be internally fully consistent. Had Relativity Theory not had the *epistemological* and ***factual*** support we saw it does, its internal logical consistency would have remained intact.

Finally, there are other figures -predominantly in the Internet- with similar opinions but, because they include in their arguments (to *validate* or *falsify* the *Theory*) political, religious, racial, and the like considerations, they do not deserve further comment in a book like this.

1.4. Relativity and Subjectivism

In Chapter 2, I said that "the desideratum in Science is that the only ***subjective*** part of it must be its practitioner (the scientist). Consequently, when we talked about the 'observer', we did not mean necessarily a human being. The 'observer' can be just a human being (but conscientious of his ***subjective*** baggage); this human being can, in turn, be assisted with the needed *instrumentation* to ***objectively*** observe a particular phenomenon; or the 'observer' can be just a ***piece of technology*** to register the phenomenon by collecting experimental data which will eventually be analyzed (it does not need to!) by a scientist and shared within the scientific community. Paraphrasing once again Hans Reichenbach, the only way to acquire ***objective** knowledge* is to be ***conscious*** of the role than *subjectivity* plays in our methods of research.

Unfortunately, as Einstein said: "The meaning of the Theory of Relativity has been misinterpreted by a large majority: philosophers play with the word relativity like a kid plays with a toy... Relativity does not mean that everything is relative in life". Summarizing: ***subjectivism*** is strange to Science, while ***relativism*** is a conspicuous part of it -- if, and only if, it can be ***objectively*** established. The confusion between *relativism* and *subjectivism*, combined with the lack of intellectual ***honesty***, has generated the irritating phenomenon of *pseudointellectualism*, and a current of thought characterized by the ***epistemic** relativism*, which has nothing to do with ***scientific** relativism*.

With respect to the scientist's awareness of his/her own ***subjectivity***, and his/her unbreakable commitment to produce ***objective** knowledge*, Jesús Zamora Bonilla shows in [2] that, contrary to what ***epistemic** relativism* affirms, even acknowledging that the scientist is a *homo economicus* with material interests, ideologies and personal passions, the *scientific community* as a whole is well-conceived and efficiently self-organized so as to pay tribute to the ***pursuit of truth*** -- creating ***knowledge*** with an indisputable ***epistemic*** value for mankind.

2. Folklore is Pseudoknowledge... and may become Pseudoscience

Along this book, starting within its title, the word 'folklore' has been ubiquitous, with its ***meaning*** being apparent from the context. Nonetheless, to be comprehensive, let us investigate what the Merriam Webster dictionary says about it. The first acceptation is well-known to everyone: "traditional customs, tales, sayings, dances, or art forms preserved among a people"; the second meaning refers to the discipline which studies it; and the third *acceptation* partially reflects my use in this book: "an often unsupported notion, story, or saying that is widely circulated". To fully define the significance of the word 'folklore' as I use it, we need to add to the third *meaning* the notion of 'tradition' included in the first, as well as to understand that the expression "unsupported notion" is synonymous with the single word 'belief'. 'Tradition' is "An inherited, established, or customary pattern of thought, action, or behavior (as a religion practice or social custom)". 'Belief' is "a state or habit of mind in which trust or confidence is placed in some person or thing".

In contrast with 'to believe' there is 'to know' which is "to be aware of the truth or factuality of" something. *Knowledge* implies the possibility of ***collective*** agreement by virtue of ***reason*** and ***facts*** -- transcending cultures and traditions. **Science** is all about to ***know***, *Folklore* is all about to ***believe***. Notwithstanding, *Science* contains *folklore*, and *folklore* carries considerable *knowledge*. *Tradition* of a human group, regardless of its nature (social, economic, scientific, cultural, religious, political, etc.), includes both ***objective knowledge*** and *beliefs*. As we saw in Chapter 1, e.g. with the word 'plummet', there are *linguistic* expressions overly effectual which, nevertheless, are based on ***false*** *factual* assumptions. This *falsity* is absolutely *innocuous*. Likewise, the scientific community is not -and there is no reason why it should be- exempt of *innocuous tradition*.

It is crystal clear, then, that what epitomizes *folklore* is the ***absence*** of the methodic use of ***reason*** and ***factual*** corroboration (the scientific method) to accept and propagate those *beliefs*. The objective of Science is, thus, to minimize that part of human traditions which are mere *beliefs*, and maximize that which corresponds to *knowledge*. Naturally, *folklore* is mostly generated outside, but it can also be created inside the scientific community. *Folklore* is *pseudoknowledge*, simply because of the way it is generated and disseminated, but it can certainly contain ***objective*** *knowledge*. The *scientific method* can convert *pseudoknowledge* into *knowledge* and, vice versa. As it was clear in this book, scientists have along history -not painlessly- thrown away as mere *beliefs* what they had cherished as ***objective*** *knowledge* for centuries.

In my Introduction to this book, I said: "By Folklore, I mean the set of popular (and scientific) beliefs, mostly erroneous, associated with Relativity Theory and with our scientific activity in general". Again: Science is not exempt of *folklore* either, but its ***method*** has been conceived to minimize it. In fact, one of the thrusts of this book has been to disclose, in the frame of Relativity Theory, both the non-scientific as well as the -less abundant though still existent- scientific *folklores*. Unfortunately, a good part of the ***non-scientific*** *folklore* is acquired under the appearance of being ***scientific***, i.e. it comes from professionals who, supposedly, practice the ***scientific method***.

Naturally, *folklore* reaches its climax within the non-scientific community but also, lamentably, in those sciences within which -due to the complexities of their subject-matters (humanities, social sciences, politics, economics, psychology, etc.)- conceiving the necessary ***experimentation*** so as to separate the ***subjective*** from the ***objective***, is extremely difficult. Because of it, those sciences are particularly vulnerable to the proliferation of *pseudointellectualism* and its associated *pseudoknowledge*.

2.1. Pseudoknowledge

As Damian Thompson suggests in [4], it is probable that Newton believed that his search for the date of the Parousia (Second Coming of Christ) in the Book of Daniel was as scientific as his *Differential and Integral Calculus*, or his *Laws of Mechanics*. In accord with his predictions, fortunately, we still have about forty to fifty years before the end of the world (?). In fact, it is estimated that about 90% of Newton's work was about *alchemy* and *mysticism*; the first, at best a proto-science[7]; the second, the paradigm of ***subjectivism***. Nonetheless, due to just 10% of his work, Newton is considered one of the greatest scientists of all history. And why is that? It is because that 10% is solidly based on ***reason*** and ***factual*** corroboration. The huge remaining 90% of his work produced or dealt with *beliefs* (*folklore*) that never were converted into ***knowledge***; the 10% that took him to universal stardom produced ***objective*** *knowledge* (science), and will be respected for ever and ever, Amen -- even after having been dethroned by Einstein.

2.1.1. The Humble Pseudoknowledge: Innocuous Folklore

Since my Grandma died and till my forties, I always had a person to continue calling 'abuela'. Reaching my forties, my Mother came to the USA from the Euclidean Necochea, and lived with me till her passing. The feature of all of them to which I primarily and subconsciously was attracted the most was their life experience. None of them had gone beyond high-school, but all of them taught me a lot -- of course not about Science, but *common sense*, *empathy*, ***realism***, *prudence*, *wisdom*, and the indispensable ***emotional*** *intelligence* to be happy and make those around me happy. Naturally, as well, that crude *wisdom* was accompanied with huge doses of *innocuous folklore*: superstitions, traditions, home medicine as useless as harmless, unfounded fears, etc.

But... there was something else that inspired me a profound respect, the higher the more time I spent surrounded with pretentious *pseudointellectuals*: my Mom and all my Grandmas ***knew*** what they did ***not*** *know*. My Mom did not philosophize about *space* and *time*, but she knew she could trust in Relativity Theory (though she would have never attempted to understand it) more than in Einstein's internal *clock* while he was chatting with Mrs. Chaplin.

2.1.2. Pseudointellectualism: Irritating and Pernicious Folklore - Pseudoscience

In 1996, Alan Sokal (1955-), Professor of Physics at the New York University and Professor of Mathematics at the University College in London, presented for publication in the prestigious cultural-studies journal *Social Text*, an article supposedly serious but that, as he declared immediately after its publication, was -in reality- a premeditated parody to an intellectual current known as 'Postmodernism'. This parodied article -after a tremendous intellectual effort on the part of Sokal- was plagued with pompous assertions without any *rational* basis or *logical* sense, ***denying*** the existence of an *external* ***objective*** world, affirming that the concepts of **reality** and **truth** were mere *linguistic constructions*, etc., and in which the only support for his perpetrated hymn to ***absurdity*** were authentic citations about physics and mathematics from French and American intellectuals who were as renowned as ignorant in those disciplines.

One thinks that the sheer title of Sokal's article -with more than 40 pages, 12 of which were references- should have been enough to raise a few eyebrows among the editors and reviewers of a magazine on Sociology. Are you ready for the title? Here it is: *Transgressing the Boundaries: Toward a Transformative Hermeneutics of Quantum Gravity*. Well... not so... but quite the contrary: the publication committee did not miss the obvious (?) relation between *Quantum*

7 A fringe science out of the mainstream though rooted in established principles.

Gravity and *Sociology*, and the article was immediately published with honors in a special edition designed to rebut the critique that some prominent scientists had previously done to *Post-modernism* and *Post-structuralism*. No one better than a real scientist (Sokal) to refute other real scientists! The so-called *Sokal scandal* had been born, and Sokal and Jean Bricmont (Professor of Theoretical Physics at the University of Louvain, Belgium), wrote a book entitled *Intellectual Impostures* [3] published in France in 1997, to complete their analysis and denouncement of what I call *pseudointellectualism*.

A predominant feature of *Post-modernism* and *Post-structuralism* is the so-called ***relativism***, which affirms that the value of **truth** or *falsity* of a proposition is **relative** to the individual or social group asserting it. If the proposition is ***factual***, the *relativism* is ***cognitive*** or ***epistemic***; if the statement is on human values, the *relativism* is ethical, moral, or aesthetic. Another trait is the indifference -I would say disdain- for ***facts*** and ***reason***, assigning a vast relevance to *subjectivity* (Solipsism). Consequently, these schools of thought consider *Science* as one more *narration*, *myth*, or *linguistic* construct among others. As an instance, Paul Feyerabend (1924-1994), a philosopher of science, referred to *Science* as "one more superstition", and some sociologists of science put *Astronomy* on the same level as *astrology*. A detailed analysis of all these phenomena and its documentation can be enjoyed in the book by Sokal and Bricmont.

With respect to *Solipsism*, I cannot help but cite Bertrand Russell's experience during his intellectual maturation:

> *... there was a curious pleasure in making oneself believe that time and space are unreal, that matter is an illusion, and that the world really consists of nothing but mind. In a rash moment, however, I turned from the disciples to the Master and found in Hegel himself a farrago of confusions and what it seemed to me little better than puns. I therefore abandoned his philosophy.*

Or, in an infinitely more tragic way... à la Borges:

> *And even so, to negate the temporal succession, to negate the self, to negate the astronomical universe, are apparent desperations and secret consolations... The world, unfortunately, is real; I, unfortunately, am Borges.*

Just for the sake of giving the reader a flavor for Sokal's denouncement of *Post-modernism* and *Post-structuralism*, so I can continue with the main thrust of this Epilogue, I transcribe below a few examples given in his book, to demonstrate these pseudointellectuals' sickening adoration for words *per se*, as well as their irresponsible disdain for ***meaning***, ***substance***, and the use of ***reason*** (emphasis is mine):

> *...This **torus** really exists, and it is exactly the structure of the neurotic...*
> Jacques Lacan (1901-1981), 1970, psychoanalyst, philosopher.

> *A literary semiotics has to be made starting from a poetic logic, in which the concept of power of the **continuum** would encompass the interval from 0 to 2, a continuum where 0 denotes and 1 is implicitly transgressed.*
> Julia Kristeva (1941-), 1969, philosopher, sociologist.

...But what does the mighty theory of ***general relativity*** *do for us except establish* ***nuclear power plants*** *and our bodily* ***inertia****, that necessary condition of life?* Luce Irigaray (1932-), 1993, philosopher, linguist, psychologist.

What is most extraordinary is that the two hypotheses, the apocalypse of real time and pure war along with the triumph of the virtual over the real, are realized at the same time, in the same ***space-time****, each in implacable pursuit of the other. It is a sign that the space of the event has become a* ***hyperspace*** *with multiple* ***refractivity****, and that the space of war has become definitively* ***non-Euclidean****.* Jean Baudrillard (1929-2007), 1995, cultural theorist, sociologist, philosopher.

First, the opinions of ***scientists*** *about* ***science*** *studies* ***are not*** *of much importance. Scientists are the informants for our investigations of science, not our judges. The vision we develop of* ***science*** *does not have to resemble what* ***scientists think*** *about* ***science****...* Bruno Latour (1947-), sociologist of science, anthropologist.

*...Transparence is no longer composed of light rays (solar or electric) but instead of elemental particles (****electrons*** *and photons) that are transmitted at the* ***speed of light****.* Paul Virilio (1932-), 1995, cultural theorist, urbanist.

...How could he [the pharaoh Ramses II who died circa 1213 BC] pass away due to a bacillus ***discovered*** *by Robert Koch in 1882?* Bruno Latour (1947-), sociologist of science, anthropologist.

When depth of ***time*** *replaces depths of sensible* ***space****; when the commutation of interface supplants the delimitation of surfaces; when transparence re-establishes appearances; then we begin to wonder whether that which we insist on calling space isn't actually light, a subliminary, para-optical light of which sunlight is only one phase or reflection.* Paul Virilio (1932-), 1984, cultural theorist, urbanist.

Evidently, as Sokal and Bricmont allegorically wrote: "the King is naked (and so is the Queen)", and if the majority of the texts these scientists denounce seem incomprehensible, "it is for the excellent reason that they mean precisely nothing" [3]. I highly recommend this book to those interested in witnessing the triumph of ***reason*** over ***inanity*** -- and this triumph is the more irrefutable and praiseworthy when we realize that the *social* and *political* leanings of the authors -as they declare them- coincide in the majority of cases with those of the renowned intellectuals they denounce. This can only indicate the victory of ***objectivity*** over *cognitive relativism* -- irrespective of the *homo economicus* and *homo politicus* nature of every human being or human agglomeration.

2.1.3. Conspiracy Theories: Highly Pernicious Folklore

A phenomenon as interesting as deleterious for humanity is the fascination that *conspiracy theories* -be their nature political, religious, racial, etc.- produce in the public. Though only rarely, we saw in Chapter 5 (The Grand Cosmic Conspiracy), that even in Science there is place for the irresistible appeal of a *conspiracy theory* -- when the inability of *rationally* explaining the ***facts***, turns into exasperation and ends imagining a villain (à la James Bond) (in this case the Cosmos) with infrahuman motives to delude a human group with little political, intellectual, religious, technological or social power (in this case the most brilliant scientists of the time as compared to Nature).

The difference between that anomalous case in the history of Science and the rest of the *conspiracy theories* is not in their ***verisimilitude***: all of them are *possible* with different degrees of *probability*; the **core** difference resides in the *intellectual attitude* of those individuals or institutions who propose them. Lorentz's proposition was simply the scientific humble assertion that he was impotent to explain the ***facts*** unless he threw away the most cherished philosophical principles of Science -- something he was not willing to do. Contrariwise, the proponents, advocates, and passive followers of *conspiracy theories* in the *non-scientific* or *pseudoscientific* realms exhibit an utterly ***opposed*** attitude: they consciously or subconsciously decide (for emotional, economical, religious, political, ethnical, social, and the like reasons) to which *belief* they will pay allegiance, granting it *-a priori-* the status of *knowledge*; they gather (if they do) all the *factual* and *non-factual* (anecdotic, political, religious, etc.) data; they select as 'evidence' those data that further their *theory*, deliberately discard those pieces of data that ***demolish*** it; they ignore the relative *weight* of each piece of 'evidence'; and there they go to sell their specious product of *pseudoknowledge*.

My first contact with this phenomenon (as I described it in Chapter 7) was in 1969 with the Moon landing presumed hoax. As James R. Hansen wrote in his book *First Man: the Life of Neil A. Armstrong*, my inanimate hero of Chapter 7 (LRRR) constitutes solid evidence in favor of the Moon landing in 1969:

> *For those few misguided souls who still cling to the belief that the Moon landings never happened, examination of the results of five*[8] *decades of LRRR experiments should evidence how delusional their rejection of the Moon landing really is.*

Even so, these LRRR experiments (Chapter 7) only prove categorically that the retro-reflector is on the Moon surface, but not strictly that Armstrong or Aldrin personally installed it. As we know, there are other reflectors that were robotically deposited. However, in 2009 the Lunar Reconnaissance Orbiter imaged the various Apollo landing sites on the surface of the moon with sufficient resolution to see the descent stages of the lunar modules, scientific instruments, and foot trails made by the astronauts.

Is it *possible* that everything was a fraud? Of course it is! Is it *probable*? Of course it isn't! And why it is **not** *probable*? Because that LRRR evidence in conjunction with mountains of other pieces of evidence -***objectively*** weighted with the anomalies- assign to the hypothesis that everything was a hoax a ***negligible*** *probability*. For the reader interested in independent evidence (outside NASA), s/he can start (with the precaution that the Internet is a superb source of *knowledge* as well as of *folklore*) in http://en.wikipedia.org/wiki/Third-party_evidence_for_Apollo_Moon_landings, and (if s/he reads Spanish) with the book *La Conspiración Lunar ¡Vaya Timo!* [12].

And why is this disdain for the rules of evidence so harmful for humanity? A detailed account of what Thompson calls *counterknowledge* and its disastrous consequences 'condemning future generations to material and intellectual

8 I assume this is a typo as only four decades have past since the Apollo 11 landing on the Moon.

poverty', can be found in his homonymous book [4]. To achieve my objective, I will only mention a few facts and statistical data cited in this and Sokal y Bricmont books which -I hope- should kindle the reader to think:

→ According to a survey by Scripps Howard in 2006, a great majority of young people, and 36% of adults in the USA ***believe*** that their government participated (actively or passively) in the terrorists attacks of September 11th, 2001 -- so that the Iraq war could be justified;

→ In 2007, it was revealed that British teachers had eliminated from their syllabus the history of the Holocaust, so as to not ***hurting*** the ***sensitivity*** of those students who denied it.

→ The bestseller *The Da Vinci Code* (which is not factual at all) has persuaded 40% of the USA public that Christian religions are hiding information on the life of Jesus Christ -- tragically confounding *beliefs* with **objective** *knowledge* which is -of course- neutrally beyond any religion;

→ In 2007, the parents of 24,000 children in Pakistan opposed to their vaccination against poliomyelitis on the basis of their *belief* in a *conspiracy* within the USA government to sterilize them;

→ In 2007, the catholic leader in Mozambique declared to the BBC that condoms made in Europe were deliberately infected with the AIDS virus with the objective of eliminating the African population;

→ A Gallup survey determined that 45% of the people living in the USA believe that their God created human beings -as we look now- about 10,000 years ago -- ignoring the ***factual*** evidence that modern humans are -anatomically- at least 100,000 years old. There is nothing inherently bad or dangerous in *believing* in God, because it is a ***belief*** immune to scientific proof/disproof (i.e. it cannot be converted into ***knowledge***), and such a ***belief*** helps many people to be content and productive. But what the Gallup survey implies is something very different: the ***denial*** of the *scientific method* by an alarming proportion of society, and that is -without doubt- very dangerous.

→ The belief that the HIV virus is innocuous, it is not sexually transmitted, and it is not the cause of AIDS, has been discarded by the scientific community, but it continues alive in the Internet. For many years, the President Thabo Mbeki of South Africa was associated with the deniers of AIDS, and in November 2008, the New York Times wrote that, due to Mbeki's support of such a belief, 365,000 people were estimated to have died in South Africa;

→ Some of the most brilliant individuals in our society, companies, and even universities are dedicated to massively commercializing *pseudoknowledge*, packaged as ***objective*** *knowledge*. These entities develop products with an incredible commercial success by purposely mixing up ***fiction*** with **reality**, i.e. combining *conspiracy theories* with ***pseudohistory***, mystery, supposed millenary secrets about the Universe, miraculous medicine, extraterrestrials, etc. This commercial success, naturally, further reinforces their motivation to continue creating more *pseudoknowledge*.

As Thompson says, gullible thinking is propagating in our society as rapidly and mutely as a virus (with an incredible ability to mutate), and nobody knows how long this epidemic will last.

2.2. How to identify Pseudoknowledge

To wrap up this book, I would like to summarize those characteristics that I have delineated along the book as useful to distinguish ***scientific*** from *pseudo-scientific* attitudes. I have to point out that, having 300 years elapsed since the *Age of Enlightenment*[9] took place, it is now very odd to find a person that, as Newton did, practices simultaneously and paradoxically ***Science*** and *pseudo-science* while favoring the latter with such a disparity. Notwithstanding, along my professional life, I have enriched my persona by interacting with intellectuals of the most diverse religions and socio-political leanings and, for many of them, it was obvious to me that their decisively *scientific* mind-sets during our professional work, instantly disappeared when the discourse changed towards social, political, or religious subject-matters. Because of that, the list that I offer below with the purpose of helping the reader to separate *Science* from *folklore*, only attempts to assess the ***objectivity*** and ***truth*** value of the statements that an individual may make, but not to pinpoint that person *per se* as *intellectual* or *pseudointellectual*. The latter purpose would be very arrogant on my part. The signs of alert, the reader is to look for when interacting with people, are:

→ General and radical ***skepticism***. To *believe* in ***nothing*** is as ridiculous as to *believe* in ***everything***. ***Reason*** and ***factual*** evidence may convert a *belief* into ***knowledge***;

→ Fascination for *beliefs* and *mysticism* as if they were ***objective*** *knowledge per se*. To *believe* in ***everything*** is as ridiculous as to *believe* in ***nothing***. ***Reason*** and ***factual*** evidence may convert a *belief* (and *knowledge*!) into ***pseudoknowledge***;

→ ***Epistemic*** *relativism*, with disdain for ***reason*** and the contrast of conclusions with ***facts***;

→ To choose a conclusion *a priori*, to select the *facts* supporting the conclusion, toss out the *facts* which disprove it, and ignoring their relative weights so as to 'logically' conclude what had already been ***emotionally*** *accepted*;

→ *Fascination* for obscure *verbiage* in which the *meaning* of the words is secondary or inexistent, with the *verbosity* designed to ***hide*** the *lack of substance*;

→ ***Abuse*** of *scientific jargon* in other fields, without explaining which is the relation with the subject-matter at hand;

→ ***Ambiguity*** that allows those who practice it, when refuted, to take refuge in alternative less radical interpretations;

→ 'Scientificism', i.e. the use of mathematics and scientific terminology to ***dress up*** their points with a tone of *objectivity* and *verisimilitude* when, in reality, the *Emperor is naked*;

→ Propensity to believe in *conspiracy theories* for emotional, political, social,

9 The Age of Enlightenment is the era in Western philosophy and intellectual, scientific, and cultural life in which reason was advocated as the primary source for legitimacy and authority.

economic, or religious causes;

→ *A priori* distrust in the intellectual and moral honesties of those intellectual or political leaders who do not share the same socio-political or religious views. From this propensity to the creation of a villain à la James Bond, there is only one step;

→ To brandish arguments of political, religious, social, linguistic, or racial nature -as well as sacred books, pseudohistory, and the like- to back up a purely ***factual*** proposition (did it or did it not occur?);

→ To resort to the *opinion* of the *majority* to establish the ***truthfulness*** of a ***factual*** proposition -- as if ***facts*** depended upon *opinions* and not all the way around;

→ To conclude, from scarce and ***anomalous*** *evidence*, the ***verisimilitude*** or ***falsehood*** of a *theory*;

→ Naïve tendency to believe that everything which is *written* can be used as solid *evidence* to prove/disprove a *theory*;

→ Repetition of phrases or ideas for which it is apparent to everyone that they are understood neither by the person who repeats them, nor by the pseudointellectual who originated them;

→ Systematic application of *euphemisms* for socio-political, economical, or religious reasons;

→ Scientists, otherwise serious, who extrapolate *theories* -without a solid basis- beyond what has been ***experimentally*** confirmed, or describe ideas in popular science books as if they were established ***fact***, weakening the seriousness of our scientific activity;

→ Distortion of ***truth*** and/or ***facts*** for political, religious, or economical reasons, or to avoid touching ethnical or social sensitivities, etc. The ***respect*** for other human beings is essential to the basic *scientific* ***belief*** in an ***external*** world (yes! it is a philosophical ***belief***); distortion of **reality** and **truth** is very dangerous: we sacrifice the future of humanity to avoid hurting the feelings of an individual or a group of individuals.

3. Why I wrote this Book and what is next

For the last time before I let you go: my objective when writing this book was not to make the reader an expert on Relativity Theory; on the contrary, I am sure you know that would take much more intellectual effort than bearing with me for about 330 pages. The basic purpose of this symbolic cavalcade was to show that the theory, solidly supported by factual evidence gathered in the last 100 years, however strange and opposed to our prejudices (disguised as they are with the 'common sense' mask) may seem to be, is ***rational***, *consistent*, and ***intelligible*** to the layperson -- as long as s/he has the audacity to understand and accept the unfounded nature of those preconceptions.

It was the desire and honor of being able to fire up this necessary *intellectual courage* on the part of the layperson, that took me to write this book and immensely enjoying the endeavor; the desire and the honor of stimulating in the youth, and resurrecting in the adult what Manuel Toharia refers to as our 'capacity to be astonished' [9], so seemingly lost in my children's generation -- who are in magical possession of technological wonders, for the reality of which our generation worked so hard. I sincerely hope my goal has been accomplished.

Another vital purpose of this book was to *demystify*, to 'defolklorize' the reader; to stimulate his getting rid of the aura of mystery and incomprehensibility that surrounds Relativity Theory, and to debunk the popular (and scientific) beliefs -mostly erroneous- that accompany Einstein's work, which has been hijacked so as to -taking advantage of the confusion between *relativity* and *subjectivity*- advance and validate totally unfounded statements in areas like psychology, morality, sociology, literature, art, etc. I am confident I have gotten this message across to the reader.

Was Einstein right with his Relativity Theory? We have seen in this book that, YES, he was right. Was Einstein right when he said that "God does not play dice"? That is the subject-matter of my next book entitled *What is Reality? - Einstein, Quantum Physics, and Folklore*. Einstein was one of the pioneers of Quantum Physics, and of the application of the concept of probability to Physics. Nonetheless, after finishing his General Relativity Theory, he spent a good part of his life trying to prove that Quantum Theory was incomplete -- in opposition to the so-called Copenhagen School led by Niels Bohr. If the reader learned a lot and was amazed more than once while reading this book, I promise that the content of my upcoming book shall be equally or more instructive and astonishing. Quantum Physics has been and continues to be an endless source of *folklore* -- inside and outside the scientific sphere.

Finally, I sincerely hope having called the attention of the reader about the catastrophic consequences of considering the concepts of **reality** and **truth** as mere *social* or *linguistic* constructs; in brief: the danger of defending the idea that 'if it is true for you, then it is true; full stop'. This *intellectual intoxication*, as Bertrand Russell wrote in his *History of Western Philosophy*, "is the greatest danger of our time, and any philosophy which, however unintentionally, contributes to it is increasing the danger of vast social disaster".

What can we do to reduce this danger? On the part of the non-scientist, to be alert of the peddlers of *pseudoscience*; I hope I have contributed somehow to kindle that ***critical*** spirit so necessary for us ***not*** to be *deceived*. On the part of the scientists, to realize the tremendous responsibility we have on our shoulders, and to vigorously take action so as to counteract the forces of *obscurantism*; popular science is a powerful way of taking action: we have to teach the layperson (*divulgare*) without *distorting* (*tergiversare*) and, by doing so, as Manuel Toharia said in [9], "attain a democracy of a better quality".

Additional Recommended Reading

[1] Bronoswki, Jacob, *The Common Sense of Science*. Cambridge: Harvard University Press, 1978.

[2] Zamora Bonilla, Jesús P., *La Lonja del Saber - Introducción a la Economía del Conocimiento Científico*. Madrid: UNED Ediciones, 2003.

[3] Sokal, Alan and Bricmont, Jean, *Intellectual Impostures*. India: Replika

Press Pvt. Ltd., 2004.

[4] Thompson, Damian, *Counterknowledge - How we surrendered to conspiracy theories, quack medicine, bogus science, and fake history*. New York: W.W. Norton & Co., 2008.

[5] Carnap, Rudolf, *An Introduction to the Philosophy of Science*. New York: Edited by Martin Garner, Dover Publications, 1996

[6] Will, Clifford M., *Was Einstein Right? - Putting General Relativity to the Test*. New York: Basic Books, 1993.

[7] Kamenov, K.G., *Space, Time, and Matter, and the Falsity of Einstein's Theory of Relativity*. Vantage Press, 2008.

[8] Crotti, Marcelo A., *La Relatividad Conceptual*. http://www.crotti.com.ar/Relatividad/.

[9] Entrevista a Manuel Toharia con motivo de la primera década del Museo de la Ciencia en la Ciudad de las Artes y de la Ciencia en Valencia, España. Vídeo en: http://www.bondiatorrent.es/index.php?option=com_content&task=view&id=139&Itemid=79.

[10] Fernández Aguilar, Eugenio, *La Conspiración Lunar ¡Vaya Timo!* Pamplona: Laetoli, 2009.